Study Guide
to accompany
University Physics
Arfken   Griffing   Kelly   Priest

Study Guide
to accompany
University Physics
Arfken Griffing Kelly Priest

T. William Houk
James Poth
John W. Snider
Miami University, Oxford, Ohio

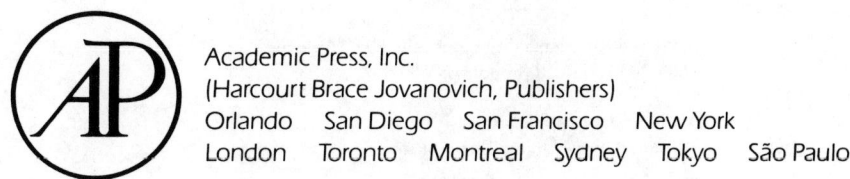

Academic Press, Inc.
(Harcourt Brace Jovanovich, Publishers)
Orlando   San Diego   San Francisco   New York
London   Toronto   Montreal   Sydney   Tokyo   São Paulo

Copyright © 1984 by Academic Press, Inc.
All rights reserved.
No part of this publication may be reproduced or transmitted in any form or by any means, electronic or mechanical, including photocopy, recording, or any information storage and retrieval system, without permission in writing from the publisher.

Academic Press, Inc.
Orlando, Florida 32887

United Kingdom edition published by Academic Press, Inc. (London) Ltd.
24/28 Oval Road, London NW1 7DX

ISBN: 0-12-059868-X
Printed in the United States of America

CONTENTS

Preface to the Student   vii
Chapter 1   General Introduction   1
Chapter 2   Vector Algebra   5
Chapter 3   Equilibrium of Rigid Bodies   11
Chapter 4   Motion in One Dimension   19
Chapter 5   Motion in a Plane   29
Chapter 6   Newton's Laws   36
Chapter 7   Work, Energy, and Power   48
Chapter 8   Conservation of Energy   58
Chapter 9   Conservation of Linear Momentum   68
Chapter 10   Many-Particle Systems   76
Chapter 11   Conservation of Angular Momentum   85
Chapter 12   Rotation of a Rigid Body   91
Chapter 13   Motion of a Rigid Body   100
Chapter 14   Oscillatory Motion   106
Chapter 15   The Mechanical Properties of Matter   116
Chapter 16   Fluid Mechanics   123
Chapter 17   Wave Kinematics   132
Chapter 18   Mechanical Waves   140
Chapter 19   Special Relativity   146
Chapter 20   Relativistic Mechanics   153
Chapter 21   Temperature and Heat   158
Chapter 22   Thermal Properties of Matter   167
Chapter 23   Heat Transfer   174
Chapter 24   The Second Law of Thermodynamics   180
Chapter 25   Kinetic Theory   188
Chapter 26   Electric Charge   194
Chapter 27   Electric Field and Gauss' Law   199
Chapter 28   Electric Potential   205
Chapter 29   Capacitance and Capacitors   212
Chapter 30   Electric Current   218
Chapter 31   Direct-Current Circuits   224
Chapter 32   Magnetism and the Magnetic Field   231
Chapter 33   Magnetic Field of Electric Current   238
Chapter 34   Electromagnetic Induction   244
Chapter 35   Inductance and Inductors   251
Chapter 36   Magnetic Properties of Matter   258
Chapter 37   Alternating Currents   265
Chapter 38   Electromagnetic Waves and Maxwell's Equations   272
Chapter 39   Reflection, Refraction, and Geometric Optics   278
Chapter 40   Physical Optics: Interference   289
Chapter 41   Physical Optics: Diffraction   296
Chapter 42   Quantum Physics, Lasers, and Squids   302
Chapter 43   Nuclear Structure and Nuclear Technology   311
Answers to Problems   321

Preface to the Student

OBJECTIVES OF THE STUDY GUIDE

This study guide is a companion volume to UNIVERSITY PHYSICS by Arfken, Griffing, Kelly, and Priest. We have adhered both to the style and format of the text. Extensive reference to the text as well as use of the same titles, section headings, and equation numbers should make the study guide simple to use as an auxiliary to the text. This study guide has been created as a supplement to the text and not as a replacement for it. We have relied on our experience to provide concise summaries of the textual material, added information, example problems, and additional exercises.

Our coverage of the material has been selective rather than exhaustive. We have stressed topics that we judge to be those that are the most difficult either conceptually or operationally for you to understand. We have provided examples which we believe illustrate the principles of the physics involved and we have worked them out fully with comments which should clarify the principles as well as the analytical and synthetic methods employed to solve problems. We have also included a number of supplementary problems which should help you to solidify your grasp of the material covered in the text and the study guide. Answers to these problems are included in an appendix at the end of the study guide so that you may check your work.

Problem solving is an important feature of the text. Accordingly, this study guide is strongly problem oriented. Perhaps the most important aspect of your introduction to physics is the development of the ability to analyze problems and synthesize their solutions. Students often feel that the answer justifies the means. Nothing is further from the truth. Although obtaining the correct answer to a problem should not be minimized, the primary goal of your efforts should be to understand the material well enough that you can be successful at developing methods to solve the problems presented to you. The outlines of various methods for attacking problems are presented in the text as well as in the study guide. These methods are standard but they cannot replace the ability to think analytically. Much factual material that you learn in this course will be repeated, albeit at a more sophisticated level, in coursework which you will take as you progress through your college careers. However the ability to use this material effectively must begin to be developed at this level.

ORGANIZATION OF THE STUDY GUIDE

Chapter titles in this study guide coincide with those in the text. Each chapter of the study guide consists of four major parts:

PREVIEW - This brief section describes the major ideas to be covered in the chapter as well as the goals of the chapter.

SUMMARY - This section presents a brief synopsis of the material presented in the text. It is broken into subsections that are numbered and titled identically to those in the text. Equations in this section are also numbered as they are in the text. Although the summaries are primarily synoptic in character, we have added material, definitions, special notes, and hints to clarify, elaborate, or emphasize the textual material.

EXAMPLE PROBLEMS - This section contains problems chosen to illustrate both the physical principles presented in the chapter and the methods used to solve these problems. We have provided complete solutions to these problems and have included extensive commentary in order that you may understand clearly <u>why</u> we have done what is done to solve the problem.

PROBLEMS - We have presented in this section a number of problems for you to work independently in addition to those which you may find in the text. We have tried to choose problems that reinforce and extend the concepts and methods in the Example Problems section.

SUGGESTIONS FOR USING THE STUDY GUIDE.

It is important that you realize that the primary source for the material which you will be studying and learning in this course is your text. Ancillaries, such as this study guide, are meant to aid you in your use of the text and in understanding what is in it. We would like to make a number of suggestions regarding your study of Physics and the appropriate use of this book.

The lecture-demonstration method is probably the most common teaching method used in physics courses. In order that you may derive the maximum benefit from the lectures which you will attend, it is imperative that you <u>read ahead</u>. Students often make the mistake of thinking that sitting and listening to a lecture over material they have not encountered before will transfer sufficient information for them to understand the material. In Physics, this is rarely the case. Lectures are meant to clarify the material you have studied and to answer questions that you may have about it. So expend every effort to keep up with your assigned reading.

The second important activity you should undertake is to carefully study the examples that are worked in your text and in the study guide. It is very important that you study them with the objective of understanding not only what operations have been done to solve the problem but why they are done.

Next, you must try to apply your new knowledge to solving problems. We suggest that you first perform the problems assigned to you from the text and then turn your attention to those in the Study Guide.

Finally, the study guide can serve as an excellent review and practice to prepare you for examinations. The chapter summaries serve as good reviews of the material covered, and the unworked problems can provide you with practice in problem solving.

Good Luck.

ACKNOWLEDGEMENTS

We express our sincere appreciation to Jane Kelly for her outstanding work in the preparation of this manuscript and thank Jeff Holtmeier of Academic Press for his help throughout the project. We are especially grateful to our families for their understanding and encouragement.

Chapter 1
General Introduction

PREVIEW

This chapter introduces the concepts upon which the quantitative nature of physics as a science depends. The types of quantities with which physics deals are defined and their nature is discussed. The concepts of units and dimensions are introduced and discussed.

SUMMARY

1.1 THE DEVELOPMENT OF SCIENCE

The primary characteristic of a science is that it seeks to discover the interrelationships which exist between quantifiable properties of systems. In some sciences the properties of the relevant systems cannot be quantified numerically, but must be quantified descriptively. Much of the social sciences as well as some branches of the natural sciences are in this category. The science of physics, which seeks to discover the relationships between physically observable quantities, is one in which numerical values can and are assigned to observed quantities. Many of the areas which physics addresses are listed in Table 1.1 of the text.

1.2 SCIENCE AND MEASUREMENT

In physics the quantities with which we most often deal are operationally defined. This means that there must be an explicit definition for each quantity which specifies how this quantity is to be measured. The stated measurement process provides an operational definition of the desired quantity.

In physics as in most of the natural sciences and mathematics there are two types of quantities.

> Definition. Fundamental quantities: those which form the primary set of quantities on which all others are based or in terms of which all others can be defined.

In order to not to have to resort to circular definitions, as does a dictionary, the actual meanings of fundamental quantities cannot be expressed in terms of other quantities but must rely on a mutually agreed upon understanding of what these quantities are.

As an example of this process we turn to plane geometry. In plane geometry one often accepts the "point" as a mutually understood fundamental quantity. From this beginning one can then construct other quantities.

> Definition. Derived quantities: those which are defined in terms of fundamental quantities.

In geometry the line defined as a collection of points is a derived quantity, assuming our choice of the point as the fundamental quantity.

The choice of a set of fundamental quantities is not unique. In our geometry we could have chosen the line as our fundamental "undefinable" quantity and then defined a point (now a derived quantity) as the intersection of two lines.

Since fundamental quantities must be agreed upon by all using them and to some extent understood without formal definition, the minimum number which can suffice to provide a complete description of an area of science is chosen. We also try to choose for that set a group which have very clear perceptual meanings.

Currently there are seven fundamental quantities employed in physics. These quantities are those defined in the System International d'Unites (SI units) for mass, length, time, temperature, number of particles in one mole of a substance, current, and luminous intensity. We consider only the first three of these at this time since they suffice for a complete set for mechanics which is the first area of physics which we will study.

1.3 LENGTH

The original metric standard of length is a platinum-iridium bar housed at the International Bureau of Weights and Measures in France. Since this standard for the meter is difficult to use and reproduce, the current definition of the SI meter is that 1 meter is the distance traveled by light in vacuum during 1/299792458 of a second (this definition was adopted on October 30, 1983). The extremely large range of many physical measurements requires the use of scientific notation and a conventional set of prefixes has been developed to be used with it. These prefixes are listed in Table 1.3 and should be committed to memory.

1.4 TIME

The SI unit of time (as well as the unit in most other systems of measurement) is the second. Currently the second is defined as the duration of 9,191,631,770 periods of the microwave radiation emitted by cesium-133.

1.5 MASS

Of the three quantities which we address in this chapter, mass is the most difficult to establish conceptually. Although it is proper to say that the mass of an object reflects the amount of matter which it contains, this is a circular definition since at this point in our development we do not have a definition of matter. A truly satisfactory development of this concept must wait until we have developed some of the tools of mechanics which will allow us to describe inertial mass and gravitational mass. Inertial mass reflects the resistance of an object to acceleration. Gravitational mass is related to the gravitational interaction between objects.

Currently the standard for mass is a platinum-iridium cylinder housed at the International Bureau of Weights and Measures. This standard kilogram is currently the only artificial non-atomic based standard in use. Although it would be preferable to have an easily accessible atomic standard for mass, the current state of technology makes such a standard difficult to achieve.

1.6 DIMENSIONS AND UNITS

Dimensions are a way of describing the qualitative nature of a physical quantity.

>**Definition.** <u>Dimension</u> or <u>dimensions</u> of a quantity impart the relationship between that quantity and the concept of the fundamental quantities from which they are derived.

As an example all quantities which measure length, such as a meter, a yard, a mile, etc. are said to have dimensions of length. Hours, seconds, years, days, on the other hand, have dimensions of time.

Examples of derived quantities' dimensions are those for velocity or speed [length/time], density [mass/length^3], force in pounds or Newtons [mass × length/time^2]. Dimensional statements are usually expressed using L, T, and M to denote length, time, and mass respectively.

Dimensions:
Length = L
Time = T
Mass = M

Thus the dimensions of force can be written as ML/T^2 and those of speed as L/T.

All relationships in physics must be stated between quantities which have the same dimensions. This is reminiscent of the statement made to you when you were learning arithmetic that you cannot add apples and pears. One way of checking all of your work in problems is to make sure that your expressions are dimensionally correct.

> **Definition.** <u>Units</u>: a concrete expression of a quantity's measure based upon a set of standards which define the system of units employed.

Examples of the units employed to measure the quantities we have discussed above are for speed: meters/sec, miles/hr, kilometers/hr; for length: meters, feet, miles, kilometers, light years; for force: pounds, newtons; and for density: kilograms/cubic meter, slugs/cubic foot.

One of the most useful techniques you should learn is the conversion of units from one set to another. This is accomplished by multiplying the original quantity by appropriate sets of conversion factors each of which is dimensionless and has magnitude one. Two examples are provided in the text and others are given below.

EXAMPLE PROBLEMS
Example 1

One of the most important quantities in physics is energy. The kinetic energy (or energy of motion) of an object is defined as $(1/2)mv^2$, where m stands for mass and v for speed.
a) Is energy a fundamental or derived quantity?   b) What are its dimensions?   c) What are its units in the SI system?

Solution

a) Since the definition of energy involves the derived quantity of speed, it is also a derived quantity. b) Using M for mass, L for length, and T for time and denoting the operation of determining the dimensions of a quantity by square brackets [ ], we see:

$$[(1/2)mv^2] = M(L/T)^2 = ML^2/T^2$$

c) In the SI system the units are:

M = kg   L = m   T = sec. (or s)

therefore the units of energy are kg m^2/s^2. Note that the dimensionless quantity 1/2 is ignored in determining the dimensions of energy.

Example 2

Poiseuille's equation for the change in pressure of a viscous fluid flowing through a pipe of radius R and length ℓ at a rate of Q ft^3/s is:

$$P = (8Q\eta\ell)/R^4$$

where η is the viscosity. What are the dimensions of viscosity?

Solution

In the United States we normally measure pressure in units of lbs/in^2. The left hand side of Poiseuille's equation has the same dimensions as Force/$Length^2$. From our discussion in the Summary the dimensions of force are ML/T^2.

$$[P] = (ML/T^2)/L^2 = M/LT^2$$

The right hand side of this equation must also have these dimensions.

$$[Q] = L^3/T \quad [\ell] = L \quad [R] = L$$

Thus the dimensions of can be calculated as:

$$[(8Q\eta\ell)/R^4] = [\eta] \times [Q\ell/R^4]$$
$$= [\eta] \times (L^3/T)(L)/L^4$$
$$M/LT^2 = [\eta] \times (1/T)$$

and therefore the dimensions of η are:

$$[\eta] = M/LT$$

Example 3

International track and field events use SI units. What is the distance of the Olympic marathon which is 40 km in miles?

Solution

40 km = (40 k̶m̶) (1000 m̶/k̶m̶) (100 c̶m̶/m̶) (1 i̶n̶/2.54 c̶m̶) (1 f̶t̶/12 i̶n̶) (1 mi/5280 f̶t̶) = 24.85 mi.

PROBLEMS

1. The force on a certain object obeys the equation F = -kx where x is the position of the object. What are the dimensions of k?

2. What are the dimensions of area and volume?

3. In most of the world land area is measured in hectares (1 hectare = 10^4 m^2). How many acres (1 acre = 43,560 ft^2) are there in a hectare?

4. A litre is 1000 cm^3 and a gallon is 231 in^3. How many litres are there in one quart?

5. The speed of light is now defined to be 299,792,458 m/s. How fast is this in mi/s and mi/hr?

Chapter 2
Vector Algebra

PREVIEW

In this chapter the concepts of <u>scalars</u> and <u>vectors</u> are introduced and the rules for performing mathematical operations such as addition, subtraction, and multiplication on vector quantities are explained. These rules are compared with the more familiar rules for operating on scalar quantities. Two different but equivalent techniques are described for adding and subtracting vectors, and situations in which each is useful are exemplified.

SUMMARY
2.1 SCALARS AND VECTORS

Of the concepts which have been developed to describe natural phenomena, some have intrinsically associated with them a direction while others do not. The former are defined as <u>vector</u> concepts or quantities and require a number (magnitude), a unit, and a direction in order to be completely described. The latter are defined as <u>scalar</u> quantities and require only a number and a unit for a complete description.

<u>Definition</u>. Vector: a quantity which has associated with it both a magnitude and a direction.

<u>Definition</u>. Scalar: a quantity which has associated with it only a magnitude.

Examples of vector quantities are velocity, force, and magnetic field, while examples of scalar quantities are mass, time, and energy. Vector quantities are identified in print by using an arrow above the symbol, or by using a wavy underline, or by **BOLDFACE** type. Thus, the equation,

$$A + B = C$$

means that the <u>vector</u> A is to be added according to the rules explained in the next paragraph, to the <u>vector</u> B to obtain the <u>vector</u> C.

2.2 ADDITION AND SUBTRACTION OF VECTORS

Vector quantities can best be visualized by representing them with <u>arrows</u>, the length of the arrow representing the magnitude of the quantity and the orientation of the arrow representing the direction associated with the quantity. Two like quantities (i.e., both forces or both velocities) may be added together by first drawing the vectors representing the two quantities with the tail end of the second

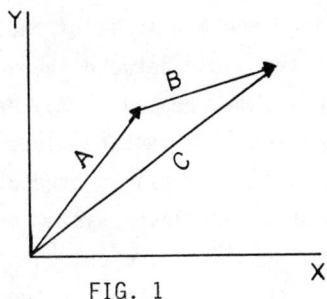

FIG. 1

arrow beginning at the arrowhead end of the first one. The sum or <u>resultant</u> of the two vectors is then represented by the arrow drawn from the tail of the first arrow to the head of the second. This rule may be extended to any number of vectors, each succeeding arrow being laid off from the head of the preceding one; the resultant of the several vectors is then represented by the arrow drawn from the tail of the first arrow to the head of the last. Fig. 1 illustrates this method of adding two vectors.

Subtraction of two or more vectors may be accomplished by simply reversing the direction of the vector(s) to be subtracted and applying the rule outlined on the previous page for addition.
$$A - B = A + (-B)$$
<u>WARNING</u>: It must be remembered that, just as in the case of scalar quantities, two or more vectors representing different concepts <u>cannot</u> be added or subtracted.

Any vector may be multipled or divided by a scalar. The only change resulting from such multiplication or division is a change in the magnitude of the vector, the direction of the vector being unaffected. As in the case of the multiplication of two scalar quantities, the vector and the scalar need not represent the same concept, and the dimensions of the resulting vector will be the same as the product of the dimensions of the two quantities which were multiplied together.

2.3 COMPONENTS

A procedure which is essentially the reverse of that outlined above for the addition of two or more vectors is called the <u>resolution</u> of a vector into its <u>components</u>. The components of a vector are those two or more vectors which, when added to each other, will yield the original vector as a resultant. While it is not a part of the above definition, it is usually convenient to require that these components be mutually perpendicular to each other so that they may be described in terms of a coordinate system with each component lying parallel to one of the axes of the coordinate system. Such a coordinate system with the three axes mutually perpendicular is called an orthogonal coordinate system.

> <u>Definition</u>. <u>Component</u>: the projection of a vector on one of the coordinate axes, there being one component corresponding to each of the three axes of the coordinate system. These components when added together vectorially yield as a resultant the original vector.

The components of a vector can be found by use of the trigonometric functions. Thus, as shown in equations (2.9a) and (2.9b) in the text, the components of a vector corresponding to any one of the axes of a coordinate system may be found by multiplying the magnitude of the vector by the cosine of the angle between the vector and that axis. This same procedure may be used to find the component of a vector along <u>any</u> direction whether that direction corresponds to one of the axes of a coordinate system or not.

Note that any of the components of a vector may be either positive or negative; the direction along a coordinate axis which is

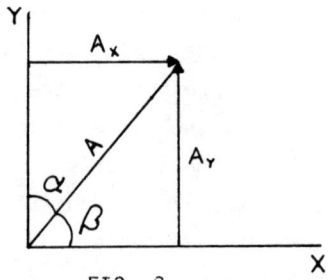

FIG. 2

termed positive is arbitrary but the sign assigned to the component must correspond to this selection. Fig. 2 shows a vector resolved into components along the X (horizontal) and Y (vertical) axes.

The utility of the vector concept in problem solving lies in the fact that, when two or more vectors are to be added or subtracted, since the corresponding components of these vectors all lie in the same direction, these components may be added or subtracted as scalars to yield the corresponding component of the resultant. Thus, if we have the vector equation,

$$A + B - C = D$$

the indicated vector operations may be replaced by the three scalar operations,

$$A_x + B_x - C_x = D_x \qquad A_y + B_y - C_y = D_y \qquad A_z + B_z - C_z = D_z$$

The magnitude of the resultant vector, D, may then be found by applying the Pythagorean Theorem, and its direction by making use of the direction cosines (see Equations 2.12 a, b, & c in the text).

A convenient device for representing vectors in component form is the unit vector. Use of unit vectors allows easy algebraic or numerical manipulation of vector quantities for problem solving.

> Definition. <u>Unit vector</u>: a dimensionless vector of unit length in a specified direction, usually a direction corresponding to one of the axes of the coordinate system being used.

Using the unit vectors, $\hat{\imath}$, $\hat{\jmath}$, and $\hat{k}$ which are used to denote unit vectors along the X, Y, and Z axes respectively, the vector, A, may be represented as

$$A = \hat{\imath}A_x + \hat{\jmath}A_y + \hat{k}A_z .$$

Since, as we have already seen, corresponding components of vectors may be added together as scalars, if the vector, C, is the sum of the vectors A and B, then

$$C_x = A_x + B_x \qquad C_y = A_y + B_y \qquad C_z = A_z + B_z$$

which may be written much more compactly as

$$C = \hat{\imath}(A_x+B_x) + \hat{\jmath}(A_y+B_y) + \hat{k}(A_z+B_z).$$

The use of unit vectors is not restricted to cartesian coordinate systems; indeed, unit vectors may be defined for any direction and two additional unit vectors are frequently used in the textbook. They are the vector, $\hat{r}$ of unit length and in the same direction as the vector to which it is applied, and the vector, $\hat{n}$ also of unit length and directed normally outward from a specified surface.

2.4 THE SCALAR, OR DOT PRODUCT

Two techniques have been found useful for multiplying two vectors together. As the name implies, in the first case, the result of the <u>scalar</u> multiplication of two vector quantities is a scalar quantity, while in the second case, the result of the <u>vector</u> multiplication of two vector quantities is a vector quantity.

> Definition. <u>Scalar product</u>: the product of the magnitudes of two vector quantities multiplied by the cosine of the angle between them.

8   Chapter 2

2.5   THE VECTOR, OR CROSS PRODUCT

Definition. Vector product: the product of the magnitudes of two vector quantities multipled by the sine of the angle between them. The direction of the vector resulting from vector multiplication is defined by the right hand rule illustrated in Fig. 2.31   in the text.

The dot and cross products of the unit vectors, $\hat{i}$, $\hat{j}$, and $\hat{k}$ have particularly simple and significant values which are listed in Table 2.1 and Table 2.2 in the text.

EXAMPLE PROBLEMS
Example 1

A baseball player has just hit a "single" and has run the 90 ft. (about 27 m) to first base. On the next play, he attempts to steal second base which is also 90 ft. away from first base. What is the player's displacement from home plate? This displacement is also the distance the catcher must throw the baseball to get the runner out.

Solution

The two portions of the runners path, from home plate to first base and from first base to second base, may each be represented by a vector 90 units long and at right angles to each other. The resultant displacement is then the vector from home plate to second base. Since the two vectors are perpendicular to each other, we may use the Pythagorean Theorem to calculate the length of the of the resultant. This calculation yields 127.3 ft. (about 39 m). The displacement being a vector quantity, we must also specify the direction (as must the catcher!) in which the ball is thrown. The angle between the resultant and the first base line is calculated as the angle whose tangent is the ratio of the length of the first-to-second base line to the length of the first base line. Since these lengths are equal, the angle is clearly <u>45 degrees</u>.

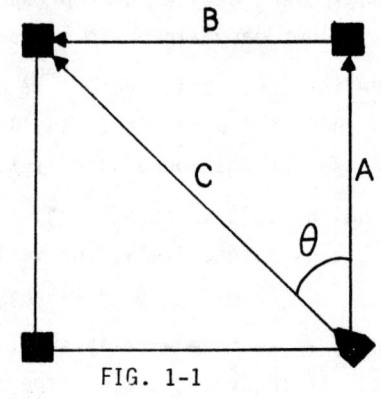

FIG. 1-1

C = A + B
Given: A = 90 ft.
        B = 90 ft.
Required: C = ?
          θ = ?

Example 2

The boom on a large derrick is 20 m long and is raised to an angle of 40 degrees above the ground. How far from the base of the derrick is the point directly beneath its upper end?

Solution

Given: ℓ = 20 m
          θ = 40°
Required: A_x
          A_x = 20 m x cos 40°
          A_x = A cos

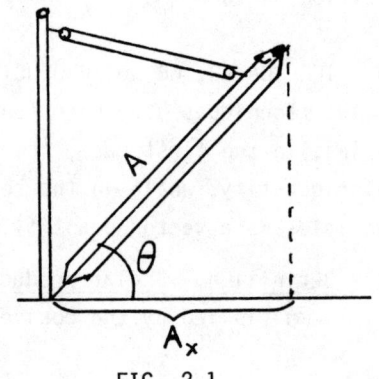

FIG. 2-1

$A_x = 15.3$ m

For this problem, it is convenient to choose a coordinate system with its origin at the base of the derrick and its X axis lying along the ground directly under the boom. The boom itself may then be represented by a vector 20 units long and the required distance may be calculated as the X component of this vector, the product of the length of the vector and the cosine of the angle between it and the X axis. This calculation yields a value of 15.3 m for the X component.

Example 3

The boom of the derrick in Problem 2 is pointed in a direction 20 degrees east of North. Describe the vector representing the boom in terms of the unit vectors, **i**, **j**, and **k** corresponding to a right handed coordinate system with the X axis directed toward the East.

Solution

If we denote the vector representing the boom by **A**, then the three components may be calculated as the product of the length of the vector and the appropriate direction cosine {see Equations 2.12, a, b, & c) in the text}. The right hand coordinate system having its X axis directed toward the East, would have its Y axis directed toward the North and its Z axis directed upward. Thus, the length of A_x is 14.4 m, the length of A_y is 5.2 m, and the length of A_z is 12.9 m, and, the vector representing the boom is

$$\mathbf{A} = \mathbf{\hat{i}}(14.4 \text{ m}) + \mathbf{\hat{j}}(5.2 \text{ m}) + \mathbf{\hat{k}}(12.9 \text{ m}).$$

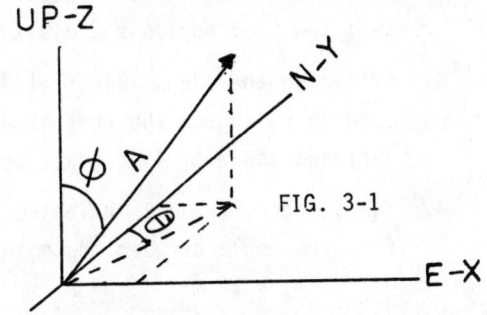

FIG. 3-1

Example 4

A carpenter is asked by a client to build a rectangular packing crate 75 cm wide by 120 cm long by 50 cm high having a volume of 450,000 cm^3. He proceeds to build the box but his carpenter's square has been damaged so that it is ten degrees off. His finished box has a bottom (and top) which is a parallelogram with an angle of 80 degrees between two contiguous edges, and the upright edge is 10 degrees from the vertical. By how much is the volume of the finished box different from the specified design?

Solution

If we define the two vectors, **A** and **B** to coincide with the two contiguous edges of the bottom of the box, and the vector, **C**, to coincide with the upright edge, a review of the definitions of the dot and cross products will show that the area of a parallelogram (the bottom of the box) may be represented by a vector which is the cross product of **A** and **B** and is directed vertically upward. The volume of the parallelepiped, which is the box, is then the dot product of this area vector with the vector **C**. Performing these two consecutive vector multiplications yields a value of

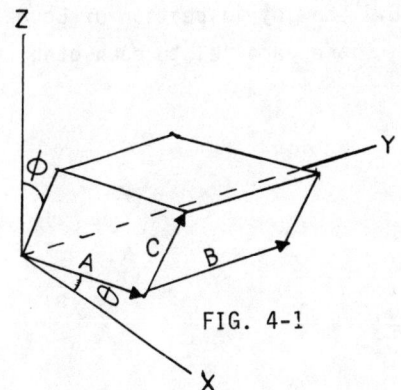

FIG. 4-1

436,430.8 cm^3 which is 13,569.2 cm^3 less than the specifications required.

PROBLEMS

1. The pilot of a small aircraft wishes to fly from Albuquerque 100 km due north to Santa Fe. Her aircraft will fly at an airspeed (speed relative to the air) of 300 km/hr. The weather bureau informs her that during the flight, the wind will be blowing at a speed of 25 km/hr from a direction 20 degrees south of West. In what direction must she point the aircraft in order to fly directly to Santa Fe? At what speed will the aircraft be flying across the ground?

2. A highway sign on the autobahn in Germany warns motorists that the road ahead has a "ten percent grade" which means that the roadway rises 10 m vertically for each 100 m of horizontal distance. Using the concept of vector components, calculate the angle at which the roadway is inclined above the horizontal and the distance measured along the roadway which corresponds to this 100 m of horizontal distance.

3. Utilizing the ideas illustrated in Example Problem 4 above, show how the cross product may be used to calculate the area of a triangle having two of its sides 30 cm and 50 cm long with an included angle of 40 degrees between them.

4. Two vectors, **A** and **B** have the components shown below. Using the idea of the dot product, find the angle between these two vectors.

    A_x = 10 m A_y = 30 m B_x = 20 m B_y = 5 m

5. Suppose that the two vectors described in Problem 4 form two sides of a triangle. Using the cross product, find the area of this triangle, and using vector addition, find the length of the third side.

6. Two vectors, **A** and **B** are described by the following equations:

    A = **î**24 + **ĵ**16 = **k̂**36   and   **B** = **î**10 - **ĵ**25 + **k̂**5

    with all units being in meters. Calculate:

    a) **A** + **B**   b) **A** - **B**   c) **A** * **B**   d) **A** X **B** .

7. Show by inspection of Equation (2.20) in the text that the dot product of two vectors which are perpendicular to each other is zero.

8. Show by inspection of Equation (2.24) in the text that the cross product of two vectors which are parallel to each other is zero.

Chapter 3
Equilibrium of Rigid Bodies

PREVIEW

In this chapter, the concepts of force and torque are introduced. These, together with Newton's First and Third Laws, are used to discuss the requirements for a rigid body to be in equilibrium, known as the First and Second Conditions of Equilibrium. The chapter also discusses the concept of center of gravity and types of equilibrium.

SUMMARY

3.1 FORCE

Force is the interaction of the environment with a body and can be regarded qualitatively as a push or a pull. An operational definition of force is that when the standard kilogram is suspended from a vertical spring, a force of 9.8 N is exerted on the spring. The unit of force is the Newton(N).

Forces can be categorized as noncontact and contact forces, and the only noncontact force considered in this chapter is gravitation. Contact forces are exerted on an object whenever it is touched by other objects, and those encountered in this chapter are tension in ropes and thrust (or normal) and tangential (or frictional) forces exerted by surfaces.

  Definition. External force: a force exerted on a body by its environment.

External forces are distinguished from internal forces, which are forces exerted on one part of a body by another part of the same body. The sum of all external forces on a body is designated by the symbol ΣF.

  Definition. Force diagram: a diagram used in problem solving in which the external
    forces acting on a body are drawn as vectors and the body is represented as either a
    point or an extended object, depending on the method of solution.

In drawing a force diagram, we include only external forces exerted on the body we have picked. The following procedure will insure that you identify these forces correctly. In addition to gravity, there is a contact force (tension, thrust or normal, tangential or frictional) wherever the body is touched by other objects. Tension can only pull (e.g., you cannot push with a rope), a smooth surface exerts a normal (or thrust) force perpendicular to the surface, and a rough surface exerts both a normal force and a frictional (or tangential) force parallel to the surface and opposed to the motion or the <u>impending</u> motion of the body. A force whose direction is unknown should be included in the force diagram with an assumed direction; an incorrect assumption is signified by a minus sign on the result.

3.2 THE FIRST CONDITION OF EQUILIBRIUM

<u>Newton's First Law</u>: every body continues in its state of rest, or of constant velocity, unless it is caused to change that state by forces exerted on it by the environment.

In Newton's first law, there is no distinction between an object being at rest or moving with constant velocity. Either state can be changed only if the sum of all external forces on the object is greater than zero.

<u>Definition</u>. <u>Equilibrium</u>: the state of a body when the forces acting on it combine to maintain the body at rest or in motion with constant velocity.

For a body to be in equilibrium, both the first condition for <u>translational</u> motion and the second condition for <u>rotational</u> motion must be satisfied. Either condition can be satisfied independently of the other, but both are required for equilibrium.

<u>First Condition of Equilibrium</u>: a body will be in translational equilibrium if and only if the vector sum of forces exerted on it by the environment equals zero:
$\Sigma F = 0$ (Eq. 3.1).

The first condition of equilibrium is most easily applied in its component form:
$$\Sigma F_x = 0, \quad \Sigma F_y = 0, \quad \Sigma F_z = 0 \tag{3.2}$$
The choice of coordinate axes is arbitrary, but we usually select them to require the resolution of as few forces as possible. When the first condition of equilibrium is applied, the body can be drawn in the force diagram as a point representing the center of gravity (defined in Sec. 3.5) and the forces can be drawn from this point.

<u>Newton's Third Law</u>: to every action force there is always opposed an equal reaction force.

The action and reaction forces described by Newton's third law must be exerted on two different bodies. Representing the two bodies by A and B, if A exerts a force F on B, then B exerts a force -F on A. The required pattern of <u>A on B</u> and <u>B on A</u> cannot be satisfied by a single body or by introducing a third body, C.

In calculating ΣF for the first condition of equilibrium, the Newton's-third-law action-reaction pair do not cancel out, because they are exerted on different bodies.

<u>Definition</u>. <u>Weight</u>: the gravitation force on a body, acting downward toward the center of the earth from the body's center of gravity (defined in Sec. 3.5).

3.3 TORQUE

<u>Definition</u>. <u>Torque</u>: the measure of the ability of a force to cause rotation, defined by $\tau = r \times F$ (Eq. 3.6), where r is the displacement from the axis of rotation of the point of application of the force F.

Calculation of a torque, therefore, always requires specification of an axis of rotation. A torque is perpendicular to the plane of r and F, and its direction is given by the right-hand rule for a vector product (Section 2.5). We use a sign convention that a positive torque causes a counterclockwise rotation and a negative torque causes a clockwise rotation when the torque is considered to be the only torque acting on a body. The unit of torque is the newton-meter (N-m).

<u>Definition</u>. <u>Moment arm</u>: the perpendicular distance from the line of action of a force to the axis of rotation, defined by $\ell = r\sin\theta$ (Eq. 3.3), where θ is the smaller angle between **F** and **r**.

The moment arm ℓ can be used to give the magnitude of the torque of a force F as $\tau = F\ell$ (Eq. 3.4).

3.4 THE SECOND CONDITION OF EQUILIBRIUM

<u>Definition</u>. <u>Rigid body</u>: a body in which the distances between every pair of points in the body are constant.

<u>Second Condition of Equilibrium</u>: a rigid body will be in rotational equilibrium if and only if the vector sum of external torques on it equals zero when taken about any axis of rotation: $\Sigma\tau = 0$ (Eq. 3.7).

The second condition of equilibrium is most easily applied in component form:
$$\Sigma\tau_x = 0, \quad \Sigma\tau_y = 0, \quad \Sigma\tau_z = 0 \tag{3.8}$$

The choice of an axis of rotation is arbitrary, but we normally select it so that as many forces as possible have no moment arms and, therefore, no torques. Once we choose the axis of rotation, it serves as a common axis for all torques acting on the body. When the second condition of equilibrium is applied, the body must be drawn in the force diagram as an extended body and the forces must be drawn from their points of application so that moment arms and torques can be calculated correctly.

3.5 CENTER OF GRAVITY

<u>Definition</u>. <u>Center of gravity</u>: the point within or near an extended body where the weight of a body may be assumed to act.

For a homogeneous, symmetric object, the center of gravity coincides with the center of symmetry. For non-symmetric objects, the center of gravity can be located experimentally by suspending the object or analytically by using the equations for its coordinates:
$$\bar{x} = \Sigma x_i W_i / W, \quad \bar{y} = \Sigma y_i W_i / W, \quad \bar{z} = \Sigma z_i W_i / W \tag{3.11}$$

<u>Definition</u>. <u>Equilibrant</u>: a force equal and opposite to the weight of a body applied at the center of gravity.

3.6 STABILITY OF EQUILIBRIUM

Types of equilibrium are classified as unstable, stable, or neutral depending on whether a small displacement from equilibrium causes the body to move away from equilibrium, to move toward equilibrium, or to assume a new equilibrium position, respectively.

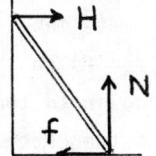

FIG. 1-1

EXAMPLE PROBLEMS

Example 1

A golf club is placed on a rough floor and leaned against a smooth wall. Identify all contact forces on the golf club

and for each give the type and body that exerts it.
Solution

The golf club is being touched by the floor and the wall, therefore, only these two bodies exert contact forces on the club, as shown in Fig. 1-1.

Since the floor is a rough surface, it exerts both a vertical normal force N and a horizontal frictional force f opposed to the impending motion of the club.

The wall is a smooth surface and exerts a normal force H perpendicular to its surface but cannot exert a force parallel to its surface.

The contact forces on the golf club are thus given by:

        N = normal force by floor
        f = frictional force by floor
        H = normal force by wall.

Example 2

A 500-N student is sitting on a 400-N box which is at rest on a classroom floor. What is the Newton's-third-law reaction force to the 900-N upward normal force of the floor on the box? By what body is this reaction force exerted?
Solution

Newton's third law requires that the action-reaction pair be exerted on two bodies, here, the box and the floor. Third bodies, such as the student and the earth, are not involved.

The action force is the 900-N upward force of the floor on the box. The reaction force is by the box on the floor. It has a magnitude of 900-N, and it is directed downwards.

Example 3

Two boys unsuccessfully attempt to move an 800-N crate that is at rest on a rough, horizontal floor. One boy is pushing horizontally on the crate with a force of 100 N, and the other boy is pulling on a horizontal rope in which the tension is 200 N. What are the horizontal and vertical forces exerted on the crate by the floor?
Solution

    Given: F = 100 N
                  T = 200 N
                  W = 800 N
    Required: N
              f

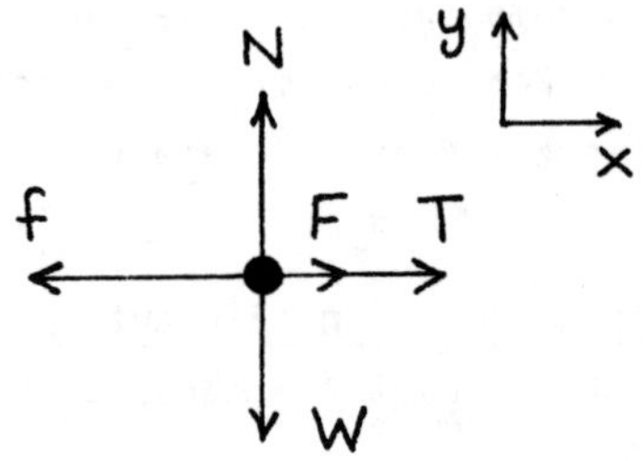

We pick the crate as the body. Since the crate is not moving, it is in translational equilibrium, and the first condition of equilibrium can be applied. The body can be drawn in the force diagram as a point representing the center of gravity, and all forces can be drawn from this point, as shown in Fig. 3-1.

FIG. 3-1

The crate makes contact with the boy, the rope, and the floor, therefore, each of these exerts contact forces on the crate. Since the floor is rough, it exerts both a normal force N upward perpendicular to the floor and a frictional force f parallel to the floor opposing the impending motion of the crate. The tension in the rope is horizontal and can only pull on the crate. The push of the boy F is also horizontal.

All forces on the crate are horizontal and vertical, therefore, if we select the coordinate axes as indicated on the force diagram none of the forces must be resolved.

Application of the first condition of equilibrium in component form (Eq. 3.2) yields:

$$\Sigma F_x = F + T - f = 0 \tag{1}$$
$$\Sigma F_y = N - W = 0 \tag{2}$$

Solving for f in Eq. (1):

$$f = F + T \tag{3}$$

Substituting in Eq. (3) for the horizontal push F and the rope tension T:

$$f = 100 \text{ N} + 200 \text{ N} = \underline{300 \text{ N}}$$

Solving for N in Eq. (2) and substituting for the weight W:

$$N = W = \underline{800 \text{ N}}$$

Example 4

What is the tension in a horizontal rope used to hold a child on a swing at an angle of 30° from the vertical? Assume the child and swing seat have a combined weight of 400 N.

Solution

Given: $\theta = 30°$
$W = 400 \text{ N}$

Required: T

We pick the combination of child and swing as the body. Since the body is being held at rest, it is in translational equilibrium, and the first condition of equilibrium can be applied. All forces on the body can be drawn from the center of gravity in the force diagram, shown in Fig. 4-1.

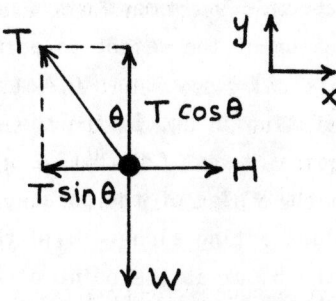

FIG. 4-1

The body makes contact only with the horizontal rope and the rope attached to the swing seat. The contact forces on the body are therefore the tensions H and T in the two ropes, and these tensions can only pull on the body. The only noncontact force on the body is its weight W. We select the coordinate axes horizontally and vertically, as indicated on the force diagram. With this choice of axes, we must resolve only the tension T, and we draw its components from its point of application.

The first condition of equilibrium applied in component form (Eq.3.2) is

$$\Sigma F_x = H - T \sin\theta = 0 \tag{1}$$
$$\Sigma F_y = T \cos\theta - W = 0 \tag{2}$$

Solving for T in Eq. (2) and substituting in Eq. (1) solved for H:

$$H = (W/\cos\theta)(\sin\theta) = W \tan\theta \tag{3}$$

Substituting in Eq. (3) for the weight W and $\tan\theta$, with $\theta = 30°$:

$$H = (400 \text{ N})(0.577) = \underline{231 \text{ N}}$$

One way to check the solution is to consider the special case of the swing hanging vertically with no tension in the horizontal rope. In this case $\theta = 0°$, and the result does indicate that the solution makes sense.

$$H = W \tan 0° = 0$$

Example 5

A block of wood of height 0.6 m and width 0.4 m weighing 200 N is at rest on a horizontal floor. What is the torque of the weight of the block about an axis of rotation which is perpendicular to a face of the block having the given dimensions and which passes through one of the lower corners of the block?

Solution

Given:
- a = 0.4 m
- b = 0.6 m
- W = 200 N
- o = axis of rotation

Required: τ_o

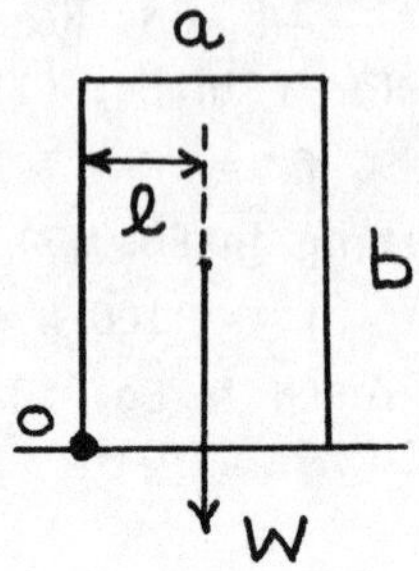

The magnitude of the torque of the weight W is given in terms of its moment arm ℓ by Eq. (3.4):

$$\tau_o = W\ell \qquad (1)$$

FIG. 5-1

The block is drawn in Fig. 5-1 as an extended body so that the moment arm can be shown directly on the diagram. The moment arm ℓ of the weight is the perpendicular distance from the vertical line of action of the weight to point o:

$$\ell = a/2 = 0.4 \text{ m}/2 = 0.2 \text{ m} \qquad (2)$$

Substituting in Eq.(1) for the weight W and the moment arm ℓ:

$$\tau_o = (200 \text{ N})(0.2 \text{ m}) = 40 \text{ N-m}$$

To find the sign of the torque, determine the direction the block will rotate under the action of the weight acting alone. Hold the block at point o to cause rotation about the axis of rotation and pull the block at the point of application of the weight at the center of gravity in the direction of its line of action. The block rotates clockwise, indicating that the sign of the torque is negative. The torque of the weight about an axis through point o is thus given by:

$$\tau_o = \underline{-40 \text{ N-m}}$$

Example 6

A 600-N person is standing upright on a rough, horizontal floor with legs spread 0.5 m apart. What is the minimum horizontal push, applied to the person's shoulder at a height of 1.5 m above the floor, required to tip the person over?

Solution

Given:
- a = 0.5 m
- h = 1.5 m
- W = 600 N
- o = axis of rotation

Required: H

Torques are involved and the second condition of equilibrium is to be applied, therefore the person must be drawn as an extended body in the force diagram, shown in Fig. 6-1. When the person is on the verge of tipping, one foot breaks contact with the floor, therefore, the contact forces on the person are the normal force N and the frictional force f of the rough floor and the horizontal push H. The

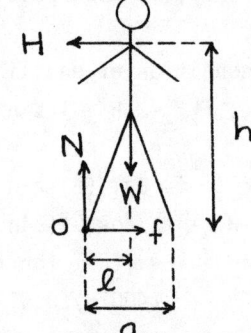

FIG. 6-1

weight W acts downward from the center of gravity.

If we select the axis of rotation through point o, the normal force N and the friction force f have moment arms equal to zero, while the unknown force H does have a moment arm h and will appear in the second condition of equilibrium. The weight W also has a torque about point o, with a moment arm ℓ .

To find the signs of the torques of the forces H and W, we determine the direction of rotation of the body when acted upon by each force individually. Holding the body at point o and pulling successively in the directions of the push H and the weight W, we find that the torque of H causes counterclockwise rotation and is positive, while the torque of W causes clockwise rotation and is negative.

Application of the second condition of equilibrium (Eq. 3.7) to the body gives:
$$\Sigma\tau_o = Hh - W\ell \qquad (1)$$
Solving Eq. (1) for H and substituting for the weight W and the moment arms h and ℓ:
$$H = W(\ell/h) = (600 \text{ N})(0.25 \text{ m}/1.5 \text{ m}) = \underline{100 \text{ N}}$$

Example 7

Weights of 2 N, 4 N, and 6 N are located along the x-axis at 1 m, 3 m, and 5 m, respectively. What is the x-coordinate of the center of gravity of the system consisting of the three weights?

Solution

Given: $W_1 = 2$ N, $x_1 = 1$ m
         $W_2 = 4$ N, $x_2 = 3$ m
         $W_3 = 6$ N, $x_3 = 5$ m

Required: $\bar{x}$

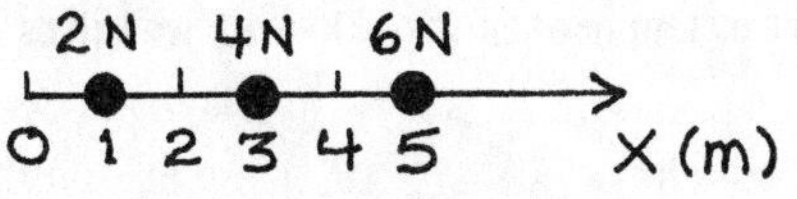

FIG. 7-1

For a system consisting of discrete weights, as opposed to a continuous distribution, the x-coordinate of the center of gravity is given by Eq. 3.11:
$$\bar{x} = \Sigma x_i W_i / W \qquad (1)$$
where W is the total weight of the system and the sum runs over the total number of discrete weights. Here, i runs from 1 to 3, and the total weight W is given by:
$$W = \Sigma W_i = 2 \text{ N} + 4 \text{ N} + 6 \text{ N} = 12 \text{ N} \qquad (2)$$
Substituting in Eq. (1) for W and calculating the summation:
$$\bar{x} = [(1 \text{ m})(2 \text{ N}) + (3 \text{ m})(4 \text{ N}) + (5 \text{ m})(6 \text{ N})]/12 \text{N} = 44 \text{ m}/12 = \underline{3.67 \text{ m}}.$$
The result of 3.67 m does seem reasonable, since the system is not symmetric and the largest weight of 6 N is located at a coordinate greater than x = 3 m.

PROBLEMS

1. A painter weighing 600 N is standing on one of the steps of a 300-N stepladder which is at rest on a horizontal floor. a) What is the Newton's-third-law reaction force to the weight of the painter? b) By what body and on what body is this reaction force exerted?

2. A 1.5-m board is leaning against a wall at a point 0.9 m above the floor. What is the minimum frictional force that will hold an 800-N block on the board without slipping?

18   Chapter 3

3. A 5-m, 80-N ladder leans against a smooth building at a point 4 m above a rough sidewalk. a) Use the first condition of equilibrium to find the normal force of the sidewalk on the ladder. b) Use the second condition of equilibrium to find the horizontal force of the wall on the ladder.

4. A 10-m, 40-N board is supported horizontally by a vertical cable at one end and a chair at the other end, and an 80-N box is placed on the board 2 m from the chair. a) Use the second condition of equilibrium to find the tension in the cable. b) Use the first condition of equilibrium to find the normal force of the chair on the board.

5. A 500-N diver stands at the free end of a 5-m, 80-N diving board that is firmly fixed at the other end and supported from below 2 m from the fixed end. What is the normal force of the support on the diving board?

6. A 2-m, 200-N beam is attached to a wall and supported in a horizontal position by a cable that connects the end of the beam to a point on the wall 1.5 m above the beam. If a 400-N sign hangs from the midpoint of the beam, find the thrust and tangential forces exerted on the beam by the wall.

7. A 2-N meter stick has small weights attached along its length as follows: 4 N at the 20-cm mark, 6 N at the 40-cm mark, and 8 N at the 80-cm mark. Where should a support be placed to balance the meter stick and weights?

Chapter 4
Motion in One Dimension

PREVIEW
   This chapter introduces the concepts and quantities required to describe motion in a quantitative and unambiguous fashion. The concept of a coordinate system, introduced in Chapter 2, is generalized to that of a frame of reference. The vector quantities displacement, velocity, and acceleration are defined, and their relationships examined with special emphasis on the case for constant acceleration. The basic kinematical relationships for the case of constant acceleration are derived. The effect of changing frames of reference on the basic kinematical quantities is also developed.

SUMMARY
4.1 FRAMES OF REFERENCE

   Definition. Event: an occurrence or happening which is specified by a position vector relative to some observer at the time the event transpires.

Whatever the event might actually be, its physical meaning is specified by its position vector and time relative to some observer.

   Definition. Displacement: the vector difference between the position vectors of two events relative to an observer. $\Delta r = r_f - r_i$ (4.1), where i = initial event and f = final event.

   Definition. Time interval: the time interval between two events is defined as:
   $\Delta t = t_f - t_i$ (4.2)

The symbol " " which indicates the difference between two quantities should always be interpreted as the final value minus the initial value of the quantity.

   Definition. Frame of reference: a set of observers who always have fixed positions relative to one another.

Since observers who "belong" to the same frame of reference always remain at fixed positions relative to one another, they can employ a common set of coordinates to describe the positions of events which occur relative to them. The location of the origin of a coordinate system used by observers in a frame of reference is arbitrary. If the frame of reference is defined by a single observer, the origin usually represents the observer. Often the term frame of reference and coordinate system can be used interchangeably.

   Definition. Inertial frame of reference: a frame of reference in which a particle initially at rest, or initially moving with a velocity v, and which is not subject to

external forces, remains at rest or moves with constant velocity v is called an
<u>inertial frame of reference</u>.

4.2 AVERAGE VELOCITY

<u>Definition</u>. <u>Average velocity</u>: the displacement which an object undergoes during some time interval divided by that time interval.

$$\bar{v} = \Delta r / \Delta t = \frac{r_f - r_i}{t_f - t_i} \tag{4.3}$$

The average velocity is a vector whose direction is the same as the <u>change in position</u>. The dimensions of velocity are L/T, the SI units of velocity are m/s.

For rectilinear motion (i.e. motion in only one dimension or direction) we may write

$$\bar{v} = \Delta x / \Delta t = (x_f - x_i)/(t_f - t_i) \tag{4.4}$$

The average velocity is defined over finite (or discrete) time intervals. Since the displacements and time intervals used to calculate average velocity are discrete, the velocities at times for which data does not exist must be estimated by interpolation or extrapolation.

The graph of position (or the component of the position vector in a given direction) versus time can be used to define the average velocity (component of the velocity) as the slope of the secant line formed between the points x_f and x_i (or r_{x_f} and r_{x_i} etc.)

4.3 INSTANTANEOUS VELOCITY

<u>Definition</u>. <u>Instantaneous velocity</u>: the limit of the average velocity as the time interval about the chosen instant is allowed to shrink to zero.

$$\bar{v} = \lim_{\Delta t \to 0} \Delta r / \Delta t = dr/dt \tag{4.6}$$

The symbol dr/dt, termed the derivative of **r** with respect to t, is merely a shorthand notation employed in calculus to indicate the limiting process in the definition.

<u>Geometrically</u> the instantaneous velocity can be interpreted as the slope of the straight line tangent to the position versus time curve at a point.

<u>Special notes</u>
1. Uniform motion (i.e. constant velocity) is represented by a straight line position versus time graph with constant slope ($v = \bar{v}$).
2. The sign of the slope of the tangent line to a curve at a point is the same as the sign of the instantaneous velocity at that time.
3. If the instantaneous position is given as an explicit function of time, the instantaneous velocity can be found by differentiation.
4. The instantaneous speed of an object is equal to the <u>magnitude</u> of the instantaneous velocity.

4.4 ACCELERATION

<u>Definition</u>. <u>Acceleration</u>: the rate of change of velocity with time is termed acceleration. The average acceleration is defined as:

$$\bar{a} = \Delta v / \Delta t = (v_f - v_i)/(t_f - t_i) \tag{4.8}$$

The average acceleration is a vector whose direction is the same as the change in velocity. The dimensions of acceleration are L/T^2, the SI units for acceleration are m/s^2.

Definition. <u>Instantaneous acceleration</u>: the limit as the time interval over which the average acceleration about a point is computed shrinks to zero.

$$\mathbf{a} = \lim_{\Delta t \to 0} \Delta \mathbf{v}/\Delta t = d\mathbf{v}/dt \qquad (4.10)$$

Geometrically the <u>average acceleration</u> is the slope of the secant connecting $\mathbf{v}_f$ and $\mathbf{v}_i$ on a velocity versus time graph. The <u>instantaneous acceleration</u> is the slope of the tangent to the velocity versus time graph at the point of interest.

An important special case is that of constant acceleration. In this case the <u>velocity</u> versus time graph is a straight line with constant slope. The slope of this line is $\mathbf{a}$ and since it is constant:

$$\bar{\mathbf{a}} = \mathbf{a}$$

4.5 THE PROGRAM OF PARTICLE KINEMATICS

Definition. <u>Kinematics</u>: the science of the <u>description</u> of motion.

Its fundamental quantities are displacement and time interval. The important derived quantities are velocity and acceleration. Kinematics deals only with the quantitative description of <u>how</u> objects move, not why.

The primary objective of kinematics is, once the acceleration of an object is known in some <u>initial configuration</u>, to find the future values of the velocity and displacement.

i.e.   given acceleration → future value of velocity → future value of displacement (or position)

4.6 LINEAR MOTION WITH CONSTANT ACCELERATION

Consider motion in only one dimension which shall arbitrarily be designated the x-direction, with constant acceleration $\bar{\mathbf{a}} = \mathbf{a} = a_x = $ constant.

The kinematical relationships which give the future values of velocity and displacement in this circumstance, with the basic assumptions that $v_i = v_0$, $x_i = 0$, and $t_i = 0$ are:

$$v = v_0 + at \qquad (4.13)$$
$$x = 1/2\,(v+v_0)t \qquad (4.12)$$
$$x = \bar{v}t$$

which can be recast in the equivalent representation

$$v^2 = v_0^2 + 2ax \qquad \text{time independent-velocity solution} \qquad (4.17)$$

$$x = v_0 t + 1/2\,at^2 \qquad \text{final velocity independent displacement solution} \qquad (4.18)$$

Solutions for the situation $\mathbf{a} = $ constant can be derived either graphically or by direct integration using the calculus. In <u>all cases</u> it should be noted that each solution requires the specification of a set of initial conditions. A given situation leading to a physical solution is specified by:

1. The value of the constant acceleration.
2. The initial conditions which specify the configuration of the system at t=0.
3. The current value of the position, time, or velocity.

4.7 RELATIVE MOTION - TWO FRAMES OF REFERENCE

If two frames of reference (coordinate systems) are in motion relative to one another the coordinate transformation between the arbitrarily chosen "rest" frame and the frame considered to be in motion is:
$$x = x' + vt \, , \quad y = y' \, , \quad z = z' \tag{4.19}$$

The assumptions on which this transformation is based are:
1. The unprimed frame is considered the rest frame.
2. The primed frame is in motion in the positive x direction.
3. v is the linear velocity of the moving frame relative to the rest frame.
4. Time intervals measured in either frame of reference are identical.

Although these assumptions may appear restrictive, the co-ordinate frames representing the two frames of reference can generally be chosen so that these conditions are met. If the relative velocity of the frames of reference with respect to one another is a constant then velocities of objects measured in the two frames are related by
$$u = u' + v \tag{4.23}$$
and the accelerations measured with respect to the two frames are identical
$$a = a' \tag{4.25}$$

Any frame of reference moving at constant velocity with respect to an inertial frame of reference is also an inertial frame of reference.

Definition. Galilean relativity: frames of reference between which the relations
$$x = x' + vt$$
$$u = u' + v$$
$$a = a'$$
hold are said to obey the principle of Galilean relativity.

EXAMPLE PROBLEMS

Example 1

The following data give the position of a car relative to a stoplight after the light changes from red to green.

x(m)	t(s)
0	0
2	1
7	2
15	3
26	4
37	5

a) Plot the position versus time on a suitably constructed set of axes.
b) Calculate the average velocities for the 1st, 4th, and 5th seconds.
c) Estimate the instantaneous velocity of 2nd and 5th seconds.
d) Estimate the time at which the velocity becomes constant.
e) What is the average acceleration of the car from t = 0 to t = 5 seconds? From t = 0 to t = 2 seconds?

Solution

a) We first construct a set of orthogonal axes and plot the points. Experience should help you to learn to choose convenient unit sizes for the axes, which should also be labeled and indicate the direction of increasing position, time, etc. The next step is to "interpolate" a curve connecting these points assuming the change in position with time is "smooth". Our guess is indicated by the solid line connecting the data points. The curve can be used to estimate positions and velocities at various points other than those for which data is given.

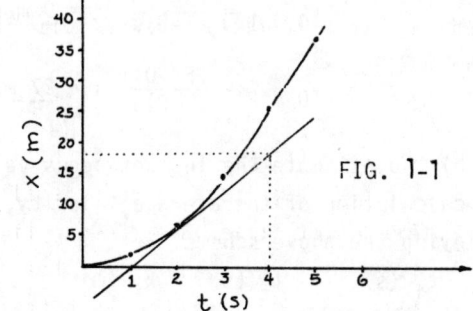

FIG. 1-1

b) The <u>1st second</u> of motion corresponds to the time interval $t_i = 0$ to $t_f = 1$. The average velocity is

$$\bar{v}_1 = \frac{x_f - x_i}{t_f - t_i} = \frac{(2-0)m}{(1-0)s} = \underline{2 \text{ m/s}} \tag{1}$$

Similarly for the 4th and 5th seconds

$$v_4 = (26-15)m/(4-3)s = \underline{11 \text{ m/s}} \; ; \; v_5 = (37-26)m/(5-4)s = \underline{11 \text{ m/s}} \tag{2}$$

c) From our previous calculation, it appears that $v_4 = v_5 =$ constant, thus one expects the instantaneous velocity at $t = 5$ s to have the approximate value of 11 m/s. To estimate the instantaneous velocity at $t = 2$ s we have drawn the <u>tangent line</u> to the curve at $t = 2$ s. It crosses the t axis at $t = 1$ s. $t = 4$ s corresponds to a distance of 18 m, thus the slope of the <u>tangent line</u> is:

$$v_{2 \text{ (estimated)}} (18-0)m/(4-1)s = 18/3 \text{ m/s} = \underline{6 \text{ m/s}} \tag{3}$$

(d) From the calculation in b) and the shape of the graph it would appear that the velocity becomes constant between 3 and 4 seconds.

(e) The initial velocity at $t = 0$ is $v_o = 0$. [We assume the car is stopped when the light is red.] From b, $v_5 = 11$ m/s thus the average acceleration is

$$a = \Delta v/\Delta t = (v_f - v_i)/(t_f - t_i) = (11 \text{ m/s}) - 0)/(5-0)s = \underline{2.2 \text{ m/s}^2} \tag{4}$$

From our estimate of the instantaneous velocity at $t = 2$ s found in part c), we calculate the average acceleration from $t = 0$ to $t = 2$

$$a = \Delta v/\Delta t = (6-0)m/s / (2-0)s = \underline{3 \text{ m/s}^2} \tag{5}$$

Note: This is a case of nonconstant acceleration.

Example 2

An object's displacement is found to obey the following relationship:

$$x = (6 \text{ m/s}^3)t^3 + 3(\text{m/s})t \tag{1}$$

a) What is the objects average velocity from $t = 0$ to $t = 1$ s and from $t = 0$ to $t = 2$ s?
b) Estimate, by numerical calculation, its instantaneous velocity at $t = 2$ s.
c) Calculate its instantaneous velocity exactly at $t = 2$ s.

Solution

a) We use the definition of average velocity: $v = \Delta x/\Delta t$

$$\text{at } t = 0 \; ; \; x = 6(0) + 3(0) = 0 \text{ m} \tag{2}$$
$$\text{at } t = 1 \text{ s} \; ; \; x = 6(1) + 3(1) = 9 \text{ m} \tag{3}$$
$$\text{at } t = 2 \text{ s} \; ; \; x = 6(8) + 3(2) = 54 \text{ m} \tag{4}$$

thus

$$v_{0\;1} = \frac{(9-0)\text{m}}{(1-0)\text{s}} = \underline{9 \text{ m/s}} \tag{5}$$

$$v_{0\;2} = \frac{(54-0)\text{m}}{(2-0)\text{s}} = \underline{27 \text{ m/s}} \tag{6}$$

b) To estimate the instantaneous velocity, and to illustrate how the limit effects the calculation of the average velocity, we compute a series of v's for decreasing time intervals employing the above scheme.

t_i (s)	t_f (s)	x_i (m)	x_f (m)	$x_f - x_i$	$t_f - t_i$	v (m/s)
0	2	0	54	54	2	27 m/s
1	2	9	54	45	1	45 m/s
1.5	2	24.75	54	29.25	0.5	58.5 m/s
1.75	2	37.41	54	16.59	0.25	66.36 m/s
1.8	2	40.39	54	13.61	0.2	68.04 m/s
1.9	2	46.85	54	7.15	0.1	71.46 m/s
1.95	2	50.34	54	3.66	0.05	73.215 m/s
1.99	2	53.25	54	0.75	0.01	75.00 m/s

Our estimate of the instantaneous velocity is therefore
$$v_2 = \underline{75 \text{ m/s}}$$

(To see how rapidly the limiting velocity is approached it might be instructive for you to calculate the successive differences in v.)

(c) By applying calculus we may readily compute the exact instantaneous velocity at t = 2 s.
$$v = dx/dt = (6t^3 + 3t) = 18t^2 + 3 \tag{7}$$
$$v = 18 t^2 + 3$$
$$v_2 = 18(2)^2 + 3 = 18.4 + 3 = \underline{75 \text{ m/s}} \tag{8}$$

Note: $x = 6t^3 + 3t$ is a fairly simple, monotonically increasing function of time. The numerical calculation performed above gives good results fairly rapidly. If $x = x(t)$ is more complicated, the numerical calculation may not appear quite so simple. (See problem below. This should demonstrate the usefulness of calculus as an analytical tool in physics.)

Example 3

A car going down a steep hill loses its brakes. The car's initial velocity is 0.5 m/s. The distance to the bottom of the hill is 0.75 km, and the car's velocity at the bottom of the hill is 77.5 m/s. What was the acceleration of the car down the hill?

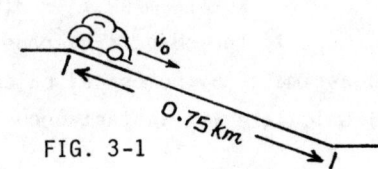

FIG. 3-1

Solution

From the <u>context</u> <u>of</u> <u>this</u> <u>example</u>, we assume that:

a) We are dealing with linear motion - down the hill.

b) The acceleration is constant.

It is certainly possible to conceive of an infinite variety of other conditions or assumptions which might affect this problem, such as wind resistance, bumps on the hill, square tires on the car, etc. From the context of this chapter, we are dealing with rectilinear motion with constant acceleration; these are the intended assumptions.

Given:
$$v_0 = 0.5 \text{ m/s}$$
$$v_f = 77.5 \text{ m/s}$$
$$x = 0.75 \text{ km}$$

Required: a

From the relationships derived in the chapter we see (Eq. 4.17) applies
$$v^2 = v_0^2 + 2ax \quad \text{or} \tag{1}$$

Therefore
$$a = [(77.5 \text{ m/s})^2 - (0.5 \text{ m/s})^2]/2(0.75 \times 10^3 \text{ m}) = \underline{4 \text{ m/s}^2}$$

(Note the square of $(0.5 \text{ m/s})^2$ is negligible.)

Example 4

A ball is thrown vertically downward from a bridge with an initial velocity of 0.3 m/s. The ball strikes the water below the bridge 3.5 s after it is released, what is the height of the bridge?

Solution

Given:
$$v_0 = 0.3 \text{ m/s}$$
$$t_f = t = 3.5 \text{ s}$$

Required: h = x

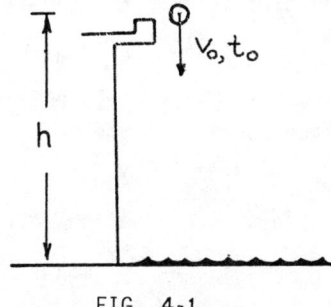

FIG. 4-1

At first glance it appears that this problem does not provide enough information to allow solution. This is typical of many problems in science - one cannot proceed immediately from the known information to the desired result. The first question that you should ask is: Is there anything I <u>know</u> about this physical situation which is not stated in the problem? The answer should be <u>yes</u>. You should remember that all objects close to the surface of the earth fall with the same acceleration, $a = 9.8 \text{ m/s}^2$. Thus we now know the acceleration. If we examine the relationships for constant acceleration, equations 4.12, 4.13, 4.14, 4.17, 4.18, we see that we can find t_f from information provided using $x = v_0 t = 1/2 \, at^2$ (Eq.4.18), the displacement time solution.

Substituting (assuming positive displacement is down)
$$x = (0.3 \text{ m/s})(3.5\text{s}) + 1/2 \,(9.8 \text{ m/s}^2)(3.5 \text{ s})^2 \tag{1}$$
$$x = 1.05 \text{ m} + 60.024 \text{ m}$$
$$x = \underline{61 \text{ m}}$$

Example 5

A train leaves a station at 2:00 pm and arrives at a station 80 km distant at 3:00 pm. The train accelerates with constant acceleration to a maximum velocity, v_{max}, and then constantly decelerates so that its velocity when it reaches the second station is zero. What is the maximum velocity of the train?

Solution

The solution of this problem requires some thought and some algebraic manipulation. Although this might be considered a "trick" problem or, in fancier terms a "pathological" case, it illustrates that the amount of information provided in a given situation determines what you are able to deduce further concerning the problem.

Given: Total distance $= x = x_1 + x_2 = 80$ km $\quad v_0 = 0$
Total time $= t = t_1 + t_2 = 1$ hr $\quad v_{1\ hr} = 0$
Accelerations are constant.

Required: v_{max}

where x_1 = distance traveled during t_1, when the train is accelerating at a_1
where x_2 = distance traveled during t_2, when the train is accelerating at a_2

We seek an expression which involves v_{max}, the distance and the time.

$$x = vt = 1/2(v_0 + v_{max})\, t \qquad (4.12, 4.14)$$

is such an expression. For the two halves of the trip

$$x_1 = 1/2(v_0 + v_{max})\, t_1 \quad ; \quad x_2 = 1/2(v_{max} + v_{1hr})\, t_2 \qquad (1)$$

(i.e., $v_{max} = v_{final}$ for the accelerating portion of the trip and $v_{max} = v_{initial}$ for the decelerating portion of the trip). But $v_0 = 0$, $v_{1\ hr} = 0$. So

$$x_1 = 1/2\, v_{max} t_1 \quad ; \quad x_2 = 1/2\, v_{max} t_2 \qquad (2)$$

Adding these two equations gives

$$x_1 + x_2 = (v_{max}/2)(t_1 + t_2) \qquad (3)$$

$$80 \text{ km} = (v_{max}/2)(t_1 + t_2)$$

$$v_{max} = \underline{160 \text{ km/hr}}$$

The interesting thing about this problem is that regardless of how hard you try, you cannot determine, independently, a_1, a_2, or t_1, t_2, or x_1 and x_2. They are related by

$$v_{max} = a_1 t_1 = a_2 t_2 \qquad (4)$$

so that $a_1/a_2 = t_2/t_1$, and these ratios must be equal. So a_1 and a_2 can be anything, but, curiously, the maximum velocity cannot exceed 160 km/hr. This is guaranteed by the fact that the ratio of time for deceleration to the time for acceleration is fixed by the above equality. Investigate this strange situation further for yourself!

Example 6

James Bond is on top of a train traveling west at 110 km/hr. An agent from SMERSH, on top of an adjacent train traveling east at 75 km/hr, fires a spear gun at 007 after he passes him. The velocity of the spear from the spear gun is 60 m/s, west, relative to the spear

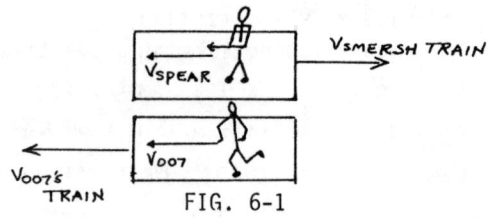

FIG. 6-1

gun. James Bond, having taken physics at Oxford realizes that his best chance for survival is to run away from the approaching spear as fast as he can. The maximum speed that 007 can sprint is 10 m/s. The burning question is: Will agent 007 survive?

Solution

First, we carefully diagram the situation, assuming the motion is essentially one dimensional, as indicated in the problem. [This is, again, a simplification, to help make the problem less complicated since we are interested in the principles involved, rather than fancy vector manipulations]. In common units the velocities are:

Velocity of 007's train relative to ground = 110 km/hr = 30.6 m/s, west
Velocity of SMERSH train relative to ground = 75 km/hr = 20.8 m/s, east
Velocity of 007 relative to his train = 10 m/s, west
Velocity of Spear relative to SMERSH train = 60 m/s, west

Let us adopt 007's position as the origin of the unprimed frame and the agent from SMERSH as the origin of the primed frame. We also adopt west as the positive x direction. Our first question is: What is the velocity of the primed frame relative to the unprimed frame?

Although one can employ formulas to answer this question, often it is best to use "common sense". The thought processes help develop "physical insight" into such situations. From our diagram it should be obvious that 007 is moving west at a velocity of 30.6 m/s + 10 m/s = 40.6 m/s, west (+x), relative to the ground. The SMERSH train is moving at 20.8 m/s, east (-x or -x' direction). Thus 007 sees the agent from SMERSH receding from him at -61.4 m/s (the minus sign indicates east, the -x or -x' direction).

Now we may immediately apply equation 4.23 for the transformation of velocities:

$$u = u' + v \quad \text{with } v = -61.4 \text{ m/s} \tag{4.23}$$

and

$$u' = v_{\text{spear relative to x' frame}} = +60 \text{ m/s}$$

$$u = +60 \text{ m/s} + (-61.4 \text{ m/s}) = \underline{-1.4 \text{ m/s}}$$

Our conclusion is that the spear is actually moving east <u>or</u> away from 007, and he cannot be hit by it as long as he keeps running.

PROBLEMS

1. The position of an object versus time is given in the table to the right.
   a) Plot the position versus time curve for this data.
   b) Construct the best continuous position versus time graph through these points that you can.
   c) Calculate the average velocity for the time intervals 0-2 s, 0-6 s, 4-6 s.
   d) Estimate the instantaneous velocity at 0, 2, 4, and 6 s.
   e) From your estimates in d) calculate

x(m)	t(s)
0	0
9.0	0.5
14.5	1.0
18.5	1.5
21.3	2.0
24.0	2.5
26.2	3.0
28.3	3.5
29.7	4.0
30.1	4.5
32.0	5.0
33	6.0

the average acceleration over the time intervals 0-2 s, 2-4 s, 4-6 s, and 0-6 s.

2. From the velocity versus time graph presented at the right, estimate the values of the acceleration at 0, 3, 5 and 7 s.

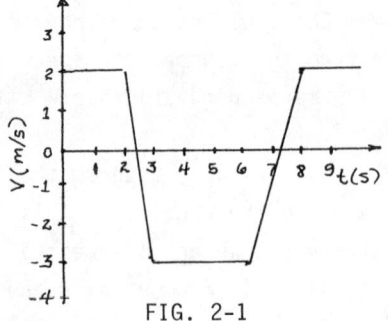
FIG. 2-1

3. The position of a particle obeys the following relationship:
$$x = t^3 - 10 t^2 - 25 t - 10$$
a) Calculate the average velocity of this particle for the following time intervals: 0-2 s, 0.5-2 s, 1.0-2 s, 1.5-2 s, 1.75-2 s. b) From your calculations in a) (plus any further calculations you feel necessary) estimate the instantaneous velocity of the particle at 2 s. c) Calculate the exact value of the instantaneous velocity at 2 s.

4. A car drives around a circular race track 5 km in length with a constant speed of 220 km/hr. a) What is the magnitude of the instantaneous velocity of the car? b) What is the direction of the instantaneous velocity of the car relative to the race track? c) What is the magnitude of the car's average acceleration as it goes one half way around the track? d) What is the magnitude of the car's average acceleration when it completes one complete revolution around the track? e) What is the <u>displacement</u> of the car if it completes one revolution of the track?

5. A baseball pitcher throws a fast ball at 150 km/hr. The ball is accelerated from rest to its final velocity over a distance of 1 m. What is the acceleration of the baseball?

6. A car traveling at 90 km/hr is 75 m from a traffic light when it turns yellow. What acceleration must the car have in order to stop before running the stoplight.

7. A golf ball is dropped from the top of the Eiffel Tower (h 300 m). a) What is its velocity as it strikes the ground? b) How long does it take for the ball to strike the ground?

8. A ball is thrown vertically upward with a velocity of 30 m/s. a) How long a time is required for it to reach a height 40 m above where it was released? b) What is its velocity at that point? c) Explain your two answers to part a).

9. A wide receiver on a certain football team can run the 40-yd. dash in 7 s. The quarterback can throw the football at a velocity of 66 ft/s. When the quarterback throws a pass to the wide receiver, what is the velocity of the football relative to the wide receiver?

10. A duck flying at 6 m/s at a height of 200 m passes directly over a hunter. The bullets from the hunters gun have a velocity of 270 m/s. What determines the direction which the hunter must aim the gun to hit the duck? (You may neglect the effect of gravity on the bullet.)

Chapter 5
Motion in a Plane

PREVIEW

This chapter generalizes the linear kinematical relationships developed in Chapter 4 to two dimensions. The case of motion of an object close to the surface of the earth is discussed as an example of constant acceleration in two dimensions. Motion in a plane circular path is presented as an example of two-dimensional motion in which the acceleration is not constant.

SUMMARY

5.1 MOTION IN A PLANE AND THE PRINCIPLE OF SUPERPOSITION

We begin by expressing the definitions of position, velocity and acceleration in the forms appropriate for two dimensions. We assume that an arbitrary Cartesian coordinate system has been chosen.

<u>Definition</u>. The position vector **r** is written as
$$\mathbf{r} = \hat{\imath}x = \hat{\jmath}y = r_x\hat{\imath} + r_y\hat{\jmath}$$

<u>Definition</u>. The instantaneous velocity
$$\bar{\mathbf{v}} = \lim_{\Delta t \to 0} \frac{\Delta \mathbf{r}}{\Delta t} = \frac{d\mathbf{r}}{dt} \tag{5.4}$$

can be written in component form as

$$v_x = \lim_{\Delta t \to 0} \frac{\Delta x}{\Delta t} = \frac{dx}{dt} = \frac{dr_x}{dt} \quad ; \quad v_y = \lim_{\Delta t \to 0} \frac{\Delta y}{\Delta t} = \frac{dy}{dt} = \frac{dr_y}{dt} \tag{5.4a}$$

<u>Definition</u>. The instantaneous acceleration
$$\mathbf{a} = \lim_{\Delta t \to 0} \frac{\Delta \mathbf{v}}{\Delta t} = \frac{d\mathbf{v}}{dt} \tag{5.5}$$

can be written in component notation as

$$\tag{5.5a}$$

$$a_x = \lim_{\Delta t \to 0} \frac{dv_x}{dt} \quad ; \quad a_y = \lim_{\Delta t \to 0} \frac{\Delta v_y}{\Delta t} = \frac{dv_y}{dt}$$

If a_x = constant and a_y = constant then the motion in the x and y directions can be solved for independently and obey the principle of superposition.

<u>Principle of Superposition</u> - The separate solutions for the x and y directions
of a particle may be linearly (vectorially) combined to form the solution for the two
dimensional motion of the particle.

In the case of motion near the surface of the earth, the acceleration of an object or particle is a constant of magnitude 9.8 m/s^2 and directed downward. The motion occurs in a vertical plane. The vertical direction is usually chosen as the y axis and the horizontal direction as the x axis. The motion in these two directions can be determined independently and the results combined by vector addition of components to describe the motion in two dimensions.

Although the vector components of the motion may be treated independently, the motions in the x and y directions <u>are</u> connected by the fact that the time, t, is a common parameter in the solutions.

5.2 MOTION IN A PLANE WITH CONSTANT ACCELERATION

For motion of an object (commonly termed a projectile) close to the surface of the earth

$$a_x = 0 \quad ; \quad a_y = \text{constant} = -9.8 \text{ m/s}^2$$

where we consider the positive y direction to be vertically upwards.

The solution to the equations of motion are

$$x = v_{xo} t \quad ; \quad v_x = v_{xo} = \text{constant} \qquad (5.7, 5.8)$$

and

$$v_y = v_{yo} + a_y t \; ; \; v_y^2 = v_{yo}^2 + 2a_y y; \; y = v_{yo} t + 1/2 \, a_y t^2 \qquad (5.9, 5.10, 5.11)$$

Note that these equations assume that the initial y position of the object is chosen as the origin of the coordinate system. If this is not the case then equation (5.11) is replaced by

$$y = y_o + v_{yo} t + 1/2 \, a_y t^2 \qquad (5.11')$$

These equations can be recast in various forms depending on the situation and information desired.

<u>Definition</u>. <u>Trajectory</u>: the path traversed by a particle in two dimensions which is described by the equation $y = y(x)$ or $x = x(y)$.

Elimination of the time from equations 5.7 and 5.11 provides the equation for the trajectory of a projectile

$$y = \left(\frac{v_{yo}}{v_{xo}}\right) x + 1/2 \left(\frac{a_y}{v_{xo}^2}\right) x^2 = x \tan \theta + \left(\frac{a_y}{2 v_o^2 \cos^2 \theta}\right) x^2 \qquad (5.12, 5.14)$$

where $v_{xo} = v_o \cos \theta$ and $v_{yo} = v_o \sin \theta$ are the magnitudes of the x and y components of the initial velocity and θ is the angle v_o makes with the x axis.

The trajectory of a particle near the surface of the earth is parabolic in shape. If the elevation of the launch position is the same as that of the impact position (i.e. $y_{initial} = y_{final}$) then the trajectory of the particle exhibits a number of special properties or symmetries:

1. The trajectory of the particle is spatially symmetric about the maximum of the curve. The time to reach the maximum height is

$$t_m = -v_{yo}/a_y \qquad (5.15)$$

The horizontal position of the maximum of the trajectory is

$$x_m = v_{xo} \, v_{yo}/a_y \qquad (5.12)$$

2. The horizontal distance covered is the range, R, and is

$$R = \frac{-2 v_o^2 \sin \theta \cos \theta}{a_y} = \frac{-v_o^2 \sin 2\theta}{a_y} \qquad (5.17)$$

The range is a maximum when $\theta = 45°$

3. The range, R, is symmetric about the launch angle of 45°. If a particle is projected at an acute angle of $45° + \theta$ or $45° - \theta$, the range will be the same.

4. The trajectories are temporally symmetric about the maximum of the trajectory. The time for the projectile to rise to y_{max} is equal to the time for it to fall from y_{max} to the earth.
5. The equations of motion of the particle are time reversal invariant. Replacing t by -t and v_0 by $-v_0$ does not change the equations of motion or their solutions.

5.3 ACCELERATION IN CIRCULAR MOTION

Since the acceleration of an object is defined as
$$\mathbf{a} = \frac{d\mathbf{v}}{dt}$$

any change in **v** will result in a nonzero acceleration. So far we have dealt only with cases where **a** = constant and therefore a_x = constant; a_y = constant. We now discuss an important case where **a** is not constant.

One important generalization can be made regarding the instantaneous velocity even for cases where **a** = constant
$$\mathbf{v} = \frac{d\mathbf{r}}{dt}$$

is always tangent to the path of the particle at each instant.

A very important special case of nonconstant acceleration is when an object or particle is constrained to move in a circular path. For this case it is most convenient to choose a moving reference frame whose origin is fixed at the object and which rotates with it. One axis is chosen tangent to the path and pointing in the direction of the instantaneous, tangential velocity vector **v**. The other axis is chosen to point radially inward toward the center of the circular path.

Since the radius vector to a point on a circle is perpendicular to the tangent to the circle at that point, these radial and tangential axes form a Cartesian coordinate system affixed to the particle.

The acceleration of the object can be considered to be composed of two components.

a_T - the tangential component, responsible for the change in magnitude of **v**.

a_R - the radial component, responsible for the change in direction of **v**.

The total acceleration is
$$a = \sqrt{a_T^2 + a_R^2} \quad ; \quad \tan\theta = \frac{a_T}{a_R} \qquad (5.22, 5.23)$$

θ is the angle which **a** makes with the radius vector.

Furthermore it can be shown that
$$a_R = \frac{v^2}{R} \qquad (5.18)$$

and the direction of a_R must be inward toward the center of the circular path. For this reason a_R is often termed the centripetal (center seeking) acceleration.

> **Definition. Uniform Circular Motion:** the motion of an object which travels in a circular path with constant <u>speed</u>, v.

For uniform circular motion

$$a_R = \frac{v^2}{R}; \quad a_T = 0$$

and since v is constant, a_R is also constant and the only acceleration is centripetal.

The fact that the <u>magnitudes</u> of **v** and **a** are constant does <u>not</u> mean that **v** and **a** are constant. They each change <u>direction</u> continuously.

EXAMPLE PROBLEMS

Example 1

A projectile is fired from a cannon at ground level with an initial velocity of 680 m/s at an angle of 36.9° above the horizontal. a) How high does the projectile rise? b) How far from the cannon does the projectile strike the ground? c) What is the time of flight of the projectile?

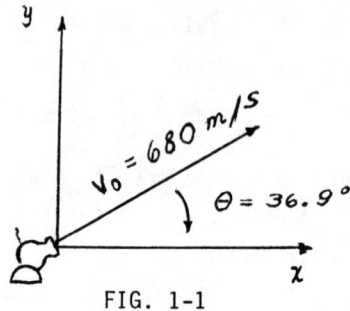

FIG. 1-1

Solution

We assume this is motion close to the surface of the earth and with constant acceleration $a_y = -9.8 \text{ m/s}^2$. We first resolve v_o into its horizontal and vertical components.

$v_{xo} = v_o \cos = v_o \cos 36.9 = (680 \text{ m/s})(0.8) = 544 \text{ m/s}$

$v_{yo} = v_o \sin = v_o \sin 36.9° = (680 \text{ m/s})(0.6) = 408 \text{ m/s}$

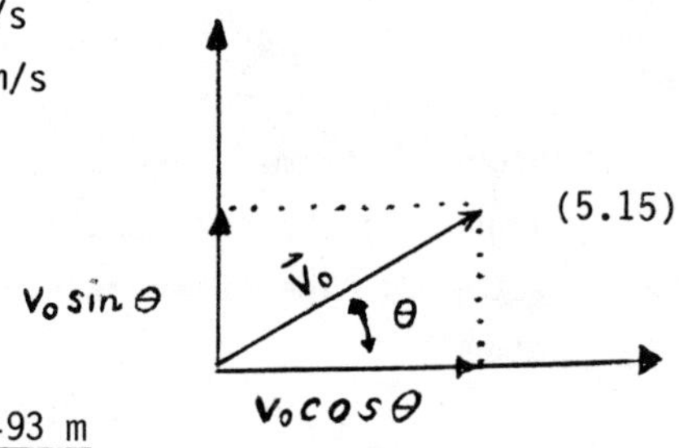

a) The maximum height is attained when

$$t = t_m = -v_{yo}/a_y$$

Substituting this into equation (5.11)

$$y_{max} = v_{yo} t_m + 1/2 \, a_y t_m^2$$

$$y_{max} = \frac{-(408 \text{ m/s})^2}{-9.8 \text{ m/s}^2} + 1/2 \left(-9.8 \text{ m/s}^2 \left(\frac{-408 \text{ m/s}}{-9.8 \text{ m/s}^2}\right)^2\right) = \underline{8493 \text{ m}}$$

(5.15)

FIG. 1-2

b) The range of the projectile is given by (5.17)

$$R = \frac{-v_o^2 \sin 2\theta}{a_y} = \frac{-(680 \text{ m/s})^2 \sin 73.8°}{-9.8 \text{ m/s}^2} = \underline{45,310 \text{ m}}$$

c) The time of flight can most easily be determined by applying the time symmetry about the maximum of the trajectory.

$$t_{flight} = 2 \, t_m = 2 \left(\frac{-v_{yo}}{a_y}\right) = \frac{-2(408 \text{ m/s})}{-9.8 \text{ m/s}^2} = \underline{83.3 \text{ s}}$$

The reader should think a bit about the solutions to this problem. Are the results above consistent with the assumptions used to solve the problem? Some interesting questions to think about are; Is a y_{max} of 17 km "close" to the earth? Can the earth be considered "flat" over a distance of 45 km? If something remains in the air for over a minute, can the rotation of the earth be neglected?

Example 2

Robin Hood and his merry men have decided to lay siege to Prince John's castle. One of their weapons is a catapult which hurls a large boulder at a velocity of 40 m/s at 45°. If the walls of the castle are 30 m high, how far from the base of the wall must Robin position his catapult so that the boulder will fall on top of the wall as it descends to earth?

Solution

There are usually a number of ways to solve any given problem. This is a good example. We could solve the quadratic expression (5.11) for the two instants of time where y = 30 m. The shorter time will

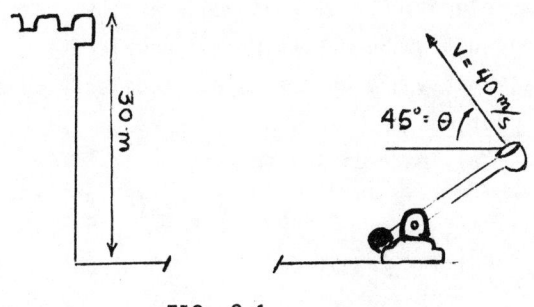

FIG. 2-1

correspond to when the boulder reaches 30 m on the way up, the longer time when it reaches 30 m on the way down. We would then substitute into equation 5.7 for the horizontal motion to determine x, which is the distance the catapult must be placed from the foot of the wall.

However, the time has already been algebraically eliminated from the equations of motion when we solved for the trajectory, thus, x can be determined directly.

$$y = x\tan\theta + \left(\frac{a_y}{2v_0^2 \cos^2\theta}\right)x^2 = x\cdot 1 + \left(\frac{-9.8 \text{ m/s}^2}{2(40 \text{ m/s})^2 \cos^2 45°}\right)x^2$$

$$30 \text{ m} = x - \frac{9.8 \, x^2}{1600}$$

using the quadratic formula we find

$$x = \underline{39.61 \text{ m}} \quad \text{or} \quad \underline{123.7 \text{ m}}$$

Thus the catapult should be placed $\underline{123.7 \text{ m}}$ from the castle wall.

Example 3

A boy swings a stone in a sling shot in a horizontal circular path of radius 0.3 m. If the stone completes one revolution each 1/4 second, a) What is the speed of the stone? b) What is the magnitude and direction of the acceleration of the stone?

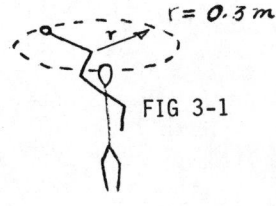

FIG 3-1

Solution

a) From (5.19)

$$v = \frac{2\pi r}{T} = \frac{2(3.1416)(0.3 \text{ m})}{0.25 \text{ s}} = \underline{7.54 \text{ m/s}}$$

b) Using (5.20) or (5.18)

$$a_R = \frac{4\pi^2 r}{T^2} = \frac{v^2}{R} = \frac{(7.54 \text{ m/s})^2}{0.3 \text{ m}} = \underline{189.5 \text{ m/s}^2}$$

The direction of a_R is inward toward the center.

Example 4

At an instant of time an object is traveling in a circular path of radius 5 m at a speed of 2 m/s. It has a tangential acceleration of 2 m/s^2.
a) What is its radial acceleration? b) What is its

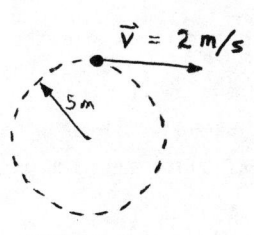

FIG. 4-1

acceleration? c) What would be its speed 2 s later?
d) What would be its radial acceleration 2 s later?
e) What would be its total acceleration 2 s later?

Solution

a) The radial acceleration is
$$a = \frac{v^2}{R} = \frac{(2 \text{ m/s})^2}{5 \text{ m}} = \frac{4}{5} \frac{\text{m}}{\text{s}^2} = 0.8 \frac{\text{m}}{\text{s}^2}, \underline{\text{inward}}$$

b) The magnitude of the total acceleration is
$$a = \sqrt{a_T^2 + a_R^2} = \sqrt{(2 \text{ m/s})^2 + (0.8 \text{ m/s}^2)^2} = \underline{2.15 \text{ m/s}^2}$$

and
$$\tan \theta = \frac{a_T}{a_R} = \frac{2 \text{ m/s}^2}{0.8 \text{ m/s}^2} = 2.5$$

$\theta = 68°$ as shown in Figure 5-4.

c) After 2 s the object's speed will be
$$v = v_i + a_T t = 2 \text{ m/s} + 2 \text{ m/s}^2 (2 \text{ s}) = 6 \text{ m/s}$$

d) The radial acceleration will be
$$a_R = \frac{v^2}{R} = \frac{(6 \text{ m/s})^2}{5 \text{ m}} = \underline{7.2 \text{ m/s}^2, \text{ inward}}$$

e) The total acceleration is
$$a = \sqrt{a_T^2 + a_R^2} = \sqrt{(2 \text{ m/s}^2)^2 + (7.2 \text{ m/s}^2)^2} = 7.47 \text{ m/s}^2$$

$$\tan \theta = \frac{a_T}{a_R} = \frac{2 \text{ m/s}^2}{7.2 \text{ m/s}^2} = 0.278$$

$\theta = \underline{15.5°}$

PROBLEMS

1. A particle's position is given by $x = 6t^2 + 2t + 6$; $y = 2t + 3$. a) What is the particle's position at $t = 0$, $t = 2$ and $t = 4$ s? b) What is the particle's velocity and acceleration at each of these times?

2. A hunter aims his gun, which has a muzzle velocity of 800 m/s, directly at a target which is at the same height as the gun and 500 m distant. By how much does he miss the bull's eye?

3. A quarterback can throw a football a maximum range of 100 m horizontally. Assuming he can throw the ball vertically upward with the same speed, how high would the ball go?

4. A baseball player hits a "line drive" home run over the centerfield fence which is 165 m from home plate and 3 m high. If the ball just clears the fence, what was the initial speed?

5. A shot putter launches his shot at a height of 2 m above the ground with an initial speed of v_o. a) At what angle relative to the horizontal should he launch his shot in order to achieve the maximum range? b) Does the launch angle depend on the initial speed? c) Does it depend on the initial height? (Note: This is an extension of example 4 from the text and requires careful thought.)

6. The moon orbits the earth at a mean distance of 248,000 miles in a period of 28 days. Assuming the moon's orbit is circular, calculate the radial acceleration of the moon.

7. A race driver rounding a curve of radius 100 m at a speed of 20 m/s. He steps on his throttle giving his car a tangential acceleration of 2 m/s^2. What is the total acceleration of the race car at this instant?

8. Show that for a particle which begins circular motion with $v_T = 0$ at $t = 0$ and constant tangential acceleration a_T, the radial acceleration is proportional to the square of the tangential acceleration.

Chapter 6
Newton's Laws

PREVIEW

The previous chapters have discussed how to describe motion; this chapter will begin the study of the interrelationship between forces, which effect motion, and the motion itself. Newton's three laws of motion are introduced and applied to a number of simple cases dealing with translational motion. The Universal Law of Gravitation is introduced along with a number of non-fundamental mechanical forces. The concepts of inertial mass and inertial reference frames and their relationship to Newton's laws are also treated.

SUMMARY

6.1 THE PROGRAM OF PARTICLE MECHANICS

The basic problem of mechanics is to study how and why objects move. More precisely, we wish to learn how to predict what will happen to a mechanical system if we know its properties at some time and the properties of its environment and how the environment and system interact with each other. In this chapter we will deal with situations which are simple enough that we need only be concerned with translational motion (we consider rotational motion later).

It is common practice to begin the study of physics with Newtonian mechanics because it deals successfully with the behavior of systems with which we have everyday experience. This provides the benefit of being able to use one's intuition and common sense to relate the quantitative statements of mechanics to familiar objects and processes. As new concepts and quantities are introduced you should take time to think about them so that you may understand how they describe things with which you are familiar.

> Definition. Net force: the vector sum of all external forces
>   (i.e., those exerted by the environment) which act on a system.

The forces which act on a system depend on the nature of the system and its environment and how they interact. These interactions are usually described in terms of force laws.

> Definition. Force law: a quantitative relationship deduced by observation
>   which provides the theoretical basis that accounts for the forces which systems
>   exert on one another due to their intrinsic properties.

The knowledge of force laws and the forces which act on a system coupled with Newton's laws of motion which relate these forces to kinematical quantities allow us to solve the basic problem of mechanics introduced above.

6.2 NEWTON'S FIRST LAW

> Newton's first law: an unaccelerated body remains unaccelerated unless it

is caused to change that state by forces exerted on it by its environment.

Often Newton's first law is stated as: objects at rest tend to remain at rest and objects in uniform motion (i.e. moving with constant velocity) tend to remain in uniform motion unless acted upon by a net external force. Although these two definitions are equivalent, the one given in the text is to be preferred because it allows us to relate this "force free" state to the concept of inertial reference frames introduced in Chapter 4.

Observers can determine whether they are in an inertial reference frame or not because objects which are initially unaccelerated in such a frame remain so indefinitely, unless there is an identifiable external force which acts on them. Conversely observers in accelerated reference frames are forced to invent fictitious forces to account for the observed motion of objects in their reference frames. An example is the tossing of a ball between two people on a merry-go-round. To the individuals on the merry-go-round the ball seems to curve away from the person to which it was thrown. To an observer on the ground, who is essentially in an inertial frame of reference, the motion of the ball is quite regular. The inertial observer sees the person to whom the ball was thrown move away from the ball's path.

Newton's first law gives an operational definition of the term zero net force.

6.3 FORCE

The intuitive definition of force as a push or pull has already been discussed in Chapter 3. We now give an operational definition of force in terms of acceleration.

> <u>Definition</u>. <u>Force in SI Units</u>: a force of 1 Newton is defined to be that push or pull exerted by a spring on a standard kilogram which gives this mass an acceleration of 1 m/s^2.

Although this definition may seem arbitrary and cumbersome, it satisfies the requirements of providing an operational method for measuring force and also defines the dimensions and units of force.

6.4 NEWTON'S SECOND LAW

Based on the definition of force given on the previous page, Newton's second law can be considered as a statement of the relationship between force, mass and acceleration of an object.

> <u>Newton's second law</u>: the acceleration of a particle is equal to the ratio of the <u>net</u> force acting on the particle to the inertial mass of the particle.

In equation form this may be written as:
$$a = F/m \quad \text{or} \quad F = ma \tag{6.1}$$

This is probably one of the two most famous equations of physics. It was not until the work of Galileo and Newton that scientists were able to abstract from experiment and observation this most important relationship. It was Newton who realized that, in the absence of friction, it was the acceleratioon of an object which was effected by the application of a net force.

It is very important to note that Newton's 2nd law:
1. is a vector relationship.
2. depends on the net force (vector sum of forces)

acting on a particle or object.
3. states that the direction of the acceleration and
the direction of the net force are the same.

Newton's 2nd law demonstrates that the larger the mass of an object the more difficult it is to accelerate. This is in agreement with Newton's 1st law which deals with the resistance or inertia of an object to change its state of motion.

> Definition. Inertial mass: the ratio of the net force exerted on an object to its acceleration.

The reason we specifically denote the mass in Newton's 2nd law as the inertial mass is that we will shortly deal with another type of mass which arises independently.

It is important to realize that Newton's 2nd law is a dynamical relationship not an identity. The forces which are involved are derived independently from force laws. Once the forces acting are identified, this equality relates those forces to the translational kinematical behavior of the system on which they act. Sometimes students make the mistake of thinking that mass x acceleration and force are the same thing. Although they are set equal to each other in Newton's 2nd law it is very important not to lose sight of the fact that they are qualitatively different.

6.5 NEWTON'S THIRD LAW

The final general observation made by Newton concerning forces has a different perspective and deals with different phenomena than the first two laws.

Newton's first two laws deal exclusively with the net force which the environment exerts on a given system and the effects those forces have on the system. In fact, Newton's 1st law is but a special case of the 2nd law. The 2nd law tells us that if the net force on an object vanishes then so will its acceleration, i.e. it will move with constant velocity. This is obviously just the 1st law.

> Newton's third law: if body A exerts a force on body B then body B exerts an equal and oppositely directed force on body A.
>
> $$F_{AB} = -F_{BA}$$

Newton's third law differs from the first two in that it is a general statement about forces and force laws rather than a dynamical statement about the effects of forces on a single system. It is very important to note that Newton's third law refers to <u>two</u> different forces acting on <u>two</u> different bodies. These two "mirror" forces are often called a Newton's third law couple or an action-reaction pair and can be identified by the symmetry which exists between them. For example, if the earth exerts a force on you (your weight), then you must exert a force on the earth. If you push on a wall, then the wall pushes on you.

Students often have a great deal of trouble with Newton's 3rd law because (as is the case for Newton's 2^{nd}d law) they make the mistake of thinking that things that are equal are identical. Certainly four quarters equal one dollar, but four quarters and a dollar bill are not exactly the same thing. Consider Fig. 1,

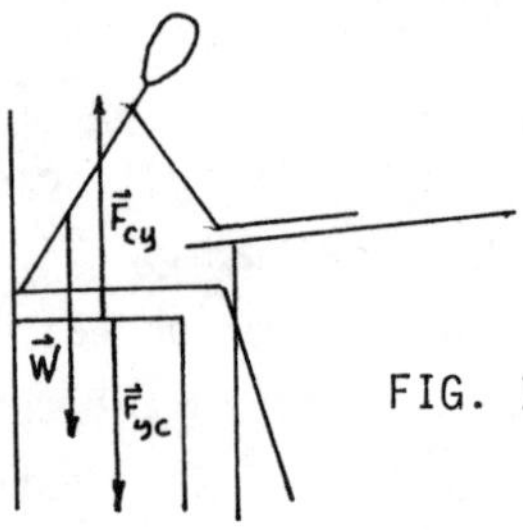

FIG. 1

which represents you sitting at your desk reading physics. In terms of Newton's 1st or 2nd laws the net force on you is zero. There are two forces acting on you, your weight (the force of the earth on you) and the contact force of the chair pushing up on you (F_{cy}). By the <u>first</u> or <u>second</u> law these forces must be of equal magnitude and oppositely directed, since the net force acting on you must vanish.

The question is, do these equal and opposite forces comprise an action-reaction pair? NO! The symmetry of Newton's statement of the third law demands that the reaction to your weight be the force which you exert on the earth (i.e. your gravitational attraction for the earth, discussed below). Likewise the reaction force to the push of the chair on you must be the force with which you push on the chair, F_{yc}. Although all these forces will have the same magnitude as your weight, they are distinct, different forces. When identifying action-reaction pairs always rely on the symmetry discussed.

6.6 THE UNIVERSAL FORCE OF GRAVITATION

In order to apply Newton's dynamical laws we must have some idea of the origin and nature of the forces which act on objects. The most familiar force, which is also a fundamental force of nature, is that of gravitation. The work of the 15th and 16th century astronomers Copernicus, Brahe, and Kepler along with the work of Galileo provided the basis which allowed Newton to formulate what is now called Newton's Universal Law of Gravitation.

> <u>Newton's Universal Law of Gravitation</u>: two objects which possess mass exert an attractive force on each other. The magnitude of the force is proportional to the product of the two masses and is inversely proportional to the square of the distance between them. The forces act along the line which connect the centers of the two masses.

Although stated in this form the law of gravitation is restricted in essence to two point masses (so that the definition of the "center" of each mass can be unambiguously defined), it can easily be generalized since the forces which act on an object obey the principle of superposition.

> <u>Definition</u>. <u>Principle of superposition of forces</u>: the gravitational force acting on an extended body can be considered to be equal to the sum of the gravitational forces acting on each mass point which comprise the body.

We have not yet defined the term mass point but the implication should be obvious. For the time being we can replace mass point by atom if we wish, since we will see that the magnitude of the gravitational force is small enough that atoms can be considered as point masses for all intents and purposes.

The proportionality constant which makes Newton's law an equality was first measured by Cavendish and is now called the universal gravitational constant and denoted by G. This is one of the fundamental constants of nature and $G = 6.67 \times 10^{-11}$ N·m^2/kg^2

In terms of G, Newton's Universal Law of Gravitation is:
$$F = Gm_1m_2/r^2$$
or in vector notation
$$F = Gm_1m_2 r_{12}/r^2$$

where r_{12} denotes a unit vector pointing from body 1 to 2 (or from 2 to 1) along the line which connects their centers.

There is, at present, no physical reason why the masses referred to in the law of gravitation, termed the gravitational mass, and the inertial mass in Newton's 2nd law should be the same. However, experimentally they cannot be distinguished from one another and it is commonly accepted that all forms of mass are identical regardless of the operational situation in which they arise.

6.7 WEIGHT AND MASS

<u>Definition</u>. <u>Weight</u>: the gravitational force exerted on an object by the earth.

If we denote the mass of the earth by M_E and the mass of the object by m, the magnitude of the weight is:
$$W = GmM_E/r^2 \qquad r > R_E$$

If the force of gravity is the <u>only</u> force acting on the object then we may substitute this force law into Newton's 2nd law giving:
$$GmM_E/r^2 = ma$$
$$GM_E/r^2 = a$$

If the motion of the object takes place close to the surface of the earth, then it can be shown that GM_E/r^2 is almost a constant as is the weight of the object (see example 5 on page 111 of the text). Since the acceleration of an object close to the surface of the earth is almost constant it is usually designated by the special letter g.
$$g = GM_E/R_E^2 = a$$

It is an interesting quirk of nature that, since inertial mass and gravitational mass are identical, the acceleration of an object acted upon only by gravitational forces does not depend on the mass of the object, but only on the mass(es) of the object(s) which are exerting the gravitational forces on it. Thus weight is a property of the environment of an object while mass is an intrinsic property of the object itself. This explains Galileo's famous observation that the acceleration of objects close to the surface of the earth does not depend on their mass (or weight).

6.8 MECHANICAL FORCE LAWS

There are only four fundamental forces of nature: gravity, electromagnetic, strong nuclear and weak nuclear, which are not derivable from other forces but are related expressly to the properties of the objects which interact by means of these forces. In studying the mechanics of everyday objects and systems, only two of these forces are of any real consequence, gravity and the electromagnetic force. Usually the electromagnetic force is manifested, at least in mechanical systems, in forms which are more easily described in terms of empirically derived mechanical force laws. We will briefly describe four such mechanical laws.

<u>Hooke's Law</u> (linear spring force): a linear spring exerts a force which is proportional to the displacement of the spring from its equilibrium position and in a direction which always tends to return the system to equilibrium:
$$F = -kx \tag{6.12}$$

This type of force is generally called a linear restoring force and occurs in cases other than just the linear spring. When written in this form the displacement x is measured from the equilibrium position of the system (i.e. $x = 0$ is the position at which the spring exerts no restoring force.

> Force of Sliding Friction: objects whose surfaces are in contact and are pressed together by a normal contact force N which are sliding past one another experience a frictional force parallel to the surfaces in the direction which opposes the sliding motion of each surface past the other. The magnitude of the force is:

$$f = \mu_k N \qquad (6.13)$$

The coefficient μ_k is called the coefficient of kinetic friction and is a property of the interacting surfaces.

> Force of Static Friction: objects whose surfaces are in contact and are pressed together by a normal contact force N which are stationary with respect to one another <u>may</u> experience a frictional force parallel to the surfaces in the direction which opposes the <u>impending</u> motion of each surface past the other. The magnitude of this force may be:

$$0 \leq f \leq \mu_s N$$

The coefficient μ_s is called the coefficient of static friction and is usually greater than the coefficient of kinetic friction. It is very important to note that the force of static friction can have any value from 0 up to a maximum of $\mu_s N$. The force of static friction is a passive force which arises as a response to some other force which is attempting to cause the surfaces to begin to slide past one another. Students often have trouble with this concept and attempt to always use $\mu_s N$ as the static force of friction. You should try to remember that this is the maximum value that this force can have. It can have any value less than this down to 0. We will work a number of examples to illustrate this force in the Example Problems section.

> Stokes' law: a spherical object moving through a viscous fluid experiences a frictional force which retards its motion that is proportional to its radius and velocity and to the frictional properties of the fluid. The magnitude of this force is:

$$F_D = 6\pi\eta rv \qquad (6.14)$$

The coefficient of fluid friction η is called the viscosity and is an intrinsic property of the fluid.

We will employ these force laws in the examples which we shall work in the Example Problems section.

6.9 APPLICATIONS OF NEWTON'S LAWS

We will not summarize this section since it is essentially material which is analogous to that in the next section. However, it is very important that you study the examples given in this section and how they are worked, especially as they relate to the Problem Solving Guide as extended on page 115 of the text.

As you begin to learn analysis of physical problems it is helpful to have a system which you can apply to analyze the problems in a step-by-step fashion. This is the purpose of the Problem Solving Guide. It may help you to make a copy of the guide on a 3 x 5 inch index card so that you may refer to it as you work problems. Although you may find some problems transparent and feel that you do not need to follow the steps in the guide, it is best to go ahead even in these cases and follow the procedure in the guide so that it becomes familiar.

EXAMPLE PROBLEMS
Example 1

A projectile of mass 0.2 kg is to be fired with an acceleration of 100 m/s^2 at an angle of 45° with respect to the horizontal. What force must the firing mechanism exert on the projectile?

Solution

Given: $m = 0.2$ kg
$g = 9.8$ m/s^2
$a = 100$ m/s^2, 45° above horizontal

Required: F_{WM}

The first step is to try to guess the answer to the problem. From Newton's 2nd law we may estimate the magnitude of the force by neglecting the effects of gravity since a is much larger than g.

$$F_M = ma = F_{net}$$
$$= (0.2 \text{ kg})(100 \text{ m/s}^2)$$
$$= 20 \text{ N} \quad (1)$$

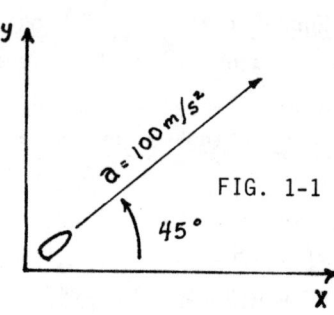

FIG. 1-1

and it must be applied at some angle greater than 45° to compensate for the weight acting downward on the projectile since F_{net} must be parallel to **a**.

The body to be chosen is obviously the projectile and the forces acting on it are shown in Fig. 1-2. Since we have no idea what angle F_M makes with the horizontal, we choose to solve for its components as the unknowns (this avoids having to solve algebraic equations with trigonometric functions in them) F_{M_y} and F_{M_x}.

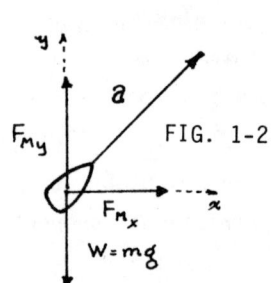

FIG. 1-2

From equation 1 the components of F_{net} are:

$$F_{net_x} = (20 \text{ N})(\cos 45°)$$
$$= 14.14 \text{ N}$$

$$F_{net_y} = (20 \text{ N})(\sin 45°)$$
$$= 14.14 \text{ N}$$

We now apply Newton's 2nd law in the x and y directions:

$$F_y = F_{net_y} = 14.14 \text{ N} = F_{M_y} - W = F_{M_y} - (0.2)(9.8)$$
$$14.14 \text{ N} = F_{M_y} - 1.96 \text{ N}$$
$$F_{M_y} = 16.18 \text{ N}$$

$$F_x = F_{net_x} = 14.14 \text{ N} = F_{M_x}$$

The force which the mechanism must exert is given by:

$$F_M = (F_{M_x})^2 + (F_{M_y})^2$$
$$= (14.14)^2 + (16.18)^2 = 21.42 \text{ N}$$
$$\tan^{-1}\theta = F_{M_y}/F_{M_x} = 16.10/14.14 = 1.14$$
$$\theta = 48.7°$$

These answers agree with our original estimates of F_M.

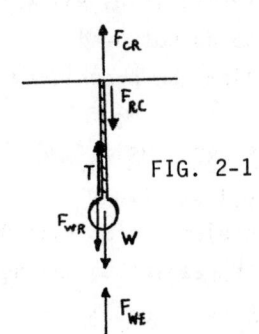

FIG. 2-1

Example 2

A weight W is suspended from the ceiling by a light, inextensible rope. Identify as many action-reaction pairs of forces as you can for this system.

Solution

Given: W

Required: Action-reaction force pairs by Newton's 3rd law.

The first action-reaction pair should be obvious. The weight of the suspended object is the force of the <u>earth</u> on the <u>object</u>. By Newton's 3rd law the <u>object</u> must exert an equal and opposite force on the <u>earth</u>. This force has magnitude W and acts at the center of the earth outward.

Since the weight is not accelerating we know by Newton's 1st or 2nd law that the net force acting on the weight must vanish. Thus the tension in the rope must be equal to W and in the opposite direction.

$$F = 0 = T - W \quad \text{(up is taken as +y)}$$

Note that T, the force of the <u>rope</u> on the <u>weight</u> is <u>not</u> the reaction force to W, the force of the <u>earth</u> on the <u>weight</u>. The equality of magnitude and oppositeness of their directions is a result of Newton's 1st or 2nd law, not the 3rd.

Since T is the force of the <u>rope</u> on the <u>weight</u>, there must be an equal and opposite force exerted by the <u>weight</u> on the <u>rope</u>.

$$F_{WR} = -T = W$$

Since the system composed of the ball and cord are not moving, Newton's 2nd law tells us that the ceiling must exert an upward force on the rope equal and opposite to the weight of the rope + weight of object. Since the rope is "light", we neglect its weight and:

$$F_{CR} = -W$$

By Newton's 3rd law the rope must exert an equal and opposite force on the ceiling.

$$F_{RC} = -F_{CR} = W$$

These action-reaction pairs are shown in Fig. 2-1.

Example 3

Two blocks, one of mass 10 kg (m_1) the other of mass 5 kg (m_2), are accelerated as a unit across a smooth floor with an acceleration of 3 m/s^2 by a horizontal force applied to the 10 kg mass toward the right. a) What is the force acting on this 15 kg system? b) What is the force which the 10 kg block exerts on the 5 kg block? c) What is the force which the 5 kg block exerts on the

10 kg block? d) What is the net force on the 10 kg block?

Solution

Given: $m_1 = 10$ kg
$m_2 = 5$ kg
$a = 3$ m/s^2 to the right

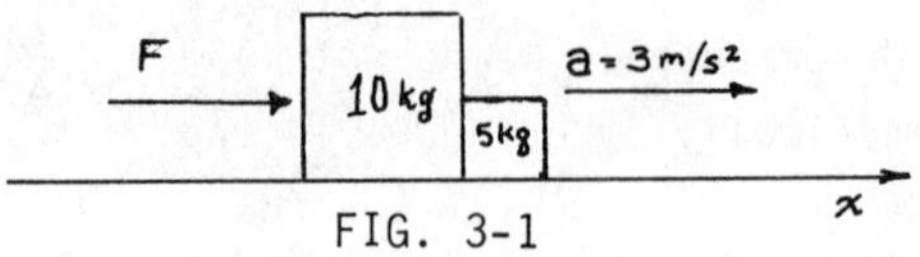

Required: F, F_{m_1,m_2}, F_{m_2,m_1}, F_{m_1}

FIG. 3-1

a) We choose the +x axis to the right as shown in Fig. 3-1. In this problem we will denote vector directions by using the unit vector $\hat{i}$. As this is a one dimensional problem we proceed immediately to apply Newton's 2nd law to the system of both bodies.

$$F = ma = (15 \text{ kg})(3\hat{i} \text{ m/s}^2) = 45\hat{i} \text{ N} \tag{1}$$

b) Since the acceleration of the 5 kg block is $3\hat{i}$ m/s^2 the net force acting on it must be:

$$F_{m_2} = ma = (5 \text{ kg})(3\hat{i} \text{ m/s}^2) = 15\hat{i} \text{ N}$$

c) By Newton's 3rd law $F_{m_2,m_1} = -F_{m_1,m_2}$ so $F_{m_2,m_1} = -15\hat{i}$ N

d) The net force acting on m_1 is by (1) and (c):

$$F_{m_1} = F + F_{m_2,m_1} = 45\hat{i} \text{ N} + (-15\hat{i} \text{ N}) = 30\hat{i} \text{ N}$$

Note that in each case the net force acting on each block is just that required to give it the common acceleration of $3\hat{i}$ m/s^2.

Example 4

A bartender slides a mug of beer down the bar to a customer who is 3 m away so that the mug just reaches the customer as it stops. The mug has a weight of 4 N and the bartender releases it with a velocity of magnitude 1.5 m/s^2. What is the coefficient of kinetic friction between the bar and the mug?

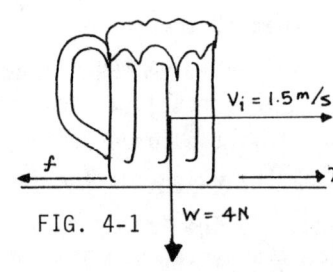

FIG. 4-1

Solution

Given: $W_{mug} = 4$ N
$v_i = 1.5$ m/s
$v_f = 0$
distance = $x = 3$ m

Required:

The mug is the system of interest and we choose our coordinate system as indicated in Fig. 4-1. Once the bartender releases the mug the only forces acting on it are W_{mug}; N, the upward normal force of the bar on the mug; and f, the force of friction opposite to the direction of motion of the mug. We assume there is no motion in the y direction:

$$F_y = 0 = N + W \quad N = -W$$

The only force acting on the mug after release in the x direction is the frictional force:

$$F_x = f = -\mu_k N$$

Applying Newton's 2nd law gives:

$$-\mu_k N \hat{i} = ma \quad (1)$$

Unfortunately we do not know **a**. However, we do have sufficient information to determine a kinematically by using Eq. 4.17

$$v_f^2 = v_i^2 + 2ax$$
$$0 = (1.5 \text{ m/s})^2 + 2(a)(3 \text{ m})$$
$$a = -0.375 \text{ m/s}^2$$

Substituting into equation 1 we have:

$$-\mu_k N \hat{i} = m(-0.375 \text{ m/s}^2)\hat{i}$$
$$\mu_k W = W/g(0.375 \text{ m/s}^2)$$
$$\mu_k = 0.375/9.8$$
$$\mu_k = 0.038$$

Example 5

A boy is pulling a trunk of mass 160 kg across a rough floor at constant velocity by means of a rope over his shoulder. The rope makes an angle of 53° with the horizontal and the coefficient of kinetic friction between the trunk and the floor is 0.5. What is the magnitude of the tension in the rope?

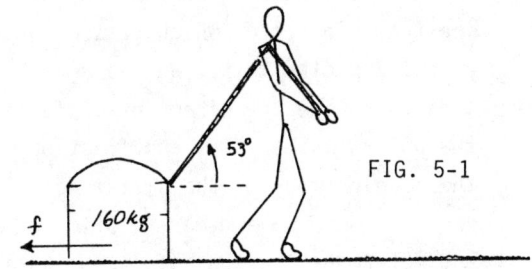

FIG. 5-1

Solution

Given: $m_{trunk} = 160$ kg
$v = $ constant; $a = 0$
$\theta = 53°$
$\mu_k = 0.5$

Required: T

Although it is often worthwhile to try to guess the answer to a problem or to estimate it, this requires some experience. At this stage it is probably best to proceed with the more methodical approach to solving this problem. We choose the trunk as the system. The forces which act on the trunk are: 1) the tension in the rope, 2) the frictional force between the floor and trunk, 3) the weight of the trunk, and 4) the normal, upward force of the floor on the trunk. These forces are diagrammed in Fig. 5-1, which also shows our choice of axes.

We must resolve T into its components.

$$T_x = T\cos 53° = (3/5)T; \quad T_t = T\sin 53° = (4/5)T$$

Since the trunk moves with constant velocity the acceleration in the x direction (as well as in the y direction) is 0. Applying Newton's laws we have:

$$F_x = (4/5)T - f = (4/5)T - \mu_k N = 0 \quad (1)$$
$$F_y = (3/5)T + N - W = (3/5)T + N - mg = 0 \quad (2)$$

Multiplying (1) by 15 and (2) by 20 gives:

$$12T - \mu_k 15 N = 0 \quad (3)$$
$$12T + 20 N - 20 mg = 0 \quad (4)$$

Subtracting (3) from (4) gives:

$$15 \mu_k N + 20 N - 20 mg = 0$$

$$N = 20\, mg/(15\, \mu_k + 20)$$

Substituting the given values for m, g, and μ_k we find:

$$N = (20)(160)(9.8)/[15(0.5) + 20] = 1140 \text{ newtons}$$

In this case the normal force exerted by the horizontal surface on the trunk is not **mg**. **T** has a vertical component which tends to "lift" the trunk off the floor, thereby decreasing the contact force **N**. Students often mistakenly think that the normal force exerted by a horizontal surface is always equal to mg. This is not true as illustrated by this example. **N** is an unknown force which must be determined from the equations generated by applying Newton's laws.

Substituting the value we have found for N into (1) yields the desired value of T.

$$4/5\, T - (0.5)(1140\, N) = 0$$
$$4/5\, T = 570\, N$$
$$T = 712.5\, N$$

Example 6

Atwood's machine consists of two masses connected by a light, inextensible cord which passes over a pulley (see problem 70, page 127 of the text). We assume the pulley has no effect other than to change the direction of force exerted by the string (i.e. the pulley is massless and frictionless). If $m_2 > m_1$, prove that the acceleration of the system is:

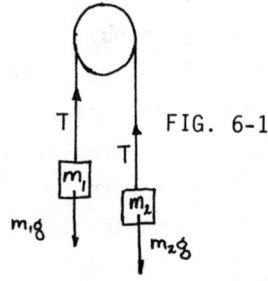

FIG. 6-1

$$a = g(m_2 - m_1)/(m_2 + m_1)$$

and determine the tension in the string.

Solution

Since each mass is acted on by the tension in the spring as well as its weight, the acceleration of the system will be less than g and m_2 will move down while m_1 will move up. A little reflection should convince you that this is actually a one dimensional problem. If mass m_2 falls a distance x then mass m_1 will rise by exactly the same amount. Only one coordinate is needed to describe the motion of the system. The fact that the pulley is assumed massless and frictionless means that T is the same throughout the cord. In fact we can redraw the system as in Fig. 6-2.

We arbitrarily choose the x axis to be positive toward the right. The next step is to choose a body or system to which we will apply Newton's laws. At this point you might pick either body 1 or 2. Since we are first interested in a, the acceleration, if

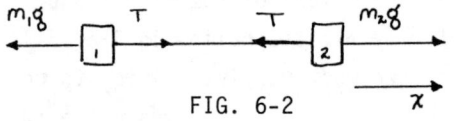

FIG. 6-2

we consider Newton's 3rd law for a moment, you should realize that the T's acting on m_1 and m_2 are equal to each other and oppositely directed (although they are not an action-reaction pair).

If we include both m_1 and m_2 in the system, the T's are internal forces and will cancel each other in the application of Newton's laws. In the x direction we then have:

$$F_x = m_2 g - m_1 g = (m_1 + m_2)a$$
$$a = g(m_2 - m_1)/(m_2 + m_1)$$

In order to find the tension in the cord we must generate an equation which contains it. Since Newton's 2nd law must apply to any body we choose m_2 and write:

$$F_x = m_2 g - T = m_2 a = m_2 g(m_2 - m_1)/(m_2 + m_1)$$

which when solved for T gives:

$$T = m_1 m_2 g/(m_2 + m_1)$$

PROBLEMS

1. What is the weight in newtons of a 5 lb bag of sugar?

2. A dragster of weight 2000 lbs covers a quarter mile in 3 s from a standing start. Assuming it has a constant acceleration, what is the force that the ground exerts on the dragster?

3. A boy swings a 110 g rock in a slingshot about his head in a horizontal circle of radius 1/2 m. The rock has a tangential velocity of 10 m/s. a) What is the magnitude of the centripetal force which is exerted by the slingshot on the rock? b) If the radius of the circular path is increased to 1 m while maintaining the same tangential velocity what is the centripetal force?

4. A car of mass 1200 kg traveling at 2.2 m/s crashes into a brick wall. It decelerates to zero velocity in 0.1 s. What is the force exerted on the car by the wall in newtons and pounds?

5. A horse is pulling a cart down a country road. By Newton's 3rd law the force exerted on the cart by the horse must be equal and opposite the force exerted by the cart on the horse. Explain why this does not mean that the horse and cart cannot move.

6. Two people are engaged in a tug-of-war. Identify the forces acting on them and the reaction forces to each.

7. If you pay $2.00 for a 5 lb bag of sugar on earth, how much should you pay for a 5 lb bag of sugar on the moon? What would be different if you were purchasing a 2 kg bag of sugar in these same two locations?

8. A block of mass 2 kg is released from a height of 4 m and slides down a rough inclined plane which makes an angle of 45° with the horizontal. The coefficient of kinetic friction is = 0.2. a) What is the acceleration of the block? b) What is the velocity of the block as it reaches the bottom of the plane?

9. In example problem 3 the force acted from left to right on the 10 kg block. Suppose the same force acted from right to left on the 5 kg block. a) Would the acceleration of the system change? How? b) What would be the magnitude of the forces F_{m_1, m_2} and F_{m_2, m_1}. c) Does this make sense and, if so, why?

Chapter 7
Work, Energy, and Power

PREVIEW

This chapter introduces the concepts of mechanical work, kinetic energy and power. The definitions of these quantities in mechanical terms provide the basic relationships which serve as the basis for the development of all succeeding descriptions of energy. The work-energy principle is derived for mechanical forces in its most general form and applied to a number of special cases including simple machines. Since the work-energy principle serves as the starting point for the development of the law of conservation of energy, which is the single most powerful conservation principle in physics, it is important to understand fully the definitions and derivations in this chapter.

SUMMARY

7.1 WORK DONE BY A CONSTANT FORCE

Definition. Work done by a constant force (1 dimensional motion):
The work done by a force which has constant magnitude and direction acting on an object which moves through a one dimensional (straight line) displacement of Δs is equal to the scalar product of F and Δs.

$$W = \mathbf{F} \cdot \Delta \mathbf{s} = F \, \Delta s \, \cos\theta \tag{7.1}$$

Work as defined mechanically in physics does not usually agree with one's preconceived meaning of the term. The limited definition given above applies only to the special case of a force that is constant which acts on an object which moves in only one direction. However, even this very special circumstance exhibits some of the differences between "physical" work and what we often think of as work.

1) In the definition of mechanical work, a force must act on an object as it undergoes some displacement. Like Newton's 2nd law we must be careful to focus our attention on the object in question and to the force in question. For example, if you do isometric exercises with a stationary barbell, the force you exert on the barbell will have done no work on the barbell because the barbell did not move. This may seem strange because you believe that you have done a great deal of "work". However the question was not what you did but what happened to the barbell. We will see in later chapters as we broaden the concept of work and energy to non-mechanical systems how to account for what happened to you. For the time being we must use our initial definition and explore its consequences.

2) If the magnitude of the force is zero then the work done is trivially zero also.

3) If the displacement of the object is zero then the force does no work.

These cases are fairly obvious from the discussion in 1).

4) If the force acts at right angles to the displacement, i.e. if the angle θ is 90°, then the force does no work. This fact often seems confusing since a force which acts at right angles to the direction of motion, such as a centripetal force, obviously affects the motion. This is a consequence of the fact that work is a <u>scalar function</u>, not a vector function.

5) Work can be positive or negative, depending on the angular relationship between the force and displacement. Usually one can think of forces which tend to decelerate objects as performing negative work and those which tend to accelerate objects as performing positive work.

6) The SI units of work are kg-m^2/s^2. The dimensions are ML^2/T^2. The SI unit of work has been given the special name of Joule.

7.2 WORK DONE BY A VARIABLE FORCE

We now generalize the definition of work given in the last section to the case where we allow the magnitude of the force to vary.

<u>Definition</u>. <u>Work done by a variable force</u> (1 dimensional motion, direction of force constant relative to displacement): The limit of the sum of the scalar product of the force acting on an object at the ith position and the finite displacement Δs_i about the ith position as $\Delta s_i \to 0$, defined over the total displacement of the object from the initial position s_i to the final position s_2.

$$W = \lim_{\Delta s_i \to 0} \sum_i F_i \cdot s_i$$

We recognize this definition as that of the definite integral. In the current situation where the motion is one dimensional and the angle between **F** and **s** is fixed, the term cos can be factored out of the sum and resulting integral to give

$$W = \lim_{\Delta s_i \to 0} \cos\theta \sum_i F_i \, \Delta s_i = \cos\theta \int_{s_1}^{s_2} F_i \, ds$$

If the force and displacement are parallel then $\cos\theta = 1$ and from this slightly more general expression we recover Eqs. 7.5 and 7.6 of the text.

$$W_{12} = \lim_{\Delta x_i \to 0} \sum F_i \cdot \Delta x_i = \int_{x_1}^{x_2} F \, dx \qquad (7.5, 7.6)$$

Furthermore, if the force is constant, it also can be factored out of the sum and integral and this definition reduces to Eq. 7.1 for the case of the work of a constant force in one dimension.

7.3 WORK IN THREE DIMENSIONS

If both the magnitude and the direction of the force vary with respect to the path of the object on which they act, the calculation of the work performed by the force must be carried out in three dimensional form. In this case the scalar product of F_i with s_i will have a different value for each interval because both F_i, the magnitude of F_i, and the angle which

F_i makes with Δs_i will change for each interval. The result is that the work in the limit as $\Delta s_i \to 0$ must be defined in terms of a <u>line</u> or <u>path</u> integral.

$$W = \int_{s_1}^{s_2} \mathbf{F} \cdot d\mathbf{s} \tag{7.7}$$

Although the notation of the path integral may not state it as explicitly as one might like, this is an integration that must be carried out along the one dimensional path specified by s. In order to evaluate such an integral the dependence of both the magnitude of F as a function of s <u>and</u> the dependence of the angle between F and ds as a function of s must be known and explicitly substituted into this expression. In general such integrals are not trivial to evaluate even if the above information is given.

If Δs_i is small, then it is essentially the same as Δr_i, the change in the position vector of the object relative to the origin of the chosen coordinate system. This is certainly the case in the limit as $\Delta s_i \to 0$. Eq. 7.7 can then be written as:

$$W = \int_C \mathbf{F} \cdot d\mathbf{r}$$

where C indicates the path along which the integral is to be evaluated. In particular, if we specify F in the form:

$$\mathbf{F} = P(x,y,z)\hat{\imath} + Q(x,y,z)\hat{\jmath} + R(x,y,z)\hat{k}$$

and r as:

$$\mathbf{r} = x\hat{\imath} + y\hat{\jmath} + z\hat{k}$$

then the line integral above takes the form:

$$\int_C \mathbf{F} \cdot d\mathbf{r} = \int_C (Pdx + Qdy + Rdz)$$

In this notation it is understood that each of the variables x, y, and z, as well as the corresponding differentials, are to be expressed in terms of an appropriate single variable which parametrically specifies the path C. The limits on the integral must also correspond to the initial and terminal points of the path C. In the next section we see that there is a possibility of determining the work performed in certain circumstances which will make the evaluation of path integrals unnecessary.

7.4 KINETIC ENERGY AND THE WORK-ENERGY PRINCIPLE

From Newton's 2nd law we know that when a net force acts on a system the system will accelerate. If the net force has a component which acts along the direction of motion it will cause the systems speed to increase or decrease doing positive or negative work on the system. The effect on the speed of a net force which performs work can be described in terms of the quantity <u>kinetic energy</u>.

> Definition. <u>Kinetic energy</u>: one half the product of the mass of a system times the square of the system's <u>speed</u>. We will use K to denote the kinetic energy.
> $$K = (1/2)mv^2 \tag{7.8}$$

We emphasize the terms net and change in the previous definition since it is important to realize that the work-energy principle as stated does not depend on the work done by the individual forces which act on a system but on the net force which acts on the system. It is also important to realize that only the change in kinetic energy is given by W_{net} not the actual value of the kinetic energy itself.

The proof of this principle or theorem depends on the application of the definition of work to Newton's 2nd law. The proof for the special case of a constant force in one dimension is found on page 137 and that for the general case of non-constant forces in three dimensions is found on pages 138-139. We will not repeat these derivations in this guide but you should study them carefully in order to convince yourself that the work-energy principle is generally valid.

As we noted in the previous section, the calculation of the work done by a general force acting along a three dimensional path can be quite complex. Although the work-energy principle does not simplify the calculation of the work done by any individual force, it greatly simplifies the calculation if we are concerned only with the effects of the net force which acts on the system. The work-energy principle is a reformulation of Newton's 2nd law in scalar form. Although some information is lost in this reformulation, a great deal is gained in ease of application.

7.5 POWER

Definition. Power: the rate at which work is performed.

$$P = dW/dt \qquad (7.15)$$

The above definition of the power is for the instantaneous rate of the performance of work. By the work-energy principle we may also state that the power is the instantaneous rate of change of energy. If an amount of work ΔW is performed in a finite time interval Δt the average power is:

$$P = \Delta W / \Delta t \qquad (7.16)$$

The SI unit of power is the joule/second or Watt. The dimensions of power are ML^2/T^3. By dividing Eq. 7.1 by Δt and taking the limit as $\Delta t \to 0$ we also find that

$$P = F \cdot v \qquad (7.17)$$

7.6 APPLICATIONS OF THE WORK-ENERGY PRINCIPLE: SIMPLE MACHINES

Definition. Simple Machine: a device which transforms an input force, displacement, or velocity into an output force, displacement, or velocity and which cannot be reduced to a combination of simpler devices.

In fact there are but four basic simple machines, the inclined plane, the lever, the pulley, and the hydraulic lift. The latter of these will be considered in another chapter. We discuss simple machines in terms of the inclined plane, the lever, and the pulley.

The purpose of a simple machine is to allow you to do something that cannot easily be accomplished otherwise. Often we use them as force multipling devices which allow us to lift, move, or change the configuration of a system which we could not do "bare-handed". They are also used to increase the range of motion of systems or the velocity with which objects move.

Definition. Ideal Machine: a device which delivers all of the work input input to it by external agents as work done against other external agents. The input work is the positive work done on the system, the output work is negative work done by the system. The net work done on an ideal machine is zero.

Ideal machines do not lose any work or energy to friction, therefore they act merely as work transmitters, and $W_{input} = W_{output}$, which allows us to write:

$$F_i \Delta s_i = F_o \Delta s_o \qquad (7.19)$$

Since the input and output displacement occur during the same time interval this is equivalent to:

$$F_i v_i = F_o v_o \qquad (7.20)$$

Definition *Ideal mechanical advantage* (I.M.A.): The ratio of the distance through which the input force acts (Δs_i) to the distance through which the output force acts (Δs_o).

From Eq. 7.20 we see that the ratio $\Delta s_i / \Delta s_o$ is equal to v_i/v_o. From Eq. 7.19 we see that for an <u>ideal</u> machine:

$$\Delta s_i / \Delta s_o = v_i/v_o = F_o/F_i \qquad (7.21)$$

For a real machine in which there are energy losses and $W_i = W_o$ the ideal mechanical advantage does not equal the ratio of the output to the input force. In this case we define a new quantity:

Definition. *Efficiency*: The ratio of the A.M.A. to the I.M.A.

The efficiency is a number < 1 which acts as a measure of the loss of energy to the machine. Ideal machines have an efficiency of 1. The symbol ε is usually used to designate the efficiency.

$$\varepsilon = A.M.A./I.M.A.$$

EXAMPLE PROBLEMS
Example 1

A child is pulling a toy box weighing 50 N with a force of 25 N at an angle of 30° above the horizontal. She pulls the box 6 m across a smooth floor and then 6 m across a carpet. The coefficient of kinetic friction between the box and carpet is $\mu_k = 0.35$. a) How much work does the force the child exerts do on the box as it slides across the smooth floor? b) the rug? c) How much work does friction do on the box as it slides across the rug? d) What is the net work done on the box as it slides across the smooth floor? e) across the rug?

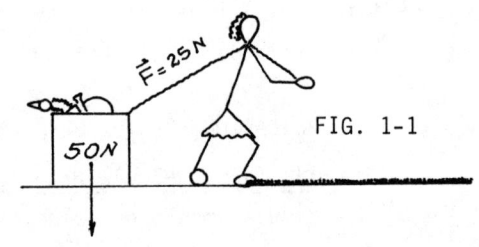

FIG. 1-1

Solution

Given: $W = 50$ N

$F = 25$ N, 30° above horizontal

$\mu_k = 0.35$

$x_{floor} = 6$ m

$x_{carpet} = 6$ m

Required: a) Work done by F on floor, b) work done by F on rug, c) work done by f across rug, d) W_{net}(floor), e) W_{net}(rug).

a) This is a case of work done by a constant force during one dimensional motion. The work done by the force the child exerts is given by Eq. (7.1):
$$W_F = F \cdot \Delta x = F \Delta x \cos 30° = (25 \text{ N})(6 \text{ m})(0.867) = \underline{130 \text{ J}}$$

b) Since the problem does not indicate that the force which the child exerts changes as the box slides onto the carpet, the work done by this force is the same for a 6 m displacement in either case. $W_F = \underline{130 \text{ J}}$

c) The fictional force is given by $f = \mu_k N$, in the direction opposite to the motion. By Newton's 2nd law the normal force in this case is $N = W - F\sin\theta$. The work performed against friction is:
$$W_f = f \cdot \Delta x = f \Delta x \cos 180° = -\mu_k (mg - F\sin\theta) \Delta x$$
$$W_f = \underline{-78.8 \text{ N}}$$

d) The only force acting on the box as it slides across the smooth floor which has a component in the direction of motion is the force exerted by the child:
$$W_{net} = \Sigma F \cdot \Delta x = 130 \text{ J}.$$

e) When the box is pulled across the carpet the child must do work against friction. The net work is
$$W_{net} = F \cdot \Delta x = F \cdot \Delta x + f \cdot \Delta x = 130 \text{ J} - 78.8 \text{ J} = \underline{51.2 \text{ J}}$$

Example 2

The force which a nonlinear spring exerts on an object is given by:
$$F = -kx^2 \qquad k = 200 \text{ N/m}^2$$

Calculate the work done by this force as the object is displaced from
a) $x = 0$ to 0.5, b) $x = 0.5$ to 1.0, c) $x = 0$ to 1.0.

Solution

The spring force is the only force acting, but since it is a variable force the work over a finite interval must be calculated using Eq. (7.6).

$$W_{1 \to 2} = \int_C F \cdot dx = -\int_C kx^2 \hat{i} \cdot \hat{i} dx = -k\int_C x^2 dx \qquad (1)$$

a) Using Eq. (1) for the interval 0 to 0.5 gives:
$$W = -k \int_0^{0.5} x^2 dx = -kx^3/3 \Big|_0^{0.5} = -(k/3)[(0.5)^3 - (0)^3] = \underline{-8.33 \text{ J}}$$

b) For the interval 0.5 to 1
$$W = -(k/3)x^3 \Big|_{0.5}^{1} = -(k/3)[(1)^3 - (0.5)^3] = \underline{-58.33 \text{ J}}$$

c) Finally for the entire interval from 0 to 1:
$$W = -(k/3)[(1)^3 - (0)^3] = \underline{-66.67 \text{ J}}$$

We see that the sum of the integrations in a) and b) is equal to c).

Example 3

A particle moves in the x-y plane subject to a force given by $F = x^2 y \hat{\imath} + y^2 \hat{\jmath}$. Calculate the work done by this force when the particle is moved from the point $(x,y) = (1,0)$ to the origin $(0,0)$ along:
a) the path $y = -x + 1$ from $(1,0)$ to $(0,1)$ and $x = 0$ from $(0,1)$ to $(0,0)$. b) the path $y = 0$ from $(1,0)$ to $(0,0)$. The two paths are diagrammed in Fig. 3-1.

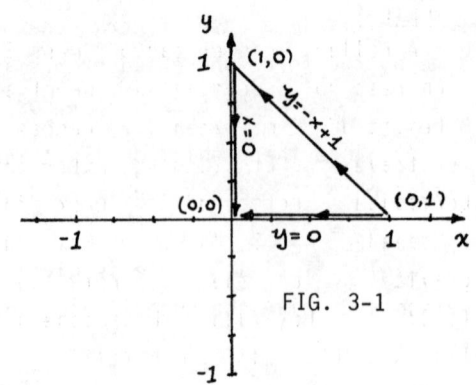

FIG. 3-1

Solution

Given: $F = x^2 y \hat{\imath} + y^2 \hat{\jmath}$

The two paths specified in Fig. 1-1.

Required: The work performed by F along each of these paths.

This problem is included in order to help you understand what a path integral is and how they must be evaluated. As we mentioned above this is not an easy topic in the calculus or its application to physics. From our discussion of path integrals we know:

$$W = \int_C F \cdot dr = \int_C x^2 y \, dx + \int_C y^2 \, dy \qquad (1)$$

a) In order to perform the integration we must substitute from the function which specifies the path $y = f(x)$ to eliminate y (or x) and dy (or dx). From $(1,0)$ to $(0,1)$ we have:

$$y = -x + 1$$
$$dy = -dx$$

Substituting into Eq. (1) gives:

$$W_{(1,0) \to (0,1)} = \int_1^0 x^2(-x+1)dx + \int_1^0 (-x+1)^2(-dx)$$

$$= \int_1^0 (-x^3 + x^2)dx - \int_1^0 (x^2 - 2x + 1)dx$$

$$= [-x^4/4 + x^3/3 - x^3/3 + x^2 - x]_1^0$$

$$= [-x^4/4 + x^2 - x]_1^0$$

$$= 0 - [-1/4 + 1 - 1]$$

$$W_{(1,0) \to (0,1)} = \underline{1/4 \text{ J}}.$$

For the path from $(0,1)$ to $(0,0)$ we have

$x = 0$, $dx = 0$ and $W_{(0,1) \to (0,0)} = \int_1^0 y^2 \, dy$

$$W_{(0,1) \to (0,0)} = [y^3/3]_1^0 = \underline{-1/3 \text{ J}}$$

The work for the entire path is:

$$W_{total} = W_{(1,0) \to (0,1)} + W_{(0,1) \to (0,0)} = 1/4 - 1/3 = -1/12 \text{ J}$$

b) For the curve $y = 0$ from $(1,0)$ to $(0,0)$ we see by inspection of Eq. (1) that the integrand in both terms vanishes identically so the work done along this path is zero.

Example 4

A roller coaster car of mass 350 kg starting from rest rolls down an incline of angle 45° from a height of 30 m. When it reaches the bottom it is travelling at 20 m/s. After rolling along a horizontal section of the track for 50 m it has a speed of 9.3 m/s. a) How much work does gravity do on the car from A to B?, b) from B to C? c) How much work is done against friction from A to B?, d) from B to C? e) What is the coefficient of friction between the car and the track.

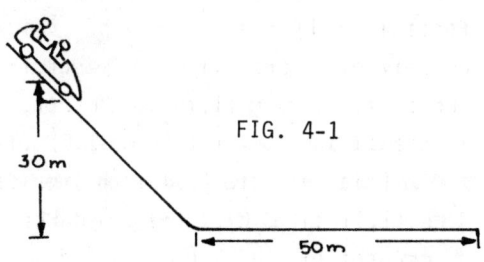

FIG. 4-1

Solution

Given: $h = 30$ m
$\theta = 45°$
$m = 350$ kg
$v_B = 20$ m/s
$v_C = 9.3$ m/s

Required: a) W_{AB}(gravity), b) W_{BC}(gravity), c) W_{AB}(friction), d) W_{BC}(friction), e) μ_k

Solution

a) From Section 7.1 and the examples in that section the work done by gravity as the car rolls down the incline is:

$$W_{AB}(\text{gravity}) = mgh = (350 \text{ kg})(9.8 \text{ m/s}^2)(30 \text{ m}) = \underline{102,900 \text{ J}}$$

b) Along the horizontal portion of the track the weight is perpendicular to the displacement and W_{BC}(gravity) is 0.

c) It is possible to calculate the work done by friction by determining the frictional force f using Newton's 2nd law and the definition of work. However, it is at this point that the power of the scalar work-energy principle becomes apparent. The net work done on the car as it rolls down the incline is the sum of the work done by gravity and that done against friction. By the work-energy principle, this net work must equal the change in kinetic energy of the car.

$$W_{AB}(\text{gravity}) + W_{AB}(\text{friction}) = \Delta K = 1/2\, mv_B^2 - 0$$

Substituting from the given values and a):

$$102,900 \text{ J} + W_{AB}(\text{friction}) = 1/2(350 \text{ kg})(20 \text{ m/s})^2$$
$$W_{AB}(\text{friction}) = \underline{-32,900 \text{ J}}$$

d) As the car rolls from B to C the only force acting in the direction of motion is the frictional force therefore only it performs any work.

$$W_{BC}(\text{friction}) = K = 1/2\, mv_C^2 - 1/2\, mv_B^2$$
$$= 1/2(350 \text{ kg})(9.3 \text{ m/s})^2 - 1/2(350 \text{ kg})(20 \text{ m/s})^2$$
$$= \underline{-54864 \text{ J}}$$

e) Now that we have the work performed by friction from A to B and from B to C we can use Newton's 2nd law and the definition of work to calculate μ_k. We choose to use the result from d) since the calculation of f along the horizontal portion of the track is simpler. For this 50 m distance, $f = \mu_k N = \mu_k mg$. Therefore:

$$W_{BC}(\text{friction}) = f\, \Delta x = -\mu_k mg \Delta x = -54864 \text{ J} = \Delta K$$

56   Chapter 7

We find upon substituting the relevant quantities that μ_k is:
$$\mu_k = (54864 \text{ J})/(350 \text{ kg})(9.8 \text{ m/s}^2)(50 \text{ m}) = \underline{0.32}$$

Example 5

A clothes basket full of clothes with a mass of 10 kg is sitting at the bottom of the basement steps. A dutiful son carries the basket up the steps for his mother and sets the basket on the floor at the top of the steps. The vertical distance from the basement floor to the next floor is 2.5 m. a) How much work does gravity do on the basket? b) What is the net work done on the basket? c) How much work was done by the boy on the basket?

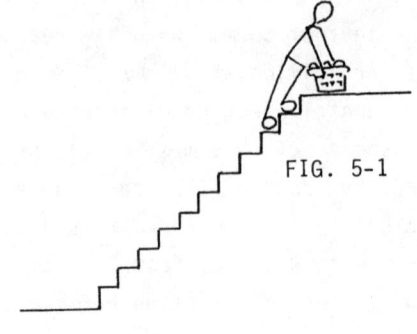

FIG. 5-1

Solution

Given: m = 10 kg
         h = 2.5 m

Required: a) $W_{gravity}$, b) W_{net}, c) W_{boy}

a) As the boy carries the basket up the stairs he moves it through a vertical displacement of 2.5 m. However, the vertical displacement is up while the force of gravity acts downward. The work done by gravity is negative. This is often stated as "The boy does work against gravity" or by saying that work was done against gravity rather than by gravity. The magnitude of the work is, of course mgh.
$$W_{gravity} = -mgh = -(10 \text{ kg})(9.8 \text{ m/s}^2)(2.5 \text{ m}) = -245 \text{ J}$$

b) From the work-energy principle we know that the net work done must equal the change in kinetic energy of the system. In this case the initial and final kinetic energies are 0, therefore so must be W_{net}.

c) The two forces acting on the basket are that of gravity and the boy. Thus $W_{net} = W_{gravity} + W_{boy} = 0$. From b) we see immediately that $W_{boy} = +245$ J.

Example 6

One of the most common levers is the first class lever where the fulcrum is placed between the input and output force. This device is usually employed as a force multiplier. A first class lever is shown in Fig. 6-1. If the angle, through which the lever moves is small then

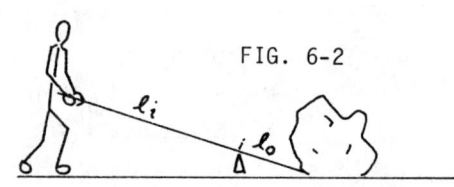

FIG. 6-2

$$\text{I.M.A.} = s_i/s_o = \ell_i/\ell_o$$

If the lever shown has ℓ_i = 1.5 m and ℓ_o = 0.5 m, what input force is required to lift an automobile weighing 4000 lbs?

Solution

Given: ℓ_{input} = 1.5 m
         ℓ_{output} = 0.5 m
         F_{output} = 4000 lbs

Required: F_{input}

First we calculate the ideal mechanical advantage for this system.

I.M.A. = ℓ_i/ℓ_o = 1.5/0.5 = 3

We treat the lever as an ideal machine, I.M.A. = A.M.A. = F_o/F_i

F_o/F_i = 4000 lbs/F_i = 3

F_i = 1333.33 lbs

PROBLEMS

1. An elevator weighing 3700 N is raised from the ground floor of a skyscraper to the 100th floor, a distance of 400 m. How much work must the motor which drives the elevator do if the elevator starts and ends at rest?

2. A man pushes a 25 kg trunk 10 m across a rough level floor at constant speed with a horizontally directed force. If the coefficient of kinetic friction is 0.2, how much work does the man do?

3. The force diagrammed in Fig. 1 acts on a particle moving in one dimension. What is the work from A B, B C, C D, D E, E F, F G, G H and the total work?

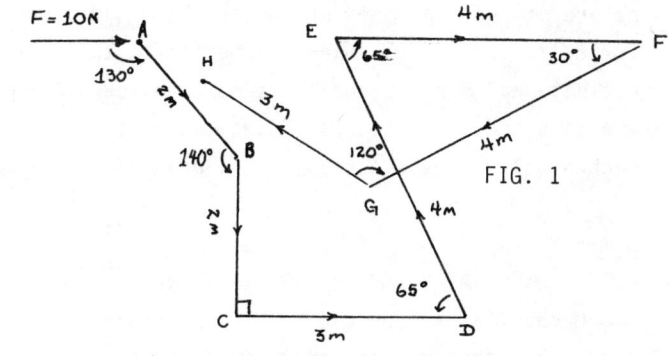

FIG. 1

4. The force acting on an object is $F = F_o(x/x_o - 1)$. What is the work done in moving the object from $x = 0$ to $3x_o$?

5. The force acting on a particle is $F = y^2 \hat{i} - x \hat{j}$. Calculate the work done as the particle moves from (0,0) to (1,2) along the path defined by $y^2 = 4x$.

6. A weight of 30 N is dragged across a table top by a string attached to a pulley on a rod at the end of the table. The pulley is 0.5 m above the table and the tension in the string is maintained at 50 N. How much work does the tension do as the weight slides from a distance 2 m away from the pulley to a distance 1 m away?

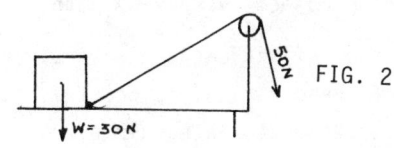

FIG. 2

7. A 10 kg block is pushed up a frictionless inclined plane of angle 37°. The force exerted on it is 100 N parallel to the plane. If the speed at the bottom of the plane is 2 m/s, what will be its speed when it has been pushed 3 m up the plane?

8. A truck with mass 20,000 kg is travelling at a speed of 100 km/hr. How fast must a car of 2000 kg be travelling to have the same kinetic energy?

9. A bullet of mass 50 g travelling 300 m/s collides with a "bullet proof" lexan window which is 2 cm thick. If the window is to just stop the bullet, what average force must it exert?

Chapter 8
Conservation of Energy

PREVIEW

Now that the concepts of work and energy have been introduced, we will begin in this chapter to explore some of the consequences of these definitions, the work-energy principle, and the special nature of certain types of forces which allow us to redefine the work performed against or by such forces in terms of an energy. We first study conservative and nonconservative forces and their definition. Next we shall see how the work of conservative forces can be expressed as potential energies and study the ramifications of this redefinition. We introduce the law of conservation of mechanical energy and study equilibria and their stability.

SUMMARY

8.1 CONSERVATIVE AND NONCONSERVATIVE FORCES

Mechanical forces can be broadly classified into two groups depending on the properties of the work which they perform or which is performed against them.

Definition. Conservative force (I): a force acting on an object which performs no work on that object when the object's total displacement forms a closed path.

A closed path in space is any locus of points which is continuous and whose initial and final points are identical. An alternative definition of a conservative force is:

Definition. Conservative force (II): a force for which the work done on an object over every path connecting two fixed points is the same.

These two definitions are entirely equivalent since either of them can be used to prove the validity of the other. This proof is given in the text on pages 154-155. The notation commonly employed to designate integrals over closed paths is an integral sign with a circle in the middle. Using this notation, the two definitions of a conservative force can be written:

$$(\mathrm{I}) \oint \mathbf{F} \, d\mathbf{s} = 0 \tag{8.1}$$

$$(\mathrm{II}) \int_{s_1}^{s_2} \mathbf{F} \, d\mathbf{s} = \text{constant} \tag{8.2}$$

The reason for giving two definitions of conservative force is that each has certain advantages depending on the situation one is considering. It should be obvious from our calculations in Example Problem 3 in Chapter 7 that not all forces are conservative. The most common nonconservative force is that of friction. In general work <u>is</u> a path dependent quantity. Conservative forces are a very special class of forces as we shall see in the next section.

8.2 POTENTIAL ENERGY

Because of its path independence, the work performed by or against a conservative force has characteristics which resemble those of kinetic energy. Like all forces, when a conservative force does positive work on an object it causes the kinetic energy of the object to increase. However, when work is done against a conservative force, i.e. when a conservative force does negative work on a system there exists the <u>potential</u> of reclaiming all of the work performed against the conservative force. As an example we consider lifting a book from the floor and placing it on a table. If the book starts at rest and ends at rest the net work done is zero by virtue of the work-energy principle. Two forces perform work on the book, the force exerted by the person lifting the book which does positive work and the force of gravity which does negative work. Even though W_{net} is zero, it is obvious to most of us that there is something different about the book after we have lifted it onto the table. We say the configuration of the system is different. If we slide the book off the edge of the table it will fall to the floor accelerating at 9.8 m/s^2 until it strikes the floor. The change in kinetic energy it experiences will be equal to the work done <u>on</u> <u>it</u> by gravity as it falls. However, this positive work will be exactly equal in magnitude to the negative work done by gravity as the book was lifted and placed on the table. From our definitions above we see that the force of gravity is conservative.

Note that the work the person did <u>against</u> gravity to lift the book on to the table can be entirely reclaimed as kinetic energy by allowing the system to return to its original configuration under the influence of the conservative force of gravity. This is the unique property of conservative forces as they have been defined above.

Another thing which we notice about conservative forces is that the work they do (positive or negative) only depends on the initial and final configuration of the system, not how the system is transformed from its initial to its final state. This follows immediately from the second definition of a conservative force. It is important to note that the change in kinetic energy has exactly the same property. We can write for a conservative:

$$W_{conservative} = \int_{s_1}^{s_2} F_{conservative} \cdot ds = f(configuration)$$

Let's consider the case where we have both nonconservative and conservative forces acting. We divide W_{net} in two parts, W_{NC} the work performed by nonconservative forces, and W_C the work done by conservative forces.

$$W_{net} = W_C + W_{NC} = \Delta K$$

We now move W_C to the right side of the work-energy relationship; <u>the side of the equation which we shall reserve for quantities which only depend on the configuration of the system.</u>

$$W_{NC} = \Delta K - W_C$$

Now path dependent quantities are on the left, non-path dependent ones on the right of the work-energy relationship.

<u>Definition</u>. <u>Change in potential energy</u>: the negative of the work done by a conservative force from the initial to final configuration of a system.

We may define the change in potential energy in this fashion because the work done by a conservative force depends only on the initial and final configuration of the system. In most cases the potential energy is a function of the position of the system.

Since the definition of potential energy depends on the definite integral and is expressed as the <u>difference</u> between the final and initial values, the potential energy function itself, U, is arbitrary to within a constant. This means that the configuration for which U = 0 is arbitrary and can be chosen to give the simplest form to U.

Since most conservative forces obey force laws which are simple functions of position, they can be substituted into Eq. 8.5 and the potential energy function determined by straight forward integration. A number of cases are discussed in the text on pages 157-160. The results are summarized in Table 8.1 of the text. You should study these examples carefully since we shall not repeat them here.

8.3 CONSERVATION OF ENERGY

In the special circumstance where there is <u>no work performed other than that by conservative forces</u>, the work-energy principle takes a special form which we term a conservation principle or law.

$$\text{If} \quad W_{NC} = 0 \quad \text{Then:} \quad 0 = \Delta K + \Delta U$$

Since this form says that the <u>sum</u> of the <u>change</u> in kinetic energy and the <u>change</u> in potential energy vanishes, it is equivalent to saying that the <u>sum</u> of the kinetic and potential energy does not change.

<u>Definition</u>. <u>Conservation law</u>: a statement that a physical quantity does not change with time, i.e. it remains constant, under specified circumstances.

In the case we are now considering the quantity which does not change with time is the sum of the kinetic and potential energy of the system.

<u>Definition</u>. <u>Mechanical energy</u>: the sum of the kinetic and potential energy (work done by conservative forces) of a system

$$\text{Mechanical Energy} = M.E. = K + U$$

With this definition we can write the law of conservation of mechanical energy as:

$$M.E._{initial} = K_i + U_i = K_f + U_f = M.E._{final}$$

It is important to understand that this is a conditional conservation principle and that it applies to the <u>sum</u> of K + U. The energy involved can change form from potential to kinetic and vice versa but the sum of the two will remain constant as long as no external forces but conservative ones do work on the system.

8.4 FORCE FROM POTENTIAL ENERGY

The definition of the change in potential energy in terms of a definite integral suggests that we should be able to calculate the force acting on a particle if the form of the potential energy is known. This can be done using the techniques of vector calculus since we are required to generate a vector, the force, from a scalar function, the potential. The general

technique employed is a bit advanced so we choose to consider a fairly simple special case at this time. In the instance where the force and motion are one dimensional, the definition of potential energy is:
$$U = -\int_{x_1}^{x_2} F(x)dx$$

For this special case the force in terms of the potential is:
$$F(x) = -dU/dx \qquad (8.27)$$

The validity of this expression can be checked by calculating the force from the potential energy function for those cases given in Table 8.1 of the text. Remember that this expression holds equally well for any one dimensional function. If F is only a function of r, you can replace x by r in Eq. 8.27.

A consequence of the relationship between force and potential energy is that if the potential energy is one dimensional and is plotted versus the independent position coordinate the resulting potential energy graph can be used to interpret the nature of the motion of the system.

Eq. 8.27 tells us that the force is the negative of the slope of the potential energy versus coordinate curve. When the slope of the potential energy curve is positive, the force will be negative and vice versa. A glance at FIG. 1 should convince you that whenever the potential energy curve has a minimum, the force tends to drive the system toward this point. The minimum itself is a point of 0 slope at which the force vanishes. Such points are called <u>equilibrium points</u>.

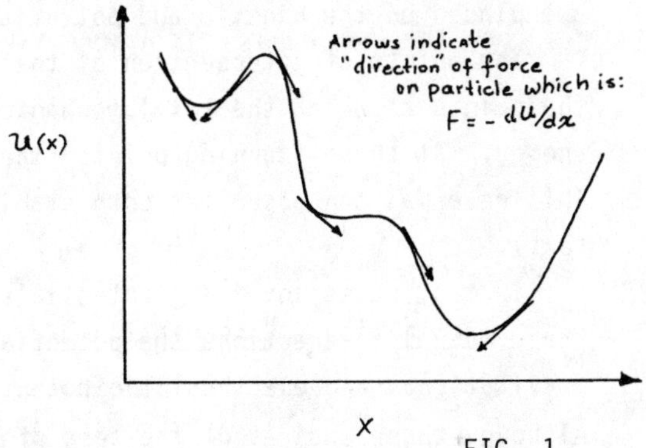

FIG. 1

8.5 STABILITY OF EQUILIBRIUM

Minimums of the potential energy curve are not the only points at which the force on a system vanishes. The slope of the potential energy curve is zero at maxima and inflection points as well. Each of these points is an equilibrium point, but each has different characteristics.

<u>Definition</u>. <u>Stable equilibrium</u>: if a system is in a state such that the net force on the system vanishes and any small displacement of the system tends to give rise to a force which tends to return the system to its zero force state, then the system is said to be in stable equilibrium.

<u>Definition</u>. <u>Unstable equilibrium</u>: if a system is in a state such that the net force on the system vanishes and any small displacement of the system tends to give rise to a force which tends to drive the system away from this zero force state, it is said to be in unstable equilibrium.

62   Chapter 8

Stable equilibrium points correspond to minima of the potential energy curve while maxima correspond to unstable equilibrium points. Inflection points correspond to a situation in which the equilibrium is bi-stable. Displacement in one direction causes a restoring force while displacement in the other direction will result in a force which drives the system away from the equilibrium configuration. In the three dimensional case inflection points are termed saddle points. A diagram of such a point is given in FIG. 8.15 in the text.

>   Definition. Neutral equilibrium: if the slope of the potential energy function is everywhere zero within a finite region of a point, the point is said to be one of neutral equilibrium.

8.6 ENERGY GRAPHS

More information can be determined if one considers the total mechanical energy plotted versus the potential energy curve. In the case where there is no work performed by nonconservative forces and the mechanical energy is constant, the mechanical energy graph depicts the limitations which are placed on the kinetic and potential energy as well as the displacement of the particle.

The points of intersection of the E = constant curve and the potential energy curve determine the points at which the total mechanical energy of the system is all in the form of potential energy. At these "turning points" the system has no kinetic energy and will reverse its motion. This reversal can occur for both stable and unstable equilibria (see FIGs. 8.16 and 8.17 of the text).

In many cases involving potentials which admit particles or systems to be bound together by their mutual interactions the potential energy is chosen to be negative. This is the case with the gravitational and electrostatic potential energies as well as the Lennard-Jones potential. Although these choices of the zero of potential energy are arbitrary, there is a good reason to choose this convention. In these cases the sign of the total energy allows one to distinguish whether the system is bound or unbound. If $E_{total} < 0$ then the system is bound. The members of the system do not have sufficient energy to escape from one another due to their mutual interaction. If $E_{total} > 0$ then the system is unbound. The components of the system will experience no "turning points" and can completely escape from one another. The limiting case is that of $E_{total} = 0$. In this situation the components of the interacting system have sufficient energy to just escape to infinite separation from one another and have zero kinetic energy left. You should study FIG. 8.18 of the text and the attendant discussion of these points.

EXAMPLE PROBLEMS
Example 1

A block is pushed up an inclined plane to a height of 5 m. The block is then released and allowed to slide down the plane. The block has a mass of 30 kg, the plane is inclined at an angle of 45°, and $\mu_k = 0.2$
a) Show that the gravitational force, mg, acting on the block is conservative by

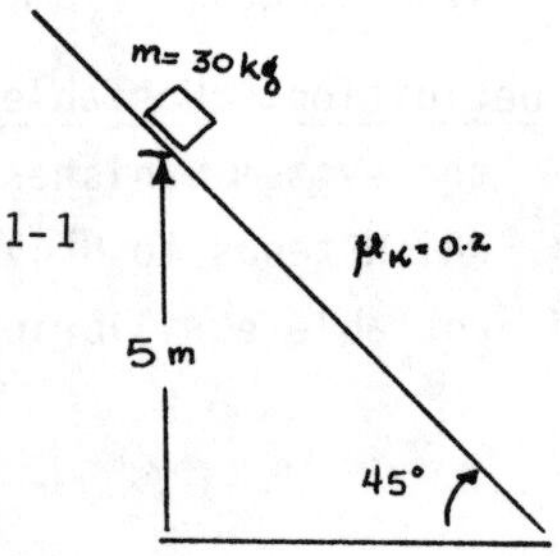

FIG. 1-1

explicitly calculating the net work done by gravity over the round trip. b) What is the net work done by friction over the round trip? c) What is the kinetic energy of the block as it reaches the bottom of the plane? d) What is its speed at the bottom?

Solution

Given: $m = 30$ kg
$\theta = 45°$
$\mu_k = 0.2$

Required: a) W_{net}(gravity), b) W_{net}(friction), c) K_A, d) v_A.

a) The vector manipulations for this problem should by now be familiar. We choose an x-y coordinate system with x positive up the plane. Since we are concerned with the work done by gravity and friction we need to know the components of W in this coordinate system, and the force of friction. Reference back to the problems in Chapter 6 shows that:

$$W_x = -W\sin\theta \,\hat{i}, \quad W_y = W\cos\theta \,\hat{j}, \quad N = -W\cos\theta \,\hat{j}$$
$$f = -\mu_k N\hat{i} \text{ when the block moves up the plane}$$
$$f = +\mu_k N\hat{i} \text{ when the block moves down the plane}$$

The work performed by the weight up and down the plane is

$$W_g = W_x \cdot x$$

Up the plane: $W_g = -W\sin\theta \,\hat{i} \cdot (5/\sin\theta)\hat{i} = -(207.8 \text{ N})(7.07 \text{ m}) = \underline{-1469 \text{ J}}$

Down the plane: $W_g = -W\sin\theta\,\hat{i} \cdot (-5/\sin\theta)\hat{i}) = \underline{+1469 \text{ J}}$

We see immediately that the net work performed by gravity is zero.

b) We can calculate the work performed by friction in exactly the same manner, the only difference being that the direction of this force always opposes the motion.

Up the plane: $W_f = -\mu_k W\cos\theta\hat{i} \cdot (5/\sin\theta)\hat{i} = 0.2(-1469) = \underline{-294 \text{ J}}$

Down the plane: $W_f = +\mu_k W\cos\theta \,\hat{i} \cdot (-5/\sin\theta)\hat{i} = 0.2(-1469) = \underline{-294 \text{ J}}$

The net work performed by friction is $W_f(net) = \underline{-588 \text{ J}}$

c) Employing the work-energy principle for the motion down the plane we can directly calculate K.

$$W_{net} = W_g + W_f = 1469 \text{ J} - 294 \text{ J} = \underline{1175 \text{ J}} = K$$

We calculate the final kinetic energy from the expression:

$$K = 1/2 \, mv_f^2 - 1/2 \, mv_i^2$$

We assume that the block starts at rest at the top of the plane so that $v_i = 0$.

$$K = 1/2 \, mv_f^2 - 0 = 1175 \text{ J}$$
$$v_f = \underline{8.85 \text{ m/s}}$$

Example 2

A ball is thrown vertically upward from the edge of a building which is 36 m high with an initial speed of 16 m/s (neglect air resistance). a) How high does it rise above the ground? b) What is its velocity as it strikes the ground?

Solution

Given: $v_i = 16$ m/s
$y_i = 36$ m

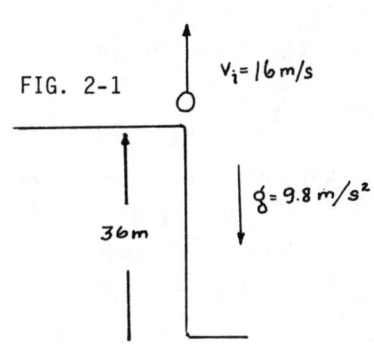

FIG. 2-1

Required: a) y_{max}, b) v at y = 0

a) Since we can neglect air resistance we may apply the law of conservation of mechanical energy. If we calculate the mechanical energy at any given point, the total energy will always remain at this value regardless of its form, kinetic, potential, or some combination of the two. Since we know the initial velocity and height of the ball we may calculate its total mechanical energy at the initial point. We use U = mgy as the gravitational potential energy with the ground taken as y = 0.

$$M.E. = 1/2\ mv^2 + mgy = 1/2\ m(16\ m/s)^2 + m(9.8\ m/s^2)(36\ m) = \underline{481m\ J}$$

When the ball reaches its maximum height, its velocity will be zero and therefore so will its kinetic energy. This is a "turning point" of the motion as we have described above. The mechanical energy at this point is all in the form of potential energy.

$$M.E. = mgy_{max} = 481m\ J$$

Substituting for g and solving we find y_{max} = $\underline{49\ m}$. Notice that we do not need to know the mass of the ball.

b) As the ball strikes the ground its height is y = 0, and its mechanical energy is then totally in the form of kinetic energy.

$$M.E. = 1/2\ mv_0^2 = 481m\ J$$

Solving for the velocity we find v_0 = $\underline{31\ m/s}$

Example 3

A mass of 5 kg resting on a frictionless horizontal surface is attached to a horizontal spring of force constant k = 200 N/m. The mass is displaced 0.3 m to the left of the equilibrium point of the spring and released (see FIG. 3-1). a) What is the potential energy of the system before the spring is released? b) How much work does the spring do on the mass from the time the mass is released until it passes through the equilibrium point for the first time? c) What is the velocity of the mass as it passes through the equilibrium point?

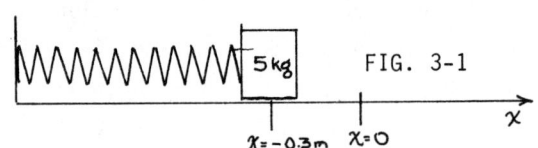

FIG. 3-1

Solution

Given: m = 5 kg
k = 200 N/m
x_i = -0.3 m

Required: a) U_i,   b) $W_{-0.30 \to 0}$,   c) v_0

a) Referring to Table 8.1 we may calculate the potential energy at the initial position.
$$U = (1/2)kx^2 = (1/2)(200\ N/m)(-0.3)^2 = \underline{9\ J}$$

We see that the potential energy <u>does not</u> depend on the sign of the displacement.

b) The work performed by the spring force is the negative of the change in potential energy.
$$W = -\Delta U = -(U_f - U_i) = -(0 - 9\ J) = \underline{+9\ J}$$

c) Since the spring force is conservative we know that the total M.E. of the system does not change during the motion. At the equilibrium point U = 0, the total M.E. is all in the form of kinetic energy and

$$(1/2)mv_o^2 = 9 \text{ J} \quad \text{which gives} \quad v_o = \underline{1.9 \text{ m/s}}$$

Example 4

An astronaut of mass m moves from a point P (at a distance r from the center of the earth) to a point Q (at a distance r + x from the center of the earth).

a) Prove that the change in potential energy of the astronaut equals

$$U = (R_E/r)^2[mgx/(1 + x/r)] \tag{1}$$

b) Show that U = mgx if $r = R_E$ and $x \ll r$. For x = 1 m, m = 80 kg, and the radius of the earth $R_E = 6.37 \times 10^6$ m, calculate U for c) $r = R_E$, d) $r = (41/40)R_E$, and e) $r = 2 R_E$.
(Note: This is problem 16, page 168 of the text. The accompanying figure is Fig. 8 page 168.)

Solution

Given: $U = -GmM_E/r$

Required: a) To show that Eq. (1) above is correct. b) e) as stated in the problem.

a) We have chosen to work this problem from the text to help you become familiar with the formal type of manipulations which are often required to put a formula into a format which will demonstrate the properties of the system in a conceptually simple (although not necessarily mathematically simple) form. $U = U_f - U_i$ with the initial radius r and final radius r +x.

$$U = -GmM_E/(r + x) - (-GmM_E/r)$$
$$= -GmM_E r/r(r + x) + GmM_E(r + x)/r(r + x)$$
$$= [-GmM_E r + GmM_E x + GmM_E r]/[r^2 + rx]$$
$$= GmM_E x/(r^2 + rx)$$

We now multiply the numerator and denominator of this fraction by R_E^2 in order that we may substitute $g = GM_E/R_E^2$.

$$U = GM_E m x R_E^2/(r^2 + rx)R_E^2 = mgx R_E^2/(r^2 + rx)$$

Factoring an r^2 out of the denominator gives the desired result

$$U = (R_E/r)^2[mgx/(1 + x/r)]$$

b) If $r = R_E$ then the squared factor in parentheses is 1. If $x \ll r$ then the term x/r in the denominator is approximately zero and Eq. (1) reduces to U = mgx.

c) The case $r = R_E$ and x = 1 m corresponds precisely to the situation in b). Therefore we may write U = mgx = (80 kg)(9.8 m/s^2)(1) = $\underline{784 \text{ J}}$.

d) For $r = (41/40)R_E$ we may still use the approximation that x/r = 0. Then:

$$U = mgx[R_E/(41/40)R_E]^2 = mgx(40/41)^2 = mgx(0.95) = \underline{746 \text{ J}}$$

e) Since $r = 2R_E$ we can again assume that x/r = 0. The square term is now $(R_E/2R_E)^2 = 1/4$. The change in potential energy for a radial displacement of 1 m at this altitude is U = (1/4)mgx = $\underline{196 \text{ J}}$.

Example 5

The hydrogen atom normally exists in its ground state which has a total mechanical energy of -2.176×10^{-18} J. The potential energy of interaction of the proton which forms the nucleus of the hydrogen atom and the electron circulating about it is given by the attractive Coulomb potential:

$$U = -ke^2/r$$

where e designates the magnitude of the charge on the electron and proton, $e = 1.6 \times 10^{-19}$ coulombs, and $k = 9 \times 10^9$ N-m^2/coulombs^2. a) Assuming that the electron moves in a circular path about the proton, show that the kinetic energy of the electron is $+(1/2)ke^2/r$. b) Show that the total mechanical energy is M.E. $= -(1/2)ke^2/r$. c) Calculate the radius of the orbit of the electron when the atom is in its ground state.

Solution

Given: $U = -ke^2r$
         M.E. $= -2.176 \times 10^{18}$ J
         $k = 9 \times 10^9$ N-m^2/coul^2
         $e = 1.6 \times 10^{-19}$ coul

Required: a) The kinetic energy as a function of r, b) the total mechanical energy as a function of r, c) the radius of the ground state orbit.

a) In order to calculate the kinetic energy as a function of r we must have some relationship that will allow us to determine the velocity of the electron in terms of its radius. Since the electron is moving in a circular orbit there must be a centripetal force acting on it. This force is provided for by the electrostatic attraction of the electron and proton. From the given form of the potential we can calculate the force on the electron:

$$F = -dU/dr = +ke^2/r^2$$

If we place this expression in Newton's second law as the net force on the electron and remember that the centripetal acceleration is equal to v^2/r we find

$$F = ke^2/r^2 = ma = mv^2/r$$

which gives:

$$ke^2r = mv^2$$

Thus we see that the kinetic energy is just one half of the above expression.

$$K = (1/2)ke^2/r$$

b) The total mechanical energy is just the sum of the kinetic and potential energies.

$$M.E. = K + U = (1/2)ke^2/r + (-ke^2/r) = -(1/2)ke^2/r$$

c) We substitute the given values for the ground state M.E., k, and e into the above expression and then solve for r.

$$M.E. = -2.176 \times 10^{-18} \text{ J} = -(1/2)(9 \times 10^9)(1.6 \times 10^{-19})^2/r$$
$$r = \underline{5.3 \times 10^{-11} \text{m}}$$

Although arrived at by purely classical considerations, this radius does have significance at the atomic level. It is the most probable radius at which the electron can be found when a hydrogen atom is in its ground state.

PROBLEMS

1. A roller coaster car has a mass of 700 kg and the coaster has a lift of 40 m. Its speed as it arrives at the end of its run is 15 m/s. If it begins with v = 0 at the top of the lift, how much work is done by friction on the car during its run?

2. A mass m is attached to a linear ideal spring of force constant k. Specify in terms of the amplitude of the system's motion the values of the position coordinate of the mass when: a) the kinetic energy is zero, b) the potential energy is zero, c) the potential energy is 1/4 the total mechanical energy, and d) the potential and kinetic energy are equal.

3. A uniform rope of mass m and length L lies on a frictionless table a height h above the floor. The end of the rope is allowed to dangle over the edge of the table and then is released. The rope slides off the table. What is the rope's velocity just as the other end of the rope slides off the table.

4. The potential energy of the center of gravity of a cola bottle might be depicted as to the right. To what positions of the bottle do each of the equilibrium points indicated on the graph correspond.

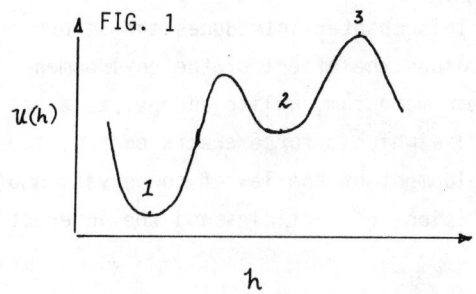

5. A potential energy function often used in molecular physics is:
$$U(r) = (1-e^{\alpha/r})$$
a) Sketch this potential energy as a function of r. b) If a particle under the influence of this force has an energy of $-E_0$, what would be the turning point radius?

6. A projectile is fired with a speed v_0 at an angle . Which of the following questions can be answered using the work-energy principle and which requires that the equations of motion from Newton's 2nd law be solved? a) What is the range of the projectile? b) What is the maximum height of the projectile? c) How long is the projectile in the air? d) With what speed does the projectile strike the ground?

7. The horizontal component of the initial velocity of a projectile is v_{xo}. The maximum height of the projectile is H and its range would be R. If the projectile explodes into two equal fragments when it is just at its maximum height, and if one of the fragments falls vertically downward, show that the other fragment strikes the ground at a distance of 3/2 R from the point the projectile was fired.

Chapter 9
Conservation of Linear Momentum

PREVIEW

This chapter introduces the second quantity derivable from Newton's laws of motion that describes the effect of the environment on a particle or system. This quantity is linear momentum. Linear momentum, unlike energy, is a vector quantity. It is introduced by means of the linear impulse which a force exerts on a system during the time that it acts. The conditions for the development of the law of conservation of momentum are developed and this law is applied to collisions of particles and the interaction of systems on which no net external impulse acts.

SUMMARY
9.1 LINEAR IMPULSE

Definition. Linear impulse (one dimension, constant force case): the product of the magnitude of the force times the time interval over which the force acts. We will employ the symbol J for impulse.

$$J = F\Delta t \tag{9.1}$$

The SI units of impulse are Newton-seconds or kg-m/s. Its dimensions are ML/T. Although it is not readily apparent why one would define this quantity, we shall see that it is particularly useful in studying cases where forces act over short periods of time. In these cases it is impractical to measure the actual force, however the impulse can be determined by measuring the change in a related quantity which we will soon define.

Definition. Linear impulse (one dimension, variable force): if the force acting on a system varies with time the linear impulse must be defined by a limiting process. It is given by:

$$J = \int_{t_i}^{t_f} F\,dt \tag{9.2}$$

Definition. Average force (one dimension, variable force): the impulse divided by the time interval over which the force acts.

$$\overline{F} = J/\Delta t \tag{9.3}$$

If we substitute Eq. 9.2 into Eq. 9.3 we find the standard definition of the time average of a quantity

$$\overline{F} = 1/(t_f - t_i) \int_{t_i}^{t_f} F\,dt \tag{9.4}$$

Let us consider Newton's second law as we have used it in the past. We see that it can be

transformed to a form which involves the linear impulse.
$$F = ma = mdv/dt$$
When the mass of the system is constant, m can be carried across the differential and this expression becomes:
$$F = d(mv)/dt \tag{9.5}$$

> Definition. Linear momentum: the product of a system's instantaneous mass times its instantaneous (vector) velocity.

The linear momentum of a system is designated by $p = mv$ (Eq. 9.6). With this definition we may recast Newton's 2nd law into a form called the impulse-momentum form.
$$F = dp/dt \tag{9.7}$$
$$Fdt = dp$$

This equation states that the instantaneous <u>net</u> external linear impulse acting on a system is equal to the instantaneous change of momentum of the system.

Our definition of linear momentum may seem more general than is justified by our derivation using the assumption of constant mass. In fact the impulse-momentum form of Newton's 2nd law is the more generally valid form. This form of the 2nd law holds for systems with non-constant mass as well as for systems moving at relativistic velocities; situations where the mass x acceleration form is not valid.

If we integrate both sides of the impulse-momentum equation we find that the change in momentum of a system over a finite time interval is the <u>net vector</u> linear impulse.

$$\int_{t_i}^{t_f} Fdt = J = \int_{p_i}^{p_f} dp = \Delta p = P_f - P_i \tag{9.9}$$

This equation is called the linear impulse-momentum principle. You should note that in this case the linear impulse is not restricted to one dimension but exhibits its full vector nature. However, it is only the <u>net</u> external force and therefore the <u>net</u> external linear impulse which satisfies Eq. 9.9.

9.3 CONSERVATION OF LINEAR MOMENTUM

If the net external linear impulse or the net external force on a system vanishes then the impulse-momentum form of Newton's 2nd law or the linear impulse-momentum principle tells us that the change in the linear momentum also vanishes. This condition gives rise to another very important conditional conservation law of nature.

> <u>Law of Conservation of Linear Momentum</u>: If the net external force or impulse acting on a system vanishes (i.e. is zero) then the linear momentum of the system is constant.

If you think about it for a while you should see that this statement encompasses Newton's 1st law. It is, in fact, more general. Although this is still a conditional conservation law, it is one of the most powerful tools available to scientists who work at the frontiers of modern physics where the conservation principles are about all we have to guide our investigations.

One of the most powerful results of the above transformation of Newton's 2nd law is for a system of particles or objects which interact with one another via internal forces obeying Newton's 3rd law. Since these forces are internal to the system, by considering the group of particles or systems as a whole, these forces have no effect on the total linear momentum of the system (the total linear momentum being defined as the sum of the individual momenta of the systems respective parts.)

Of course if you wish to know how the individual parts of the system's momenta change with time, then you must consider the "external" forces acting on that part of the system. These will include the forces that other parts of the large system exert on the part in question. What were formally internal forces with matching reaction forces are now external forces with respect to the part of the system you are considering.

Again we emphasize that the law of conservation of momentum is an extremely important and powerful physical principle which you should try to understand fully.

9.4 TWO-PARTICLE COLLISIONS

One of the most fruitful applications of the law of conservation of linear momentum is to the study of collisions. In many instances, especially at the atomic and subatomic level, the external forces acting on a pair of colliding objects is zero or can be considered negligible with respect to the forces with which the objects interact. You might think that this is only true at the atomic level of nature but if you ever see two tractor-trailer rigs collide you would certainly change your opinion. We will discuss collisions of two particles in one dimension first and then generalize to the multidimensional case.

> Definition. Perfectly elastic collision: if the total kinetic energy of the colliding particles or systems is the same after the collision as before (i.e. the total kinetic energy is conserved) then the collision is perfectly elastic.

In a perfectly elastic collision no energy is lost to friction, heat, or deformation of the bodies involved. The individual participants in the collision may each have different values of the kinetic energy before and after the collision but the sum total of all their kinetic energies remains unchanged.

The actual value of the momenta and kinetic energies of the participants in the collision can be determined by application of the law of conservation of momentum and the criterion that the total kinetic energy is conserved. The case for the collision of two particles in one dimension is treated in the text. The general result for the velocities of the two particles after the collision are given in Eqs. (9.21) and (9.22).

$$v'_1 = \frac{m_1 - m_2}{m_1 + m_2} v_1 + \frac{2 m_2}{m_1 + m_2} v_2 \qquad (9.21)$$

$$v'_2 = \frac{2 m_1}{m_1 + m_2} v_i - \frac{m_1 - m_2}{m_1 + m_2} v_2 \qquad (9.22)$$

The effect of the relative masses of the colliding particles is shown in Table 9.1 page 183. This table should be studied to help give you a "feel" for what happens in a collision between two particles. Especially to note is the fact that the maximum transfer of kinetic energy from one particle to another occurs when the particles have the same mass.

Definition. <u>Perfectly inelastic collision</u>: If when two bodies collide their relative velocity after the collision is zero, the collision is perfectly inelastic.

In a perfectly inelastic collision the two objects actually stick together and become one object whose mass is the sum of the original masses and whose velocity is common to both of the masses. The equation for a perfectly inelastic collision in one dimension comes directly from the law of conservation of momentum.

$$m_1 v_1 + m_2 v_2 = (m_1 + m_2) v \qquad (9.26)$$

These are the only simple cases which can be handled analytically. If the collision is intermediate between these two extremes, some other information must be given or the final velocities actually measured to determine what has occurred. Do not forget, however, that as long as the external forces vanish the law of conservation of momentum applies. Difficulty arises with the work done on the objects by internal forces which cause the loss of kinetic energy. These internal forces do not effect the momentum conservation as noted above.

In two or more dimensions the law of conservation of momentum may be treated in terms of its vector components. Just as kinematics, forces, and Newton's laws obey the principle of superposition, so does the law of conservation of linear momentum. The conservation conditions can be applied independently to the orthogonal components of the various momentum vectors involved.

EXAMPLE PROBLEMS
Example 1

The force exerted by a tennis racket on a tennis ball of mass 30 g during a volley is depicted in Fig. 1-1. a) What is the impulse delivered to the tennis ball? b) What is the average force which acts on the ball? c) If the ball rebounds off the racket with a speed 3/2 as large as that which it had before it struck the racket, what are its initial and final speeds? d) How much work did the racket do on the ball?

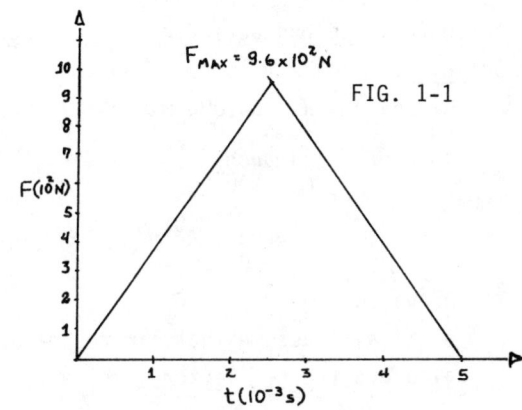

FIG. 1-1

Solution

Given: $F(t)$
$F_{max} = 9.6 \times 10^2$ N
$\Delta t = 5 \times 10^{-3}$ s
$m = 0.03$ kg
$v_f = 3/2 \; v_i$

Required: a) J, b) F, c) v_i, v_f d) W_{net}

a) The force exerted on the tennis ball by the racket is much greater than its weight (0.294 N). We can, therefore, neglect the force of gravity and treat this as one dimensional motion. The linear impulse is given by Eq. 9.9

$$J = \int_{t_i}^{t_f} F \, dt = \text{area under F vs t curve} = 2(1/2)(2.5 \times 10^{-3} \text{s})(9.6 \times 10^2 \text{ N}).$$

$$J = 2.4 \text{ N-s}$$

The direction of J is the same as that of the rebounding ball.

b) The average force is $F = J/\Delta t = 2.4 \text{ N-s}/5 \times 10^{-3} \text{ s} = \underline{4.8 \times 10^2 \text{ N}}$

c) From the linear impulse-momentum principle

$$J = \Delta p = p_f - p_i$$

We choose the direction of motion of the rebounding ball as the +x direction.

$$p_f = mv_f \hat{i} = 3/2 \, mv_i \hat{i} \quad ; \quad p_i = mv_i(-\hat{i}) = -mv_i \hat{i}$$

$$J = 2.4\hat{i} \text{ N-s} = p_f - p_i = 3/2 \, mv_i \hat{i} - (-mv_2 \hat{i})$$
$$= 5/2 \, mv_i \hat{i} = 5/2 \, (0.03 \text{ kg}) v_i \hat{i}$$

Solving for v_i

$$v_i = (2.4 \text{ N-s})/(5/2)(0.03 \text{ kg}) = \underline{32 \text{ m/s}}$$
$$v_f = (3/2) v_i = \underline{48 \text{ m/s}}$$

d) The net work done may be determined by the work-energy principle

$$W_{net} = K = 1/2 \, mv_f^2 - 1/2 \, mv_i^2$$
$$= (1/2)(0.03 \text{ kg})(48 \text{ m/s})^2 - 2/1(0.03 \text{ kg})(32 \text{ m/s})^2$$
$$= 34.56 \text{ J} - 15.36 \text{ J}$$
$$W_{net} = \underline{19.2 \text{ J}}$$

Example 2

The battleship New Jersey can fire a shell whose mass is 900 kg a maximum distance of 38 km. Assume the shell is fired at an angle of 45°. a) What impulse must the gun provide? b) If the force from the gun acts for 0.01 s, what is its average value?

Solution

Given: R = 38,000 m

θ_{launch} = 45°
m = 900 kg
Δt_{gun} = 0.01 s
v_i = 0

Required: a) J , b) F

a) We will assume that air resistance, curvature of the earth, etc. are negligible. In order to calculate the net vector impulse we must apply the linear impulse-momentum principle.

$$J = \Delta p$$

We calculate Δp by first determining the muzzle velocity of the projectile using the maximum range Eq. 5.17.

$$R = -2 v_0^2 \sin\theta \cos\theta / a_y = -v_0^2 \sin 2\theta / a_y = -v_0^2 \sin 90°/(-9.8 \text{ m/s}^2)$$

$$= v_0^2 \sin 90°/9.8 \text{ m/s}^2 = 38,000 \text{ m}$$

which gives:

$$v_0 = 610 \text{ m/s}$$

The initial momentum is zero. The final momentum is:

$$p_f = mv_f = mv_0 \cos\theta \hat{i} + mv_0 \sin\theta \hat{j} = (900 \text{ kg})(610 \text{ m/s})(.707)\hat{i}$$
$$+ (900 \text{ kg})(610 \text{ m/s})(.707)\hat{j}$$
$$p_f = 388143 \text{ N-s } \hat{i} + 388143 \text{ N-s } \hat{j}$$

By the linear impulse-momentum principle

$$J = \Delta p = p_f - 0 = \underline{388143 \text{ N-s } \hat{\imath} + 388143 \text{ N-s } \hat{\jmath}}$$

b) The average force is

$$F = J/\Delta t = (388143 \text{ N-s } \hat{\imath} + 388143 \text{ N-s } \hat{\jmath})/0.01 \text{ s}$$

$$F = 38,814,300 \text{ N } \hat{\imath} + 38,814,300 \text{ N } \hat{\jmath}$$

The magnitude of $F = \sqrt{F_x^2 + F_y^2} = 54,891,709$ N, not an inconsiderable force.

Example 3

A 300 gram cue ball traveling at 2 m/s suffers a perfectly elastic head on collision with an 8 kg bowling ball which is at rest. What is the final velocity of a) the cue ball? b) the bowling ball?

Solution

Given: $m_{CB} = 0.3$ kg
$m_{BB} = 8$ kg
$v_{CB} = +2$ m/s
$v_{BB} = 0$

FIG. 3-1

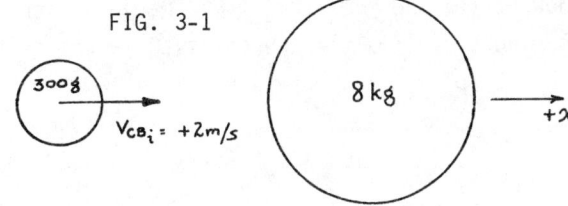

Required: a) v'_{CB}, b) v'_{BB}

a) This is a perfectly elastic collision in one dimension with the target at rest. We may employ Eqs. 9.23 and 9.24 from the text with $v_1 = v_{CB}$, $v_2 = v_{BB} = 0$, $m_1 = m_{CB}$, and $m_2 = m_{BB}$. The positive x direction is chosen as the direction in which the cue ball is initially traveling. Using Eq. 9.23 we find:

$$v'_{CB} = (m_{CB} - m_{BB}) v_{CB}/(m_{CB} + m_{BB}) = (0.3-8)\text{kg} (2 \text{ m/s})/(0.3+8)\text{kg}$$

$$v'_{CB} = -1.85 \text{ m/s}$$

We see that the cue ball recoils in the negative x direction after the collision.

b) For the velocity of the bowling ball after the collision we employ Eq. (9.24)

$$v'_{BB} = 2 m_{CB} v_{CB}/(m_{CB}+m_{BB}) = (0.6 \text{ kg})(2 \text{ m/s})/(0.3 + 8) \text{ kg} = 0.144 \text{ m/s}$$

Example 4

An automobile with a mass of 2000 kg traveling east at 60 m/s collides with a truck of mass 10,000 kg traveling north at 45 m/s. The two vehicles stick together when they collide. a) What is their resultant velocity just after the collision? b) How much kinetic energy is lost during the collision?

FIG. 4-1

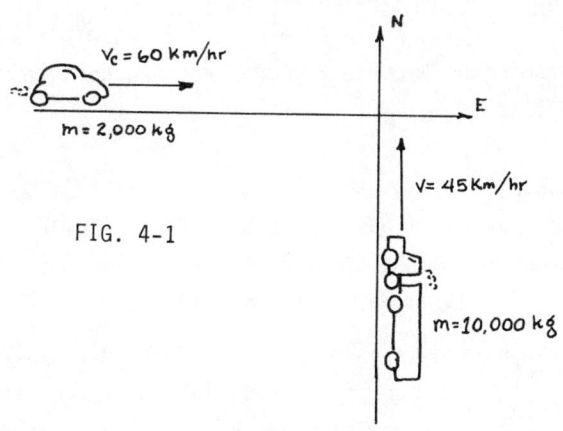

Solution

Given: $m_C = 2,000$ kg
$m_T = 10,000$ kg

v_C = 60 m/s, east
v_T = 40 m/s, north

Required: a) **v** after the collision, b) ΔK

This is a two dimensional perfectly inelastic collision. We assume that the external forces acting on the car and truck are negligible compared to the interaction forces. We apply the law of conservation of momentum to each component of the momentum. We choose the positive x axis east and the positive y axis north.

The momentum before the collision is:

$$p_x = m_c v_c = (2000)(60\ \hat{i})\text{kg-m/s} = 120,000\ \hat{i}\ \text{kg-m/s}$$

$$p_y = m_T v_T = (10,000)(40\ \hat{j})\text{kg-m/s} = 400,000\ \hat{j}\ \text{kg-m/s}$$

The components of the momentum after the collision must equal those before the collision. The mass of the moving object is now the combined masses of the car and truck.

$$p' = (12,000)(v_x \hat{i})\text{kg-m/s} = 120,000\ \hat{i}\ \text{kg-m/s}$$

$$v_x = \underline{10\ \text{m/s}}$$

$$p' = (12,000)(v_y \hat{j})\text{kg-m/s} = 400,000\ \hat{j}\ \text{kg-m/s}$$

$$v_y = \underline{33.33\ \text{m/s}}$$

$$\mathbf{v} = \underline{(10\ \hat{i} + 33.33\ \hat{j})\ \text{m/s}}$$

b) The initial kinetic energy is the sum of the kinetic energies of the car and truck before the collision.

$$K = 1/2\ m_c v_c^2 + 1/2\ m_T v_T^2$$

$$= 1/2(10,000\ \text{kg})(40\ \text{m/s})^2 + 1/2(2,000\ \text{kg})(60\ \text{m/s})^2$$

$$= 8 \times 10^6 \text{J} + 3.6 \times 10^6 \text{J}$$

$$K = 1.16 \times 10^7 \text{J}$$

After the collision the magnitude of the velocity is

$$v = \sqrt{(10\ \text{m/s})^2 + (33.33\ \text{m/s})^2} = 34.8\ \text{m/s}$$

and the final kinetic energy is

$$K' = (1/2)(m_c + m_T)v^2 = (1/2)(12,000\ \text{kg})(34.8\ \text{m/s})^2 = 7.27 \times 10^6 \text{J}$$

The change in kinetic energy is

$$K = K_f - K_i = 7.27 \times 10^6 \text{J} - 1.16 \times 10^7 \text{J} = -4.33 \times 10^6 \text{J}$$

This is the amount of energy which goes into deformation and heating of the two vehicles during the collision.

PROBLEMS

1. A baseball has a mass of 0.2 kg. The ball is thrown with a velocity of 30 m/s and after being batted has a speed of 50 m/s in the opposite direction. a) What is the change in momentum of the ball? b) What is the impulse?

2. A spring is used to fire a dart of mass 20 g from a child's toy gun. The dart leaves the gun with a speed of 1.2 m/s. The force from the spring acts on the dart for 0.04 s. a) What is the change in momentum of the dart? b) What is the impulse? c) What average force acts on the dart? d) How much work is done on the dart by the spring?

3. The force which a bow exerts on an arrow is given by:
$$F = 60 - [1.5 \times 10^5 \text{ m/s}^2]t^2 \text{ N}$$
What is the impulse this force provides on the arrow during the time from which the bow is released (t = 0) until the force becomes zero?

4. A particle of mass m undergoes a perfectly elastic collision with a larger mass M. If the initial kinetic energy of the smaller mass is E_o, show that the maximum kinetic energy it can <u>lose</u> during the collision is $4mME_o/(M + m)^2$.

5. A 10 kg mass sliding across a smooth surface collides completely inelastically with another block of mass 12 kg traveling in the opposite direction at 1.5 m/s. The final velocity of the system is 2 m/s in the direction which the 10 kg block was originally moving. What was the initial velocity of the 10 kg block?

6. The ballistic pendulum is a device used to measure the velocities of small projectiles. A bullet makes a completely inelastic collision into a block of material which is allowed to swing freely as a pendulum. Immediately after the collision the momentum of the bullet and block equals the momentum of the bullet before it struck the block. The kinetic energy of the block is then converted to gravitational potential energy. If the mass of the bullet is m and it has a velocity v and the mass of the block is M, show that the velocity of the bullet is:
$$v = \frac{m + M}{m}\sqrt{2gy}$$

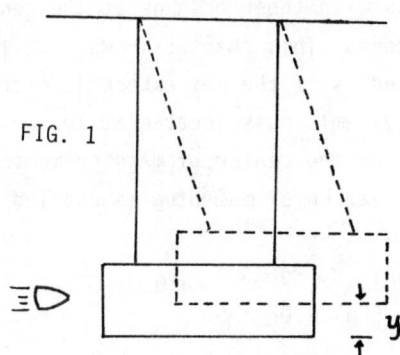

FIG. 1

where y is the maximum height to which the block rises.

7. A bullet weighing 5 g passes through a wooden block of mass 0.5 kg attached to a string 2 m long which hangs from the ceiling. The center of gravity of the block rises 0.3 m due to the impact of the bullet. What is the speed of the bullet as it emerges from the block if its original speed is 500 m/s? (Assume the bullet spends a negligible amount of time in the block.)

8. Two blocks, one having twice the mass of the other, have a spring compressed between them. The potential energy stored in the spring is 90 J. What is the kinetic energy of each block if the blocks are released?

9. A billiard ball rests on a smooth table top. A cue ball of the same mass traveling at 10 m/s strikes the billiard ball and is deflected 30° from its original direction. The billiard ball acquires a velocity directed 45° with respect to the original velocity. a) Compute the speed of each ball after the collision. b) Is the collision perfectly elastic? c) If not, what fraction of the original kinetic energy is lost?

Chapter 10
Many-Particle Systems

PREVIEW
Most of the mechanics we have studied so far involves particles which have no extension in space, or systems which behave like particles. In many cases we have used examples where the system is not a particle, but we have largely ignored any complications which might arise because of this. This chapter introduces the concept of the center of mass. Using the center of mass or coordinate frames with their origins at the center of mass allows us to justify the results of earlier chapters. This chapter shows that the <u>translational</u> motion of a system of particles can be treated as if the net external forces acting on the system act on a single particle whose mass is the system's mass located at the center of mass. We also treat classical relativity again and consider the center of mass or momentum reference frame. The last section of the chapter deals with conservation of momentum as applied to a special case of nonconstant mass.

SUMMARY
10.1 CENTER OF MASS

<u>Definition</u>. <u>Center of mass</u>: in a collection of particles or a continuous mass distribution, the point defined by:

$$r_{cm} = (1/M) \sum_i m_i r_i \quad \text{(discrete masses)} \tag{10.1}$$

or

$$r_{cm} = (1/M) \int r\, dm \quad \text{(continuous mass)} \tag{10.6}$$

is called the center of mass of the system. M is the total mass of the system.

Alternatively we could define the components of the center of mass but the definition is not independent of that above. The components of the center of mass can be determined in any orthogonal coordinate system but they are especially easy to compute in Cartesian coordinates.

$$\begin{aligned} x_{cm} &= (1/M) \sum m_i x_i \quad \text{or} \quad (1/M) \int x\, dm \\ y_{cm} &= (1/M) \sum m_i y_i \quad \text{or} \quad (1/M) \int y\, dm \\ z_{cm} &= (1/M) \sum m_i z_i \quad \text{or} \quad (1/M) \int z\, dm \end{aligned} \tag{10.3, 10.8}$$

The center of mass of a system is obviously a function of its geometry. Although the position vector of the center of mass in space as expressed in an arbitrary reference frame depends on the choice of the origin, <u>the position of the center of mass relative to the system is fixed</u>. This means that the center of mass of a system is a real physical quantity which "belongs" to the system and is determined only by the system's configuration. For example, the center of mass of an equilateral triangle made of uniform material is in the center of the triangle, i.e. at the point where the bisectors of the three angles cross. Depending on our choice of reference frames we may label this point with different $\vec{r}$'s, but they will all point to the same location relative to the system.

In a uniform gravitational field the center of mass and gravity are identical. In nonuniform gravitational fields the center of mass and the center of gravity may vary substantially from one another.

In the definition above for continuous mass distributions, the integral over dm must be transformed into an integral over the volume of the object using the density.

10.2 DYNAMICS OF THE CENTER OF MASS

To completely determine the motion of a system of particles one would be required to specify the position vector of each particle (or point within the body) as a functin of time. Even though this is theoretically possible in the framework of Newtonian mechanics, it is practically impossible if the number of particles becomes very large. Furthermore, this approach actually makes the physics of the motion inscrutable because of the large number of variables which one must keep track of. To use a trite expression, you can't see the forest for the trees. Fortunately there is a simple reduction of the motion of a system of particles which can be performed which provides a much more intuitive conceptualization. We have taken the first step in this reduction by defining the center of mass. We now show that the translational dynamics of a system can be treated as if the system consisted of a single particle of mass M, the mass of the system, concentrated at the center of mass.

> **Definition.** **Center-of-mass velocity:** the time rate of change of the center-of-mass position vector r_{cm}. The center-of-mass velocity is equal to the center-of-mass weighted sum of the velocities of the individual particles.
>
> $$v_{cm} = dr_{cm}/dt = (1/M) \Sigma m_i v_i \qquad (10.9, 10.10)$$

Although this definition seems trivial it has far reaching consequences. If we write down the total momentum of the system of particles, we can see immediately that it can be expressed in terms of the velocity of the center of mass.

$$P_{total} = P_i = \Sigma m_i v_i = M(1/M) \Sigma m_i v_i = M v_{cm}$$

Thus the total linear momentum of the system can be written as if it were the momentum of a single particle of mass M moving with v_{cm}. We could proceed as in the text to define a_{cm} and apply Newton's 2nd law in the form F = ma to achieve the result we desire, but let's get there directly by using the impulse-momentum form of the 2nd law. We have the total momentum; its time rate of change is the <u>net external</u> force which acts on the system. We may write immediately:

$$dp_{total}/dt = \Sigma F_{ext} = M dv_{cm}/dt = M a_{cm} \qquad (10.16)$$

And we see that, as was mentioned above, the translational motion of the <u>system</u> can be treated as if we had a single particle of mass M at the center of mass. This result deserves some more comment. It tells us that whatever the relative motion of the particles which make up the system might be, the response of the total system to the net external force acting on it cause the whole system to accelerate and move through space as if it were a point particle of mass M. It is also important to note that if one sums all the forces acting on all the particles of the system, the result is the net external force acting on the system because the internal forces sum to zero by Newton's 3rd law.

If the net external force acting on the system vanishes then we know from the last chapter that the total linear momentum of the system is conserved. This implies that the <u>sum</u> of the individual linear momenta of all the particles must remain constant also.

The total kinetic energy of the system does not depend on just the external work done, but on the work done by internal forces also. The kinetic energy of the system cannot in general be expressed as some simple function of the center of mass parameters. Therefore, it need not be conserved even when the net external force on the system is zero, i.e. when the total linear momentum is conserved.

10.3 GALILEAN RELATIVITY

We have already dealt with inertial reference frames in Chapters 4 and 5. Most simply stated, inertial reference frames are those which move with constant relative velocity. If the velocities are not large, then the coordinate and velocity transformations between inertial reference frames are the Galilean transformations derived in Chapter 4. If we consider the primed frame to be moving in the direction of the positive x axis in the unprimed (rest) frame with velocity v, then the transformations are:

$$x = x' + vt \tag{10.22}$$

$$x' = x - vt \tag{10.23}$$

$$u = u' + vt \tag{10.24}$$

Since v is a constant we see that:

$$d^2x'/dt^2 = a' = d^2x/dt^2 = a$$

Since observers in the two frames of reference measure the same accelerations, Newton's laws will have the same form in both frames. The actual observation of positions and velocities may differ in the two frames but the observed forces and accelerations will not. We recall these developments at this time in order to have them fresh in our minds for the next item of discussion.

10.4 CENTER-OF-MOMENTUM REFERENCE FRAME

In many cases, especially collisions in which there are no net external forces acting, and linear momentum is conserved, it is useful to transform coordinates from a laboratory fixed frame of reference to one which is moving with the center of momentum (or mass).

<u>Definition</u>. <u>Center-of-momentum reference frame</u>: any frame of reference whose origin is such that the total linear momentum **p** of the system is zero.

In Newtonian mechanics (i.e. nonrelativistic mechanics) $p_{total} = \Sigma p_i = \Sigma m_i v_i = M v_{cm}$. Therefore, we see that in the center-of-momentum frame of reference $v_{cm} = 0$. The above definition does not specify the origin of this reference frame. It is usually most convenient to choose the origin to coincide with the center of mass so that $r'_{cm} = 0$. This special reference frame is called the center-of-mass reference frame. If $r'_{cm} = 0$ then obviously $v'_{cm} = 0$ and the center-of-mass reference frame is also a center-of-momentum reference frame.

We may transform velocities from a laboratory fixed reference frame to the center of mass reference frame using the Galilean velocity transformation.

$$v = v' + v_{cm} \tag{10.31}$$

In collisions where p_{total} it conserved, transformation of the velocities of the colliding particles from the laboratory frame of reference to the center-of-mass frame reveals that in the center-of-mass frame the velocities of the two particles are colinear and oppositely directed both before and after the collision. The initial and final center-of-mass velocities for perfectly elastic and inelastic one dimensional collisions are derived on page 201 of the text.

The initial velocities in both cases are:

$$v'_{1_i} = m_2(v_{1_i} - v_{2_i})/(m_1 + m_2) \quad (10.32)$$

$$v'_{2_i} = m_1(v_{2_i} - v_{1_i})/(m_1 + m_2) \quad (10.33)$$

For a totally inelastic collision the two particles stick together and must have a common velocity. Obviously this must be v_{cm}, so the final velocities of the two particles in the center-of-mass reference frame are zero. In the case of a totally elastic collision the initial velocities and final velocities have the same magnitude but opposite directions (see Eqs. 10.34 and 10.35).

The statements made in the above paragraph hold true in collisions in higher dimensions as well. A two dimensional collision can be analyzed in the center-of-mass frame of reference much more easily than in a laboratory frame. We consider the case of a particle with velocity v_1 and mass m_1 incident on a particle at rest with mass m_2. Since the orientation and origin of the laboratory frame are arbitrary, we choose the x axis to be along the line connecting the incident and stationary particles, and choose the origin at the position of the stationary particle, or target as it is usually called. The center-of-mass frame will also have its x axis oriented along

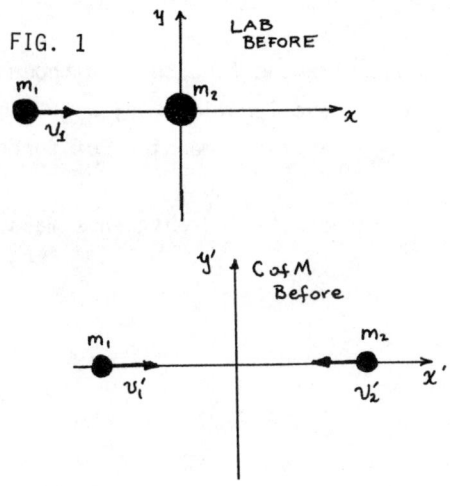

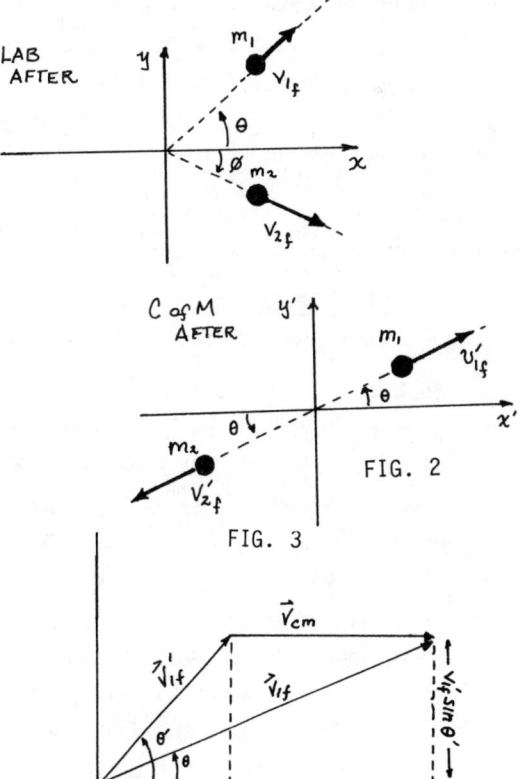

the line connecting the particles. The two frames of reference and the velocities one sees before the collision are shown in FIG. 1. Note that with this choice of orientation the axes of the two reference frames are parallel and v_{cm} lies along the x axis.

After the collision the two particles move off at different angles with respect to the laboratory system. We will call the angle which the incident particle makes with the x axis the scattering angle. This angle is chosen since the target particle is rarely observed. Since momentum is assumed conserved in the collision, v_{cm} will still point along the x axis after the collision.

In the center-of-mass reference frame the target and incident particle move away from each other along colinear paths which make an angle θ with the x axis. Note that in the center-of-mass frame there is no ambiguity about what to call the scattering angle. The particles after the collision as viewed from the two reference frames are shown in FIG. 2.

Since $\vec{v} = \vec{v}' + \vec{v}_{cm}$ and since we know $\vec{v}_{cm}$ is parallel to the x axis we can consider the graphical vector addition of these quantities. This is shown in FIG. 3 which corresponds to FIG. 10.24 in the book. The appropriate trigonometry is shown which gives the relationship between the scattering angles in the two frames of reference.

$$\tan\theta = v'_{1_f} \sin\theta' / (v'_{1_f} \cos\theta' + v_{cm}) \tag{10.42}$$

10.5 ROCKET PROPULSION

Newton's 2nd law in the mass x acceleration form cannot be applied to systems whose mass changes with time. In most cases such problems must be handled by using the impulse-momentum form of the 2nd law. If the net external force on the system vanishes, then conservation of momentum considerations can be employed. A rocket in a force free environment is an example of a system with no net external force but which exhibits the results of work performed by internal forces.

If the rocket ejects mass from its engines with a velocity u_o relative to the rocket at a rate $\alpha = -dm/dt$ then the instantaneous mass of the rocket is related to its instantaneous acceleration by:

$$u_o \alpha = m(t)a(t) \tag{10.48}$$

The functional reference to time has been added to emphasize the fact that, although the left hand side of this equation is constant, both the mass and the acceleration vary with time. As the mass decreases the acceleration will increase. The term $u_o \alpha$ has the dimensions of force and is called the thrust of the rocket engine.

The velocity of the rocket can be expressed as a function of its instantaneous mass:

$$v = v_o + u_o \ln(m_o/m) \tag{10.50}$$

EXAMPLE PROBLEMS
Example 1

Three equal masses are arranged in a plane so that they form an equilateral triangle of side length a. a) Using the coordinate system shown in FIG. 1-1 calculate r_{cm}. b) Show that the same result for r_{cm} is obtained by first calculating the center of mass of two of the particles and then calculating the center of mass of the third particle and a fictitious particle of mass 2m located at the position of the center of mass of the other two.

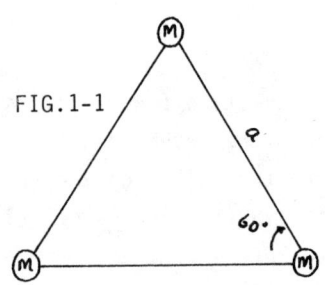

FIG. 1-1

Solution

Given: $m_1 = m_2 = m_3 = m$

Required: r_{cm}

a) First we write down the position vectors of the three masses in the chosen coordinate frame.

$$r_1 = 0\hat{i} + 0\hat{j}, \quad r_2 = a\hat{i} + 0\hat{j},$$
$$r_3 = 0.5\, a\hat{i} + a\sin 60°\, \hat{j} = 0.5\, a\hat{i} + 0.867\, a\hat{j}$$

We calculate the components of the center of mass using Eq. (10.3).

$$x_{cm} = (mr_{1x} + mr_{2x} + mr_{3x})/3m = (a + 0.5a)/3 = 0.5a$$
$$y_{cm} = (mr_{1y} + mr_{2y} + mr_{3y})/3m = 0.867a/3 = 0.289a$$

Thus $r_{cm} = \underline{0.5a\,\hat{i} + 0.289a\,\hat{j}}$

b) For simplicity we choose to first calculate the center of mass of the two particles lying on the x axis. By symmetry it should be evident that $r_{12cm} = \underline{0.5\, a\hat{i}}$. We proceed with the prescribed calculation.

$$r_{12} = 0.5a\,\hat{i} + 0\hat{j}, \quad m_{12} = 2m\,; \quad r_3 = 0.5a\,\hat{i} + 0.867a\,\hat{j}$$

We calculate the center of mass in vector format this time so that you may be familiar with it.

$$r_{cm} = [2m(0.5a\,\hat{i} + 0\hat{j}) + m(0.5a\,\hat{i} + 0.867a\,\hat{j})]/(2m + m)$$

$$r_{cm} = (1.5a\,\hat{i} + 0.867a\,\hat{j})/3 = \underline{0.5a\,\hat{i} + 0.289a\,\hat{j}}$$

and the results are identical. This is a general property of the additive nature of the definition of center of mass.

Example 2

A brass cylinder (density = ρ) of height h and radius R has a right circular conical cavity of depth h/2 and base radius R machined out of one end. The z axis indicated in FIG. 2-1 is the symmetry axis. Where is the center of mass of this object?

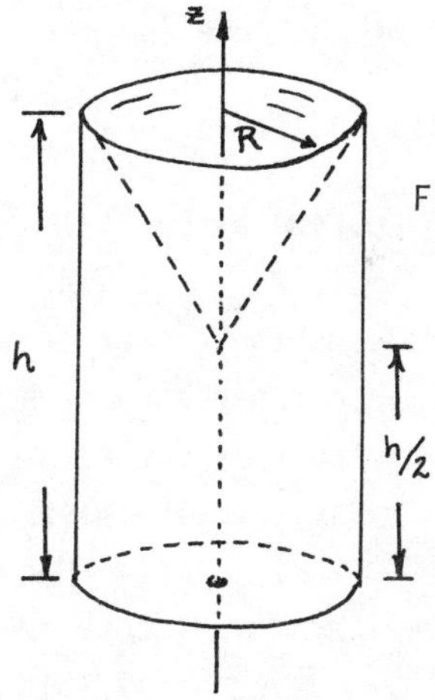

FIG. 2-1

Solution

Given: R = radius of cylinder, base of cone
h = height of cylinder
h/2 = "height" of cone
ρ = density of brass

Required: r_{cm}

One could try to perform an integration over the volume of this object, but a more clever (and much easier) way to approach this problem is to employ the result of Example Problem 1. Symmetry demands that the center of mass lie on the z axis, we only need determine its z coordinate. Let us call M the mass of the solid brass cylinder and m_{cone} the mass of a solid brass cone of the same dimensions as the cavity. Then $M_o = M - m_{cone}$. We may calculate Z_{cm} of the solid brass cylinder using the result of problem 1 as

$$Z_{cm}^{solid} = [(M - m_{cone})\, Z_{cm}^o + m_{cone}\, Z_{cm}^{cone}]/M$$

where Z_{cm}^o is the z component of the center of mass of the machined cylinder. From 1 we may write, after a little algebra:

$$Z^0_{cm} = (MZ^{solid}_{cm} - m_{cone}Z^{cone}_{cm})/(M - m_{cone}) \tag{1}$$

The z component of the center of mass of a solid brass cylinder would be located at h/2, again by symmetry. The mass of a solid cylinder would be $\pi R^2 \rho h$.

We need calculate only the mass and center of mass of a solid cone which we "subtract" from the solid cylinder. In order that the integration not be complicated by our coordinate system, we calculate the center of mass of a right circular cylinder of base radius R and height a using the geometry of FIG. 2-2. The z coordinate of the center of mass is:

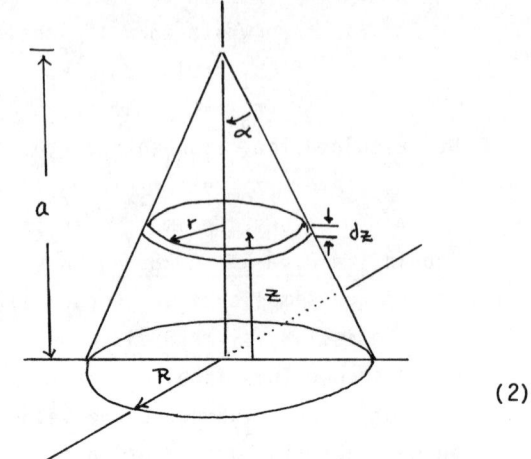

FIG. 2-2

$$Z_{cm} = (1/m_{cone})\int_0^a zdm \tag{2}$$

$dm = \pi(r')^2 \rho dz$ and $r' = (a-z)\tan \alpha = (a-z)(R/a)$.
Substituting into Eq. (2) gives:

$$Z_{cm} = (\pi\rho R^2/a^2 m_{cone}) \int_0^a (a-z)^2 zdz$$

$$= (\pi\rho R^2/a^2 m_{cone})a^4(1/2 - 2/3 + 1/4) = (\pi\rho R^2 a^2)/12 \, m_{cone}) \tag{3}$$

The mass of the cone is

$$m_{cone} = \rho\int_0^a \pi r'^2 dz = \pi\rho R^2/a^2 \int^a (a-z)^2 dz = (1/3) \pi\rho R^2 a \tag{4}$$

Dividing Eq. (3) by Eq. (4) gives the center of mass of the cone.

$$Z_{cm} = (1/4)a \tag{5}$$

which tells us the center of mass lies on the z axis one fourth of the height of the cone from the bottom. In our geometry $a = h/2$ and the center of mass of the cone would be located $(1/8)h$ from the top of the cylinder or at $(7/8)h$ from the bottom. We can now calculate r_{cm} for our machined cylinder. The mass of the cone is $(1/6)\pi\rho R^2 h$

$$Z_{cm} = [(\pi R^2\rho h)(h/2) - (\pi R^2\rho h/6)(7h/8)]/(\pi R^2\rho h - \pi R^2 \rho h/6)$$

after some tedious algebra we find

$$Z_{cm} = \underline{(17/40)h}$$

Example 3

A block of mass 2 kg traveling at 10 m/s is approaching a 5 kg block at rest. They undergo a completely inelastic collision. a) What is the final laboratory velocity of the blocks? b) What is the center-of-mass velocity? c) What are the velocities of the two blocks in the center-of-mass frame before the collision? d) What is the center-of-mass total kinetic energy before the collision? e) What is the difference in the

initial and final laboratory total kinetic energy?
Solution
Given: $m_1 = 2$ kg
$m_2 = 5$ kg
$v_{1i} = 10$ m/s
$v_{2i} = 0$

Required: a) v_{final}, b) v_{cm}, c) v'_{1i}, v'_{2i}, d) K'_{cm}, e) K

a) This question is similar to problems we have worked before in Chapter 9. Since the final and initial momenta must be the same:
$$p_i = m_1 v_{1i} = p_f = (m_1 + m_2) v_f = (2 \text{ kg})(10 \text{ m/s}) = (7 \text{ kg})(v_f)$$
$$v_f = \underline{2.86 \text{ m/s}}$$

b) Since this is a perfectly inelastic collision we know that both blocks move together after the collision. Therefore the velocity of the center of mass must be 2.86 m/s. This can be checked by using Eq. 10.10:
$$v_{cm} = \frac{m_1 v_{1i} + m_2 v_{2i}}{m_1 + m_2} = \frac{(2 \text{ kg})(10 \text{ m/s})}{7 \text{ kg}} = \underline{2.86 \text{ m/s}}$$

c) In order to calculate the velocities of the blocks in the center-of-mass frame we employ Eq. 10.31:

$v_{1i} = v'_{1i} + 2.86$ m/s 10 m/s $= v'_{1i} + 2.86$ m/s v'_{1i} $\underline{7.14 \text{ m/s}}$

$v_{2i} = v'_{2i} + 2.86$ m/s $0 = v'_{2i} + 2.86$ m/s $v'_{2i} = \underline{-2.86 \text{ m/s}}$

d) The total kinetic energy in the center-of-mass frame before the collision is
$$K' = 1/2 \, m_1 (v'_{1i})^2 + 1/2 \, m_2 (v'_{2i})^2$$
$$K' = 1/2 \, (2 \text{ kg})(7.14 \text{ m/s})^2 + 1/2 \, (5 \text{ kg})(-2.86 \text{ m/s})^2 = \underline{71.4 \text{ J}}$$

e) In the laboratory frame before the collision
$$K_i = (1/2)(2 \text{ kg})(10 \text{ m/s})^2 = 100 \text{ J}$$
Afterward:
$$K_f = (1/2)(7 \text{ kg})(2.86 \text{ m/s})^2 = 28.6 \text{ J}$$
The change in kinetic energy is $\Delta K = K_f - K_i = -71.4$ J. The amount of energy available to perform internal work on the systems during an inelastic collision is the amount of kinetic energy in the center-of-mass frame.

Example 4
A miner opens a coal chute and begins depositing coal on a long horizontal conveyor belt. The motor which drives the conveyor belt operates at constant speed so that the belt moves with speed v_o. Mass is deposited onto the belt at a rate dm/dt. a) Assuming no loss of mass from the conveyor belt, what is the force the motor must exert?
b) What power must the motor deliver to the belt?

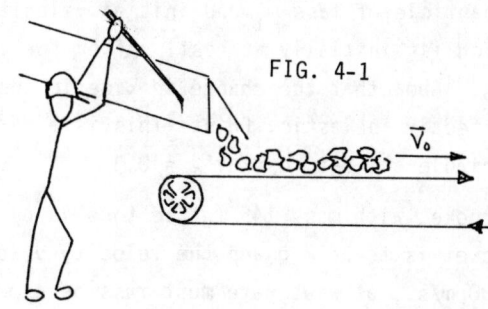
FIG. 4-1

a) Since the mass is not constant we employ the impulse-momentum form of Newton's 2nd law.
$$F = dp/dt = d/dt(MV) = MdV/dt + VdM/dt \qquad (1)$$
since $v = v_0$ is a constant and since $dM/dt = dm/dt$ Eq. (1) becomes
$$F = v_0 dm/dt \qquad (2)$$
b) The power delivered by the motor is
$$P = dW/dt = FV = v_0^2 \, dm/dt$$

Since v^2 is a constant it can be taken across the derivative. If we then multiply and divide by 2 we find
$$P = 2 \, d(1/2 \, MV^2)/dt = 2 \, dK/dt.$$
The motor does work at twice the rate the kinetic energy increases.

PROBLEMS

1. A mobile is to be made of a uniform thin rod 26 cm long of mass 15 g. Objects of mass 2, 3, and 5 g are hung at the left end, 8 cm from the left end, and at the right end respectively. Where must the supporting string be placed in order that the mobile will be in balance?

2. A buoy consists of two hollow cones of base radius 12 cm, height 30 cm, and mass 3 kg attached, one to each end, of a hollow cylinder of radius 12 cm, length 2 m, and mass 12 kg. In order that the buoy will float upright in the water, one cone is filled with cement of mass 36 kg. Where is the center of mass located?

3. A barbell consists of 2 spherical 2 kg masses connected by a light rod of length 20 cm. One of the masses is connected to a spring of rest length 15 cm, k = 30 N/cm which is anchored to the ground. The other mass is attached to a spring of rest length 20 cm, k = 20 N/cm also attached to the ground. The barbell is lifted to a height of 30 cm, held horizontal and released. What is the initial acceleration of the center of mass of this system?

4. In one high energy accelerator, protons are accelerated to a speed of 2.9×10^8 m/s and impinge on a stationary target of protons. Another accelerator accelerates protons to a velocity of 2×10^8 m/s. The protons in this second accelerator are split into two beams which then collide head on with one another. Neglecting relativistic effects a) which accelerator has the most kinetic energy available to provide to inelastic collisions between the protons? b) How much more?

5. A particle of mass m_1 and initial velocity v_1 collides head on with a particle of mass m_2 which is initially at rest. After the collision the particles have velocities v'_1 and v'_2. Show that the change in kinetic energy $\Delta K = K' - K$ will be a maximum if the collision is perfectly inelastic. [Hint: This is a maximum-minimum problem. Choose v'_1 as the independent variable and set $dK/dv'_1 = 0$.]

6. A rocket with mass 145 metric tons is on its launching pad. If the initial acceleration of the rocket is to be 2 g and the velocity u_0 of the exhaust gases relative to the rocket is 5000 m/s., at what rate must mass be expelled from the rocket?

Chapter 11
Conservation of Angular Momentum

PREVIEW

In the past few chapters we have studied the translational dynamics of systems. These considerations have focused on the response of particles and systems to net external applied forces. We begin in this chapter to describe the rotational analogues of translational dynamics. We start with the definition of angular momentum of a particle and generate from this definition the angular analogue of Newton's 2nd law in the impulse-momentum form. The definition of angular momentum is generalized to a system of particles. Finally we study a useful separation of the total angular momentum of a system into two parts, the orbital and spin angular momentum. These qualitatively different angular momenta have a somewhat more intuitive meaning for a system of interacting particles than does the total angular momentum of the system.

SUMMARY

11.1 ANGULAR MOMENTUM OF A PARTICLE

> Definition. Angular momentum (single point particle): the vector cross product of the position vector of the particle relative to the chosen coordinate system and the linear momentum of the particle. The angular momentum will be designated by L.

Since the angular momentum is defined as a cross product of two vectors $L = r \times p$ it has the following properties:
1. L is perpendicular to the plane formed by r and p.
2. The actual value of the angular momentum of a particle depends on the coordinate system chosen.
3. The magnitude of L is $L = mrv\sin\theta$, where θ is the smallest angle between r and v.
4. A particle undergoing uniform motion (v = constant) has a constant angular momentum about any origin chosen in the plane formed by r and v. The magnitude of L is $L = mvb$, where b is the perpendicular distance from the origin to the axis colinear with v. b is called the impact parameter and is an important quantity when dealing with collisions.
5. A particle which moves in uniform circular motion (v_t = constant) has constant angular momentum relative to an axis passing through the center of its circular path. The magnitude of L about this axis is $L = mv_t r$.

By differentiating the definition of L with respect to time it can be readily seen that the time rate of change of L is:

$$dL/dt = r \times F = \tau_{external} \qquad (11.8)$$

where τ_{ext} is the torque exerted on the particle. This expression is the rotational analogue of Newton's 2nd law in the impulse-momentum form. We see that there is a formal analogy between

86   Chapter 11

torque and force, and between angular momentum and linear momentum.

>   **Law of Conservation of Angular Momentum**: if the net external torque on a particle vanishes (i.e. $\tau_{ext} = 0$) then the angular momentum of the particle is constant.

If the forces which act on the particle are only central forces then the angular momentum is conserved since the force and position vector from the center of force are colinear and their cross product is identically zero.

11.2 ANGULAR MOMENTUM OF A SYSTEM OF PARTICLES

>   **Definition.** <u>Angular momentum of a system of particles</u>: the sum of the individual angular momentum vectors of the particles which comprise the system taken about a common axis.

$$L = \sum_i L_i = \sum_i r_i \times p_i \tag{11.13}$$

The rotational form of Newton's 2nd law can be developed from this definition if the internal forces between the particles in the system obey what is often termed the strong form of Newton's 3rd law. The internal forces must be equal in magnitude, oppositely directed, and <u>must be colinear</u>. Central forces such as gravity and the electrostatic force obey the strong form of Newton's 3rd law but not all forces between particles are so nicely behaved. The electromagnetic forces which arise between charged particles in motion are equal and opposite but do not act along the line connecting the particles. The rotational form of Newton's 2nd law for systems of particles which is derived in the text cannot be applied directly to such systems. If the internal forces do obey the strong form of Newton's 3rd law, then it can be readily shown that the sum of the torques due to the internal forces vanishes and:

$$\tau_{external} = dL/dt \tag{11.8}$$

which is the same equation as we developed for a single particle. In this circumstance, which is by far the most common that we encounter in mechanics, the law of conservation of angular momentum also holds in the same form as for a single particle. If the net external torque on a system vanishes the angular momentum will be conserved.

$$\tau_{external} = 0 \rightarrow dL/dt = 0 \rightarrow L = \text{constant} \tag{11.16}$$

11.3 SPIN AND ORBITAL ANGULAR MOMENTUM

If the total angular momentum of a system of particles (or a system of extended bodies) is referred to center of mass coordinates rather than laboratory coordinates, a natural division of the total angular momentum into two qualitatively distinct parts occurs. This transformation is detailed clearly on page 221 of the text and results in the following definitions.

>   **Definition.** <u>Orbital angular momentum</u>: the angular momentum of the center of mass relative to the origin of the (inertial) laboratory reference frame.

$$L_{cm} = L_o = Mr_{cm} \times v_{cm} \tag{11.25}$$

>   **Definition.** <u>Spin angular momentum</u>: the sum of the angular momenta of the individual particles of the system relative to the center of mass.

$$L_s = \sum_i r_i \times p_i \tag{11.26}$$

With these two definitions the total angular momentum can be written as:
$$L = L_o + L_s \qquad (11.17)$$
This division allows us to view the rotational motion as the superposition of two qualitatively different types of angular motion. The orbital angular momentum corresponds to the revolution of the system as a whole about the origin of the coordinate system. The spin angular momentum corresponds to the rotation or spin of the system about its center of mass. In many cases the center of mass and one of the particles are almost coincident. An example of this is the solar system where the planets revolve about the sun. The angular momentum associated with this motion is the planet's orbital angular momentum. On the other hand, most of the planets as well as the sun rotate about an axis through their centers of mass. The angular momentum associated with this motion is their spin angular momentum. This is one of the reasons that common usage distinguishes between the terms revolve and rotate. In terms of their physical properties as rotations, there is no real difference between these. Both of these are angular momenta.

From the above definition we should note that in classical mechanics if we choose to locate the origin of our coordinate frame at the center of mass of the system, the orbital angular momentum vanishes.

Finally we note that if the net external torque on the system is zero then the <u>total</u> angular momentum is conserved and
$$L_o = - L_s$$

EXAMPLE PROBLEMS
Example 1

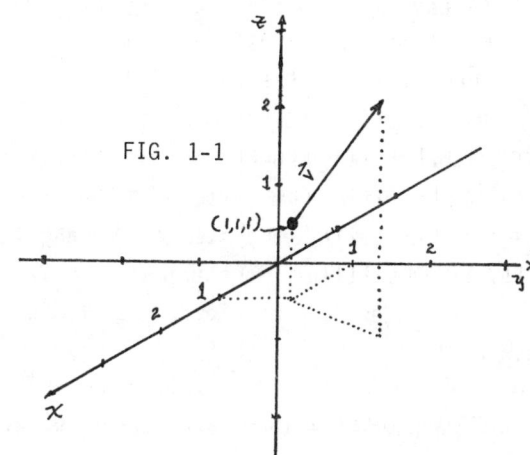

FIG. 1-1

A particle of mass 2 kg has a velocity $v = (3\hat{i} + 4\hat{j} + 5\hat{k})$ m/s. a) What is the angular momentum about the origin when it passes through the point $r = (\hat{i} + \hat{j} + \hat{k})$ m ? b) Show that L is a constant by calculating it at some other point on the path of the particle.

Solution
Given: $m = 2$ kg
   $v = 3\hat{i} + 4\hat{j} + 5\hat{k}$
   $r = \hat{i} + \hat{j} + \hat{k}$

Required: a) L_o when particle is at $\hat{i} + \hat{j} + \hat{k} = r$
b) L_o when $r =$ some other point on the trajectory of the particle.

a) We employ the definition of L and the vector cross product table for the unit vector (Table 2.2).

$$L = r \times p = (2 \text{ kg})(r \times v) = (2 \text{ kg})(\hat{i}+\hat{j}+\hat{k})\text{m} \times (3\hat{i}+4\hat{j}+5\hat{k})\text{m}^2/\text{s}$$
$$= [3(\hat{i}\times\hat{i}) + 4(\hat{i}\times\hat{j}) + 5(\hat{i}\times\hat{k}) + 3(\hat{j}\times\hat{i}) + 4(\hat{j}\times\hat{j}) + 5(\hat{j}\times\hat{k})$$
$$+ 3(\hat{k}\times\hat{i}) + 4(\hat{k}\times\hat{j}) + 5(\hat{k}\times\hat{k})] (2) \text{ kg-m}^2\text{s}$$
$$= (4\hat{k} - 5\hat{j} - 3\hat{k} + 5\hat{i} + 3\hat{j} - 4\hat{i}) (2) \text{ kg-m}^2\text{s}$$
$$L = \underline{(2\hat{i} - 4\hat{j} + 2\hat{k})} \text{ kg-m}^2/\text{s}$$

88   Chapter 11

b) In order to locate another point through which the particle passes we integrate
$$v = dr/dt = 3\hat{i} + 4\hat{j} + 5\hat{k}$$
from $r_i = \hat{i} + \hat{j} + \hat{k}$ at $t = 0$ to r_f at $t = 1$ (The choice of $t = 0$ at r_1 is arbitrary but makes the evaluation of the integral simple. The choice of $t = 1$ for the "other" point is also not necessary but convenient).

$$\int_{r_i}^{r_f} dr = \int_{t=0}^{t=1} (3\hat{i} + 4\hat{j} + 5\hat{k})$$

$$r_f - (\hat{i} + \hat{j} + \hat{k})m = 3t\hat{i} + 4t\hat{j} + 5t\hat{k} \Big|_{t=0}^{t=1} = (3\hat{i} + 4\hat{j} + 5\hat{k})m$$

$$r_f = (4\hat{i} + 5\hat{j} + 6\hat{k})m$$

Now, recalculating L in the same manner as before
$$L = r \times mv = (4\hat{i} + 5\hat{j} + 6\hat{k}) \times (3\hat{i} + 4\hat{j} + 5\hat{k})2 \text{ kg-m}^2\text{s}$$
$$= (+16\hat{k} - 20\hat{j} - 15\hat{k} + 25\hat{i} + 18\hat{j} - 24\hat{i})2 \text{ kg-m}^2/\text{s} = \underline{(2\hat{i} - 4\hat{j} + 2\hat{k})\text{kg-m}^2/\text{s}}$$

which is identical to a).

Example 2

Assume that the earth orbits the sun in a circular orbit. Neglecting the earth's rotation about its axis, what is its angular momentum about the sun?

Solution

Given: $M_E = 5.974 \times 10^{24}$ kg
         $R = 1.496 \times 10^{11}$ m
         $T = 1$ year

Required: L about the sun.

We assume the earth undergoes uniform circular motion about the sun. In order that L be strictly constant we choose the origin of our coordinate system at the sun (essentially at the center of mass of the earth-sun system.) The angular momentum of the earth is then $L = mrv_T$ (Eq. 11.5). We must just calculate v_T.

The tangential speed of the earth is the circumference of the earth's orbit divided by the period.
$$v_T = 2\pi R/T \cong 2\pi(1.496 \times 10^{11} \text{ m})/\pi \times 10^7 \text{s} = \underline{3 \times 10^4 \text{ m/s}}$$

where we have used the useful approximation 1 year $\cong \pi \times 10^7$ s, a bit of physics trivia! We now calculate L
$$L = mrv_t \quad (6 \times 10^{24} \text{ kg})(1.5 \times 10^{11} \text{ m})(3 \times 10^4 \text{ m/s})$$
$$= \underline{2.7 \times 10^{40} \text{ kg-m}^2/\text{s}}$$

The velocities and angular momenta involved in planetary motion are truly astronomical. The earth's velocity in its orbit is roughly 70,000 miles/hr.

Example 3

In the text it is stated that the angular momentum of a particle undergoing uniform circular motion is constant relative to an origin at the center of the circular path. Consider a particle of mass m undergoing uniform circular motion with constant speed v_T.

Consider an origin a distance 2 R from
the center of the circle. Show by simple
vector arguments that the instantaneous
angular momentum about this axis is <u>not</u>
constant. Explain why this is so?

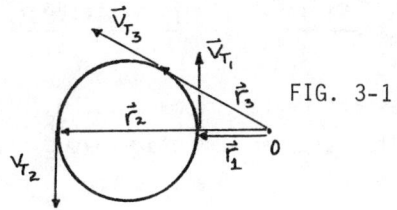
FIG. 3-1

Solution

    Given: m

               v_T = constant, uniform circular motion

              origin = 2 R from center of circular path.

    Required: Show L $\neq$ constant.

Since the origin must lie on an extended diameter of the circle we consider the angular momentum when the particle is at the three points shown in FIG. 3-1. Note that $r_1 \neq r_2 \neq r_3$ and that, even though the angle between r_1 and v_{T_1} and r_2 and v_{T_2} is 90°, by the right hand screw rule the angular momentum about O for these two positions is oppositely directed and is three times as large for r_2 as for r_1. Obviously the angular momentum at r_3 is zero since r_3 and v_T are parallel at this point.

    This may seem a bit confusing, but all is well because the centripetal force which keeps the particle in its circular orbit exerts a torque on the particle about the axis we have chosen. This can be seen at r_3 since $F_3 || v_T$ it is $\perp$ to F_c and $|\tau_0| = |r \times F| = rF \sin 90°$ = rF. Note that at r_1 and r_2 the torque vanishes. Thus the torque and angular momentum taken about O change in concert to maintain the particle in its circular path.

Example 4

    Amusement park rides often take advantage of the fact that the total angular momentum of a system is the sum of its orbital and spin angular momentum. One of our favorites is the scrambler. We model it as two masses connected by a light rigid rod of length 1.5 m. The masses are 27 kg (my daughter) and 80 kg (me). The rod is connected to the center of the scrambler by a light rod of length 5 m. Assume the two masses rotate about a pivot with tangential speed of 5 m/s. The orbital speed of the end of the rod connecting the masses to the center of the scrambler is also 5 m/s. Calculate the approximate magnitudes of the orbital, spin, and total angular momentum.

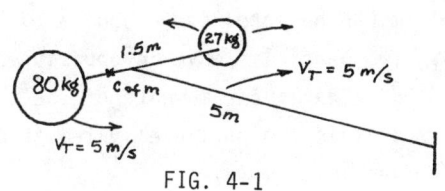
FIG. 4-1

    Given: m_1 = 27 kg

               m_2 = 80 kg

               v_T = 5 m/s for both the masses on the carriage and the carriage

               r_{12} = 1.5 m

               r_0 = 5 m

    Required: L_0, L_s, $L = L_0 + L_s$

The center of mass of the carriage is not located at the pivot. If we calculate it for the "carriage" alone taking the pivot to be the origin at the center of the rod which connects the masses we find

$$r_{cm} = \frac{(m_1 r_1 + m_2 r_2)}{m_1 + m_2} = \frac{(27)(.75) + 80(-.75)}{107} = \frac{20.25 - 60}{107} = \frac{-39.75}{107} = -0.37 \text{ m}$$

which places the center of mass .37 m from the center toward the 80 kg mass. For the calculation of the orbital angular momentum this distance is relatively insignificant compared to the 5 m length of the center connecting rod. Thus

$$L_o \simeq M r_{cm} v_{cm} \quad \text{(assuming sin } \theta = \sin 90° = 1\text{)}$$
$$= (107 \text{ kg})(5 \text{ m})(5 \text{ m/s}) \simeq \underline{2675} \text{ kg-m}^2/\text{s}$$

The spin angular momentum is

$$L = \Sigma m r'_i v'_i \sin \theta_i = m_1 r'_1 v'_1 + m_2 r'_2 v'_2$$
$$= (27 \text{ kg})(1.13 \text{ m})(5 \text{ m/s}) + (80 \text{ kg})(0.37 \text{ m})(5 \text{ m/s})$$
$$= 152 + 148$$
$$= \underline{300} \text{ kg-m/s}^2$$

The total angular momentum depends on whether the "carriage" spins in the same direction as the rotation of the carriage about the center of the ride or in the opposite direction.

$$L_s \parallel L_o \quad L = (2675 + 300) \text{kg-m}^2/\text{s} = 2975 \text{ kg0m}^2/\text{s}$$
$$L_s \nparallel L_o \quad L = (2675 - 300) \text{kg-m}^2/\text{s} = 2975 \text{ kg-m}^2/\text{s}$$

Although this is a simplified calculation of the properties of this amusement park ride, think about it the next time you ride one - it's fun!

PROBLEMS

1. A particle of mass 1.5 kg has velocity $v = 2i + 6j$ when it passes through the point $r = 2i + 6j$. What is its angular momentum about the origin at this point?

2. A car of mass 2000 kg is traveling at 90 km/hr when it enters a circular curve of radius 50 m. What is its angular momentum about the center of the curve?

3. A communications satellite of mass 800 kg is to be placed in a synchronous orbit. (A satellite in a synchronous orbit maintains a fixed position above a point on the earth's surface)
   a) What distance from the center of the earth must the satellite be placed into orbit? b) What must the satellite's orbital velocity be? c) What is the satellite's orbital angular momentum?

4. Two particles move in opposite directions along parallel straight paths separated by a distance d. The linear momenta of the particles are equal in magnitude. Show that the angular momentum of this two particle system is constant for any choice of the origin in the plane in which the particles move.

5. A small bead of mass m is attached to a light inextensible string which passes through a hole in a frictionless surface. The mass is initially moving in a circular path with tangential speed v_o and radius R. The tension in the string is increased until the radius of the circular path is decreased to R/2. a) What is the ratio of the final tangential speed to the initial tangential speed? b) What is the ratio of the final tension to the initial tension?

6. The moon revolves around the earth in its orbit once every 28 days. Relative to the sun as origin and neglecting the spin of the moon and earth about their axes, calculate a) the orbital angular momentum of the earth-moon system, b) the spin orbital angular momentum of the earth-moon system, c) assuming the spin and orbital angular momentum vectors are parallel, what is the total angular momentum of the earth-moon system about the sun?

Chapter 12
Rotation of a Rigid Body

PREVIEW

In the last chapter we began the development of the mechanics of rotating systems. The definitions of angular momentum for particles and systems given are completely general as is the rotational form of Newton's 2nd law in terms of torque and the time rate of change of angular momentum. In the present chapter we focus on the rotational motion of a certain class of systems, rigid bodies, which are constrained to rotate about an axis which has a fixed direction. For this special case it is possible to derive a rotational analogue of the mass times acceleration form of Newton's 2nd law. The quantities used to describe angular motion in kinematical form are first introduced. Based on these definitions, the rotational equivalent of mass is determined for a rigid body rotating about a fixed axis. Rotational kinetic energy is defined and the dynamics of rigid bodies is addressed.

SUMMARY

12.1 ANGULAR KINEMATICS

Before we can study the rotational dynamics of rigid bodies, we must develop an appropriate set of displacements, velocities, accelerations, etc. to describe rotational motion. The appropriate quantities are angular measures. Most of us are familiar with angular measurement from geometry and trigonometry. In these branches of mathematics angles are often measured in degrees. In dealing with the physics of rotating objects it is advantageous to deal with an angular measure which is more directly related to linear measure. In fact, you are already familiar with this measure. When we use $\pi = 3.14159...$ to calculate the circumference or area of a circle, we are actually making use of the type of angular measure we now introduce.

> Definition. One radian: the angle subtended at the center of a circle of radius r by an arc of length r laid off along the circumference of the circle.

This is in fact the measure of angle developed by the ancient geometers, such as Euclid, who discovered that the radius of a circle stands in a constant ratio to its circumference. This ratio is 2π radians. From the above definition we see that the angle in radians is the arc length s divided by the radius of the circle, $\theta = s/r$ or $s = r\theta$ (Eq. 12.1). This relationship between arc length, a linear measure of displacement, and the angle measured in radians is very convenient. The radian is a dimensionless quantity since it is the ratio of two lengths.

> Definition. Angular displacement: the change in angle, measured in radians, through which an object rotates about a specified axis. The vector direction of angular displacement is perpendicular to the plane of rotation and points in the direction specified by the right hand rule (given below).

In order to give the most general definition of angular quantities they must be vectors. The following rule specifies how the direction of angular kinematical quantities is defined.

<u>Right hand rule</u>. The fingers of the right hand are curled in such a fashion that the fingers point in the direction of increasing θ, then the thumb of the right, hand when held perpendicular to the fingers, points in the direction of the angular displacement.

Often it is not necessary to specify the complete vector direction of kinematical quantities. However, even in "one dimensional" situations positive and negative displacements must be specified. The above rule and definition are consistent with the assignment of counterclockwise angular displacements being positive and clockwise one being negative. The zero of angular displacement is usually chosen to be along the x axis of the cartesian coordinate system taken so the z axis is defined by the direction of the vector angular displacement. The axis which coincides with the direction of the angular displacement is called the axis of rotation.

We denote the angular displacement as θ. With the above definitions we may formally define the quantities analogous to linear velocity and acceleration.

<u>Definition</u>. <u>Angular velocity</u>: the time rate of change of angular displacement.
$$\omega = \lim_{\Delta t \to 0} \Delta\theta/\Delta t = d\theta/dt \tag{12.4}$$

The units of ω are rad/s or 1/s; its dimensions are $1/T$.

<u>Definition</u> <u>Angular acceleration</u>: the time rate of change of the angular velocity
$$\alpha = \lim_{\Delta t \to 0} \Delta\omega/\Delta t = d\omega/dt \tag{12.7}$$

The units of α are rad/s^2 or 1/s^2; its dimensions are $1/T^2$.

These definitions are general. If the rotation is confined to a plane then θ, ω, α will all have directions which are positive or negative. Also if the motion is planar we can derive a relationship between tangential linear quantities and angular quantities by differentiating Eq. 12.1
$$v_T = r\omega \tag{12.6}$$
$$a_T = r\alpha \tag{12.9}$$

If the vector angular velocity, position vector in the plane of rotation, and the vector tangential velocity are to be employed Eq. 12.6 becomes:
$$\mathbf{v}_T = \mathbf{r} \times \boldsymbol{\omega} \tag{12.13}$$

The <u>mathematical form</u> of the definitions of ω and α are identical to those of **v** and **a** of linear kinematics. In particular any kinematical equation derived in the linear variables **r**, **v**, and **a** can be immediately transposed into angular form by replacing the linear quantity by its rotational analogue. As an example, the equations for constant angular acceleration in a plane are:
$$\omega = \omega_0 + \alpha t \tag{12.10}$$
$$\theta = \theta_0 + \omega_0 t + (1/2)\alpha t^2 \tag{12.11}$$
$$\omega^2 = \omega_0^2 + 2\theta\alpha \tag{12.12}$$

12.2 ANGULAR MOMENTUM AND ROTATIONAL KINETIC ENERGY

The angular momentum about the origin of an arbitrary coordinate system defined in Chapter 11 is quite general $\mathbf{L} = \mathbf{r} \times \mathbf{p}$. Unfortunately the treatment of the general rotational motion of a system of particles or a rigid body employing this definition is a very complicated problem in mechanics. Those of you who continue in physics and engineering will no doubt have to come to grips with this knotty problem at some future time. It is much too involved a problem to consider in an introductory text.

We will confine ourselves primarily to the study of the rotation of a rigid body about a fixed axis. A rigid body is one in which every element of mass comprising the body is always at rest relative to the other elements of mass of the body. If the rigid body is further constrained to rotate about a fixed axis in space, every point in the body executes circular motion in some plane perpendicular to the axis about that axis. We consider the rigid body to be composed of points of mass m_i.

> **Definition.** <u>Angular momentum about an axis of a point mass.</u> The angular momentum of a mass point m_i in a rigid body about an axis (usually chosen to be the z axis) is the cross product of the vector $\mathbf{R}_i$ perpendicular to the axis of rotation which points from the axis to the position of the mass point and the linear momentum of the mass point.

This is actually the component of the regular angular momentum along the axis of rotation. The particle in the rigid body executes circular motion in the plane defined by $\mathbf{R}_i$ and $\mathbf{p}_i$. Its angular velocity points along the axis of rotation. The magnitude of the angular momentum about the axis is:

$$L_{z_i} = R_i p_i \sin 90° = R_i (m_i v_i) = R_i m_i (R_i \omega)$$
$$L_{z_i} = m_i R_i^2 \omega \tag{12.15}$$

The total angular momentum along the axis of rotation is the sum of the individual angular momenta of all the point masses which we consider the body composed of

$$L_z = \Sigma L_{z_i} = \Sigma m_i R_i^2 \omega = (\Sigma m_i R_i^2) \omega$$

By analogy with linear quantities we see that the quantity in parentheses must be the rotational analogue of the mass in linear dynamics. The fact that it is not a simple quantity should not be surprising.

> **Definition.** <u>Moment of inertia</u>: the rotational analogue of mass is defined by the relationships
> $$I = \Sigma m_i R_i^2 = \int_M R^2 dm \tag{12.18, 12.23}$$

The moment of inertia of a rigid body rotating about a fixed axis depends on a) the axis of rotation, b) the distribution of mass in the body. The same rigid body will have different moments of inertia depending on the axis about which it is constrained to rotate. For example a circular hoop has all of its mass at a constant radius R. If it rotates about an axis through the middle of the hoop perpendicular to the plane of the hoop it has a moment of inertia $I = MR^2$. If it rotates about an axis which is along one of its diameters, i.e. through the center of the hoop but in the plane of the hoop, its moment of inertia is $I = (1/2)MR^2$. We mention in passing that the moment of inertia of a rigid body which is not constrained to rotate about a fixed axis is a second rank tensor, one of the reasons not to attack general rotations at this point.

Definition. <u>Rotational kinetic energy</u>. For a rigid body constrained to rotate about a fixed axis the rotational kinetic energy of the body is (by analogy with the corresponding linear quantity)

$$K_R = (1/2)I\omega^2 \tag{12.20}$$

One general theorem which we can prove about a rigid body that is not constrained to rotate about a fixed axis is that its total kinetic energy can be expressed as

$$K_{total} = (1/2)Mv_{cm}^2 + (1/2)I_c\omega_c^2 \tag{12.21}$$

Where I_c and ω_c are the moment of inertia and the angular velocity about an axis passing through the center of mass of the rigid body. This theorem helps to confirm the statements we have made in previous chapters justifying the development of center-of-mass coordinates. It confirms that in many cases a body's motion can be thought of as a translation of the center of mass plus a rotation about that center of mass.

12.3 CALCULATION OF THE MOMENT OF INERTIA

The calculation of the moment of inertia of a rigid body entails the evaluation of the integral

$$I = \int_M R^2 dm = \int_M R^2(x,y,z)\,\rho dV \tag{12.27}$$

We have indicated explicitly the fact that the radius of the mass element dm depends on its location relative to the axis of rotation. In general this is a three dimensional integration, but it can often be done as a one dimensional integral using the same techniques (or tricks) developed for calculating the center of mass of an object in Chapter 10. If you have not covered this topic in calculus yet, you will. It is one of the favorite applications in multidimensional calculus. We will not cover this specific topic in the summary, but two items in this section deserve mention here.

Definition. <u>Radius of gyration</u>: that point at which the total mass of a rigid body could be concentrated to give the same moment of inertia that the body exhibits about a given axis.

The radius of gyration, like the moment of inertia, depends on the axis of rotation and the geometry of the rigid body. Its defining relationship is:

$$s^2 = (1/M)\int_M R^2 dm \tag{12.25}$$

<u>Parallel axis theorem</u>. The moment of inertia of a rigid body about any axis is equal to the sum of the moment of inertia about a parallel axis through the center of mass of the rigid body and the mass of the body times the perpendicular distance between the two axes squared. $I = I_c + Mh^2$ (Eq. 12.29).

This theorem allows the calculation of the moment of inertia of a rigid body about any axis parallel to an axis about which the moment of inertia has been determined. (Provided one knows where the center of mass of the rigid body is located.)

12.4 DYNAMICS OF RIGID BODY MOTION

We are now in a position to convert the angular momentum form of Newton's 2nd law into a relationship between torque and the kinematical quantity angular acceleration. It is important to remember that this is only possible in the simple form which we derive because we are considering the body in question to rotate about an axis fixed in space. In fact, it is not necessary that the moment of inertia remain fixed in time as long as the axis of rotation remains fixed.

We consider here the case where I remains constant in time. Eq. 11.12 states that $\tau_{ext} = dL/dt$. In our circumstances we may write

$$\tau_{ext} = dL/dt = d(I\omega)/dt = I\alpha \tag{12.32}$$

Based on the analogies we have developed we see this is F = ma in rotational form.

One of the interesting applications of these properties is to the angular momentum of systems which can change their moments of inertia. Since the moment of inertia of a body depends on the distribution of mass about the axis of rotation, if this distribution is altered somehow, the rotational properties of the body and its rotational motion are altered. Examples of this are the spin of divers and ice skaters. This effect also has other nonathletic applications.

EXAMPLE PROBLEMS

Example 1

A 33 1/3 rpm record is approximately 30 cm in diameter. When the turntable on which the record is placed is turned on, it reaches its final angular velocity in 0.1 s. a) What is the final angular velocity of the record in rad/s? b) What is the angular acceleration of the turntable (assume α = constant)? c) Through what angular displacement does the record turn before it reaches its final angular velocity? d) What is the tangential velocity of a point on the rim of the record?

Solution

Given: $\omega_i = 0$
ω_f = 33 1/3 rpm
r = 15 cm
t = 0.1 s

Required: a) ω_f in rad/s, b) α, c) $\Delta\theta$, d) v_{T_f}

This is a fairly simple problem in angular kinematics. If you have difficulty dealing with or thinking about angular variables, try recasting the problem in terms of linear displacements and velocities. A analogous problem can be written in the linear case. Can you do it?

a) There are 2π radians in one revolution or cycle. This is just a conversion of units from rpm to rad/s

$$\omega_f = (33 \ 1/3 \ \text{rev/min})(2\pi \ \text{rad/rev})(1 \ \text{min}/60 \ \text{s}) = 3.5 \ \text{rad/s}$$

b) If the angular acceleration is constant we may use Eq. (12.10)

$$\omega_f = \omega_0 + \alpha\Delta t \rightarrow 3.5 \ \text{rad/s} = 0 + \alpha(0.1 \ \text{s}) \rightarrow \alpha = \underline{35 \ \text{rad/s}^2}$$

c) To find θ we employ Eq. (12.12) $\omega_f^2 = \omega_0^2 + 2\alpha\theta$. Substituting we have:

$$(3.5 \ \text{rad/s})^2 = (0)^2 + 2(35 \ \text{rad/s}^2)\theta \rightarrow \theta = \underline{0.175 \ \text{rad}}$$

d) Since we have expressed our angular variable in terms of radians, the conversion to linear speed is simple. We employ Eq. (12.6)

$$v_T = r\omega \rightarrow v_T = (0.15 \ \text{m})(3.5 \ \text{rad/s}) = \underline{0.525 \ \text{m/s}}$$

Example 2

A solid disk of mass M radius R starting at rest rolls a distance L down a plane inclined at an angle θ with the horizontal without slipping ($I = 1/2\, MR^2$). a) What is the linear acceleration and the force of friction? b) What is the torque exerted by f? c) What is the angular acceleration of the disk? d) What is the final translational kinetic energy? e) What is the final rotational kinetic energy? f) What is the final total kinetic energy?

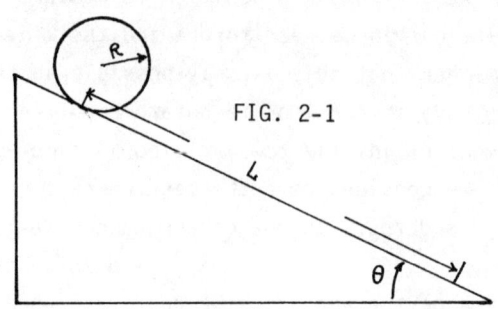

FIG. 2-1

Solution

Given: R, θ, L, I, M

Required: a) a_x, b) τ_f, c) α, d) $K_T = 1/2\, Mv_f^2$, e) $K_R = 1/2\, I\omega_f^2$, f) K_{total}

a) The force diagram for this situation is shown in FIG. 2-1. For rolling without slipping the arc length Rdθ through which the disk rolls must equal the linear distance the disk travels down the plane dx_{cm}. Thus $dx_{cm} = Rd\theta$ which leads directly to $a_{cm} = R\alpha$. (Rolling without slipping is thoroughly discussed in Chapter 13.) We apply Newton's 2nd law in both the translational and rotational form:

$$F_x = Mg\sin\theta - f = Ma_x; \quad \Sigma\tau = fR = I\alpha = 1/2\, MR^2\alpha \qquad (1,2)$$

We now substitute $a = R\alpha$ into Eq. (2)

$$fR = 1/2\, MRa_x \quad 2f = Ma_x$$

Substituting into Eq. (1) we find

$$Mg\sin\theta - f = 2f \rightarrow f = \underline{1/3\, Mg\sin\theta} \qquad (3)$$

and

$$a = 2f/M = 2/3\, g\sin\theta \qquad (4)$$

We see that the linear acceleration is 2/3 of the value it would be if the disk slid down a frictionless plane.

b) From Eqs. (2) and (3) the torque is $\underline{fR = (1/3)MgR\sin\theta}$

c) Since $a = R\alpha$; $\alpha = a/R = \underline{[(2/3)g\sin\theta]/R}$

d) The disk begins at rest and rolls a distance L down the plane. The final linear velocity squared is:

$$V_f^2 = V_0^2 + 2aL = (4/3)gL\sin\theta$$

The final translational kinetic energy is $(1/2)MV_f^2 = \underline{(2/3)MgL\sin\theta}$

e) If the disk has rolled without slipping, L is also the arc length through which the disk has turned. The angular displacement is θ = L/R. The final angular velocity squared is

$$\omega_f^2 = \omega_0^2 + 2\alpha\theta = 2[(2/3)g\sin\theta]/[L/R^2] = (4/3R^2)gL\sin\theta$$

The final rotational kinetic energy is

$$(1/2)I\omega^2 = (1/2)[(1/2)MR^2](4/3R^2)gL\sin\theta$$
$$= \underline{1/3\, MgL\sin\theta}$$

f) The total final kinetic energy is $K_f = K_T + K_R = \underline{MgL\sin\theta = Mgh}$

Thus the work done by gravity, mgh, still appears as kinetic energy, but in two different forms.

Example 3

Consider Atwood's machine with masses $m_1 = 10$ kg, $m_2 = 8$ kg. The masses are connected by a light inextensible cord passing over the pulley without slipping. The pulley has a moment of inertia of 0.5 kg-m^2 about the axis of rotation and a radius of 0.1 m. a) Find the linear acceleration of the masses. b) What is the angular acceleration of the pulley? c) What are the tensions T_1 and T_2?

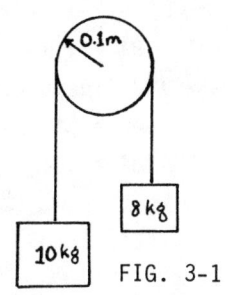

FIG. 3-1

Solution

Given:
$m_1 = 10$ kg
$m_2 = 8$ kg
$I = 0.5$ kg-m^2
$R = 0.1$ m

Required: a) a b) α, c) T_1 and T_2

a) In this case $T_1 \neq T_2$. The difference between T_1 and T_2 supplies the torque which accelerates the pulley. We choose the positive linear direction in the direction of motion of m_1, since common sense tells us that this is the direction the masses will move. We apply Newton's 2nd law to the translational motion of the masses and the rotational form to the rotation of the pulley

$$F_{linear}^{(1)} = m_1 g - T_1 = m_1 a \longrightarrow 98 \text{ N} - T_1 = 10 \text{ a} \tag{1}$$

$$F_{linear}^{(2)} = T_2 - m_2 g = m_2 a \longrightarrow T_2 - 78.4 \text{ N} = 8 \text{ a} \tag{2}$$

$$\Sigma\tau = T_1 R - T_2 R = I\alpha \longrightarrow 0.1 T_1 - 0.1 T_2 = (0.5 \text{ kg-m}^2) = (0.5 \text{ kg-m}^2) a/0.1 \tag{3}$$

Where we have used the fact, again, that $a = \alpha R$ since the cord does not slip on the pulley. These equations when solved simultaneously (we don't show the work here since it is straightforward but lengthy) gives a = <u>0.288 m/s^2</u>

b) The tangential acceleration of the wheel is = a/R = <u>2.88 rad/s^2</u>

c) The tension in each part of the string may be figured from Eqs. (1) and (2)

$T_1 = 98N - 10a = 98 - 2.88 = $ <u>95.12N</u>
$T_2 = 8a + 78.4N = 2.3 + 78.4 = $ <u>80.7N</u>

Example 4

Calculate the moment of inertia of a solid disk about an axis through the center of the disk parallel to the plane of the disk. The disk has radius R, mass M, and thickness x.

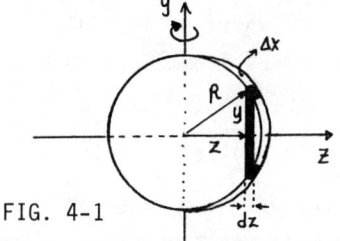

FIG. 4-1

Solution

Given: R = radius of disk
M = mass of disk
Δx = thickness of disk
ρ = constant (assumed) rotation
about axis through disk

We choose the axis of rotation to be the y axis and calculate the moment of inertia using a strip at z of height y and width dz. The contribution of this strip to the moment of inertia about the y axis is

$$dI = z^2 dm$$

in this case the linear strip's mass can be written as

$$dm = \rho dV = \rho(2y dz \Delta x)$$

Thus
$$dI = 2\Delta x \, \rho z^2 y \, dz$$

and
$$I = 2\Delta x \rho \int_{-R}^{R} z^2 y \, dz$$

we use $y = \sqrt{R^2 - z^2}$ to find

$$I = 2\Delta x \rho \int_{-R}^{R} z^2 (R^2-z^2)^{1/2} dz$$

$$= 4\Delta x \rho \int_{0}^{R} z^2 (R^2-z^2)^{1/2} dz$$

$$I = \frac{\Delta x \rho \pi R^4}{4}$$

Using $\Delta x \rho \pi R^2 = M$ we find

$$I = (1/4)MR^2$$

The integral performed in this calculation is not quite as simple as in the text and actually requires some pretty fancy footwork. We suggest at this stage in your career you consult a table of integrals when you are faced with such a task.

PROBLEMS

1. A car with 14 inch diameter wheels accelerates from 0 to 60 mph in 11 s. What is the angular acceleration of the wheels?

2. The flywheel in a car slows down from 3000 rpm to 1700 rpm in 0.3 s. a) What is its angular acceleration? b) Through what angle does it rotate while slowing down?

3. An ultracentrifuge generates a force equivalent to 180,000 times the earth's gravitational force. (i.e. the acceleration is 180,000 g) If the mean radius of the centrifuge tubes is 4.7 cm, what must be the rotational velocity in rad/s and rpm?

4. The flywheel of an engine is required to provide 1000 J of energy when its angular velocity decreases from 70 to 60 rad/s. What must be its moment of inertia?

5. A vertical pole of length L is released and allowed to topple under its own weight. Show that the tangential velocity with which the end of the pole strikes the ground is $v_T = \sqrt{3gL}$.

6. A grindstone of mass 10 kg is in the shape of a solid uniform disk of radius 15 cm. The normal operating speed of the stone is 2000 rpm. It is turned off and a lathe bit is pressed normally against the rim with a force of 20 N. The coefficient of friction is 0.7. a) What is the initial rotational kinetic energy of the grindstone? b) What torque does the force of friction exert on the stone? c) How long does it take the for the grindstone to stop?

7. A light inextensible string is wrapped around a solid drum of radius 0.25 m and mass 128 kg. A constant tangential force of 100 N is applied to the drum by pulling on the string. a) What is the angular acceleration of the drum. b) After 10 m of string have been unwound what is the angular velocity of the drum. c) Rather than pulling on the cord a 100 N weight is hung on the cord of the string. Calculate the angular acceleration of the drum. Explain why it is not the same as in case a).

8. Calculate the moment of inertia for a uniform disk of mass M, radius R and thickness Δx about an axis through its center of mass lying in the plane of the loop (same case as example problem 4) by choosing the z axis in FIG. 4-1 as the axis of rotation. [Hint: consider the moment of inertia of a thin rod about its center.]

9. Calculate the moment of inertia of a uniform annular ring of inner radius R_1 and outer radius R_2 about an axis through the center of the ring perpendicular to the plane of the ring.

10. A skater with her arms extended is spinning with an angular velocity of 2 rad/s. Her arms each have a mass of 6 kg and their center of mass is 0.6 m from the axis of rotation. Her body's moment of inertia about the spin axis is 4 kg-m^2 (excluding her arms). She draws her arms in so that their centers of mass are 0.07 m from the axis of rotation. a) What is her new angular velocity? b) What is the change in her rotational kinetic energy. c) Where does this energy come from?

Chapter 13
Motion of a Rigid Body

PREVIEW

This chapter extends the developments of the preceding chapters to two special cases where the axis of rotation of a rigid body is not fixed in space. First the rotational forms of work and power for a rigid body with a fixed axis of rotation are developed. This completes the formal treatment of rotational motion for simple rigid bodies rotating about a space fixed axis. Next the case of rolling without slipping is discussed. This case can be treated easily because, although the rotational axis of the body through its center of mass moves through space, the axis remains fixed relative to the body. This allows the problem to be reformulated in terms of a pure rotational motion, at least on an instantaneous basis. The final topic is the case of a rapidly spinning top. Although more complicated than rolling without slipping, this subject can still be handled by elementary means.

SUMMARY

13.1 WORK, ENERGY AND POWER IN ROTATION

For a rigid body rotating about a fixed axis the rotational forms of Newton's 2nd law can be converted into a rotational work energy relationship in exactly the same manner as was the case in linear motion.

> Definition. Rotational work: the scalar product of the net external torque exerted on a system and the angular displacement through which the torque acts.

We state this definition in vector terms in order that it be as general as possible. For a body with one axis fixed in space, only the components of the external torque parallel to that axis have any effect in changing the rotational state of the system. These components are either parallel or anti parallel to $\Delta\theta$. Thus the above definition reduces to

$$W_{ext} = \int \tau_{ext_z} \, d\theta$$

By expressing the right hand side of this equation as an integral over the time rate of change of angular momentum using the rotational form of Newton's 2nd law we derive the rotational work-energy principle:

$$W_{ext} = K_{rot} = (1/2)I\omega_f^2 - (1/2)I\omega_i^2 \tag{13.3}$$

Analogously the rotational power defined as the time rate of change of rotational energy is

$$P = \tau\omega \tag{13.4}$$

13.2 ROTATION WITH TRANSLATION

As we mentioned in the last chapter, the general case of rigid body rotation is very complicated. There are, however, a few cases where the axis of rotation of a rigid body is not fixed in space that can be handled in a fairly straightforward and elementary manner. Of these, the simplest is the case of a body which rolls across a surface without slipping.

When an object rolls in this fashion there is a one to one relationship between the angular displacement of the body and its linear translation. Although the axis through the center of mass moves translationally through space, the points in the rigid body which lie on this axis are fixed, i.e. the axis of rotation always passes through the same locus of points relative to the body.

The contact point of a body rolling without slipping on the surface is instantaneously at rest with respect to the surface. The center of mass on the other hand moves with the translational velocity of the object. The interesting fact, and the one which provides the clue as to how this motion may be alternatively described, is that each point moves with a velocity that is linearly related to the distance of the point from the contact point. For a circular object, the point diametrically opposite the contact point moves with velocity twice that of the center of mass. If we draw the diameter of the contact point to the opposite side of the circular object then the instantaneous velocity of each point along this diameter is $r(v_{cm}/R)$. It is fairly obvious that the term in parentheses is

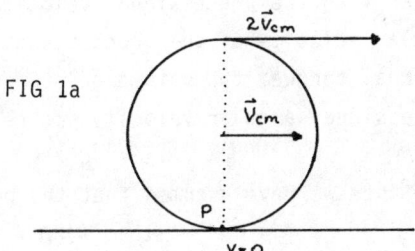

FIG 1a

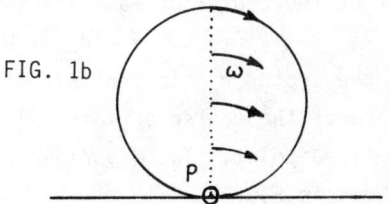

FIG. 1b

the angular velocity of the circular object about the center of mass. Thus we can consider the object to be <u>instantaneously</u> rotating about an axis through the contact point with angular velocity $\omega = (v_{cm}/R)$. This is the simplification alluded to before.

In this case the kinetic energy of the object
$$K = (1/2)Mv_{cm}^2 + (1/2)I_c\omega_c^2 \qquad (12.21)$$
which we derived in Chapter 12 may be rewritten using the parallel axis theorem as:
$$K = (1/2)(MR^2 + I_c)\omega^2 = (1/2)I_p\omega^2 \qquad (13.5)$$
where I_p indicates the moment of inertia about an axis passing through the point of contact.

13.3 PRECESSION

Another case of rigid body rotation in which the axis of rotation does not remain fixed but can still be treated in a simple manner is that of a rapidly rotating symmetric top with one point fixed. The word top is used in physics as a generic term for a spinning body. Symmetric tops include things like bicycle wheels, flywheels, gyroscopes, etc. In this case the top is spinning rapidly (we will define rapidly later) about its axis of symmetry and one point on this spin axis is fixed. If we think of a toy gyroscope, usually the point on the bottom of the gyroscope rests on the ground or a stand. This is the fixed point which lies on the axis of rotation.

We refer to the spin angular velocity as ω and the spin angular momentum as L. If an external torque is applied to a rapidly spinning top such as we have described the angular momentum is not conserved since we know that τ_{ext} = dL/dt. If the torque has a component along the direction of L, this component will cause the magnitude of L to increase. A component perpendicular to L will alter the direction of L. If the torque exerted on a rapidly spinning top is perpendicular to L the motion of the top can be decomposed into essentially two separate motions, the spin of the top about its symmetry axis and a <u>precession</u> of the spin axis of the top in space.

This precession of the spin axis is a general phenomenon that occurs with all rotating bodies, but the division which we have indicated above is only useful if the spin angular velocity is much larger than the precessional velocity. The conventional choice is that $\omega > 10 \, \Omega$, where Ω is the angular velocity of the precessional motion. The extensive discussion in the text shows that the external torque, the spin angular momentum (which points in the direction of ω) and the precessional angular velocity are related by:

$$\vec{\tau} = \vec{\Omega} \times \vec{L} \qquad (13.11)$$

Since we have assumed that the body is rigid we can write this expression as:

$$\vec{\tau} = I_c (\vec{\Omega} \times \vec{\omega})$$

Furthermore, if the torque on the object is exerted by gravity, the weight may be considered to act at the center of mass and the torque is:

$$\vec{\tau} = \vec{r}_{cm} \times \vec{W}$$

where the vector $\vec{r}_{cm}$ is the instantaneous position of the center of mass relative to the top's fixed point. These various relationships are shown in FIG. 2.

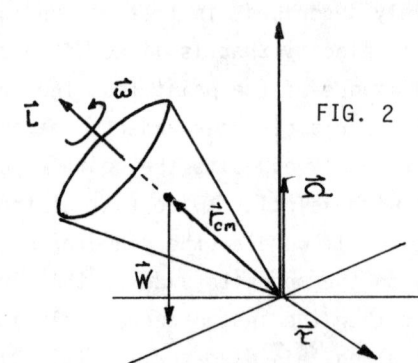

FIG. 2

EXAMPLE PROBLEMS
Example 1

A car of mass 2000 kg accelerates from 0 to 20 m/s in 12 s. The wheels of the car roll without slipping and each has a moment of inertia of 1.5 kg-m^2 and a radius of 0.3 m. a) What is the total final kinetic energy of the car? b) How much power is required to accelerate the car. c) How much of the power is supplied as rotational kinetic energy of the wheels?

Solution

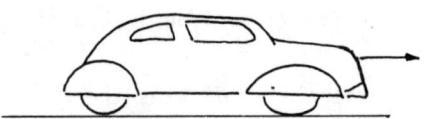

Given: m = 2000 kg,
         v_o = 0,
         v_f = 20 m/s,
         t = 12 s
         I = 1.5 kg-m^2,
         R_w = 0.3 m

Required: a) K_{total}, b) P, c) K_w

a) The final velocity of the car is 20 m/s. This must also be the velocity of the center of mass of the wheels. Since the wheels roll without slipping the angular velocity of the wheels is $\omega = v_{cm}/R$. The total kinetic energy is:

$$K = (1/2)Mv_{cm}^2 + 4(1/2)I\omega^2 = (1/2)Mv_{cm}^2 + 4(1/2)(I)(v_{cm}/R)^2$$
$$K = (1/2)(2000 \text{ kg})(20 \text{ m/s})^2 + 4(1/2)(1.5 \text{ kg-m}^2)[(20 \text{ m/s})/.3]^2$$
$$K = 400{,}000 \text{ J} + 13{,}333 \text{ J} = \underline{413{,}333 \text{ J}}$$

b) The power is
$$p = \Delta w/\Delta t = 413{,}333 \text{ J}/12 \text{ s} = 34{,}444 \text{ watts}$$

c) The power supplied for rotational motion is
$$p = \Delta w_{rot}/\Delta t = \frac{13{,}333 \text{ J}}{12 \text{ s}} = 1111 \text{ watts}$$

Example 2

A uniform rod of mass M and length L has a small, heavy ball of mass M attached at its end. The rod is pivoted at one end and held horizontal. It is then released. What is the linear velocity and momentum of the ball as it passes through its lowest point.

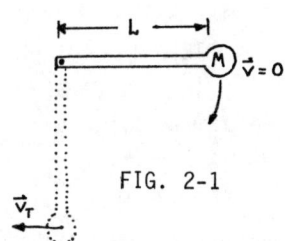

FIG. 2-1

Given: L, M = mass of rod, M = mass of ball.
Required: v_T, p_T of ball when the rod is vertical.

This problem can be most easily solved by using conservation of mechanical energy in a rotational format. We choose the position of the center of mass of the system when the rod is vertical as the zero of gravitational potential energy. Neglecting the size of the ball, the center of mass of the system is (Rod center of mass is 1/2 L).
$$r_{cm} = [(1/2)ML + ML]/2M = (3/2)L/2 = (3/4)L$$
The potential energy of the system when it is horizontal is then
$$u = (3/4L)Mg$$
When the rod swings through the vertical its energy is all kinetic and $K = (1/2)I\omega^2$. To calculate I we use the additive property of the moment of inertia
$$I_1 = \Sigma m_i R_i^2 = \underbrace{\int R^2 dm}_{(rod)} + \underbrace{ML^2}_{(ball)} = 1/3 \, ML^2 + ML^2 = 4/3 \, ML^2$$

By conservation of mechanical energy
$$U = \Delta K = (3/4L)Mg = (1/2)(4/3 \, ML^2)\omega^2$$
$$\omega = \sqrt{9g/8L}$$
The tangential velocity of the ball is
$$v_T = R\omega = L\sqrt{9g/8L} = \sqrt{9gL/8}$$
Its linear momentum is
$$p_T = Mv_T = M\sqrt{9gL/8}$$

Example 3

A yo-yo consists of a solid disk of radius R and mass M with a string wound around it. The end of the string is held fixed. a) What is the acceleration of the center of mass? b) What is the tension in the string? c) What is the velocity of the center of mass in terms of h, the distance the yo-yo has fallen?

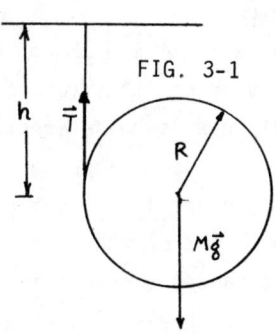

FIG. 3-1

Solution

   Given: M, R, h, I = 1/2 MR^2

   Required: a) T, b) a_{cm}, c) v_{cm}

a) In order to find the acceleration of the center of mass and T we first apply Newton's 2nd law for the translational motion. We take the positive direction down, since it is obvious that this will be the direction of a_{cm}.

$$-T + Mg = Ma_{cm} \qquad (1)$$

In order to generate another equation for T and a_{cm} we apply the rotational form of Newton's 2nd law. We choose the center-of-mass rotational axis. The only force which exerts a torque about this axis is T.

   TR = Iα Rα = a_{cm} (since we assume the string does not slip)
   TR = (1/2)MR^2 a_{cm}/R

$$T = (1/2)M\, a_{cm} \qquad (2)$$

Substituting into Eq. (1)

$$-(1/2)M\, a_{cm} + Mg = M\, a_{cm} \rightarrow a_{cm} = \underline{2/3\ g} \qquad (3)$$

   b) From Eqs. (2) and (3) we see T = $\underline{1/3\ Mg}$

   c) Since the acceleration of the center of mass is constant we may employ the kinematical relation $v_f^2 = v_0^2 + 2ah$ (Eq. 4.17).

$$v_f^2 = 0 + 2(2/3\ g)h$$

$$v_f = \underline{\sqrt{(4/3)gh}}$$

This problem can also be solved by considering the axis of rotation to be the point of contact of the string with the disk and employing the parallel axis theorem. Can you verify the result to c) using energy considerations?

Example 4

The front wheel of a bicycle has a radius R = 35 cm and mass M. If the rider leans the bicycle over 10° when he is traveling 5 m/s, a) What is the direction of the change in angular momentum of the bicycle wheel? b) What is the precessional angular velocity of the wheel? (Consider only the wheel in this calculation)

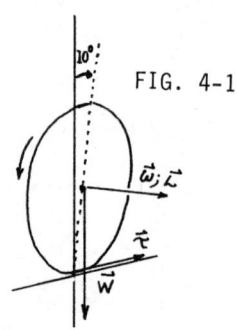

FIG. 4-1

Solution

   Given: R = 35 cm, M, v_{cm} = 5 m/s, θ = 10°

   Required: a) L (direction), b) Ω

a) This is an interesting example of both a "fast top" and rolling without slipping. When the wheel of the bicycle is leaning over, its weight exerts a torque about an axis through P, the point of contact of the wheel with the ground. This torque causes a change in the angular momentum of the wheel. Since the wheel rolls without slipping, it is most convenient to choose P as the point through which the instantaneous angular momentum L and the torque pass. We see from FIG. 4-2

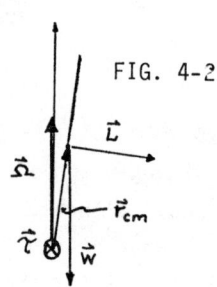

FIG. 4-2

that $r_{cm} \times W = \tau$ points into the paper. Since $\tau = dL/dt$, ΔL must point in this direction also. Thus, as in the case of the gyroscope, the wheel will precess about the vertical axis. In this case the wheel turns to the rider's left, the same direction in which he leans. Those of you experienced in riding a bicycle without holding the handle bars should appreciate this explanation!

b) To calculate the precessional frequency we make use of Eq. 13.11. The magnitude of in this case is given by

$$\tau = \Omega L \sin\theta = \Omega L \sin 100° \quad \text{(note, } \Omega \text{ vertical)} \tag{1}$$

Now the torque about P is $\tau = MgR \sin 170°$, the moment of inertia of the wheel can be taken as MR^2, and the angular velocity of the wheel is $= v_{cm}/R$. Solving (1) for Ω and substituting gives

$$\Omega = \frac{\tau}{L\sin 100°} = \frac{MgR\sin 170°}{I\omega\sin 100°} = \frac{MgR(0.17)}{MR^2(v_{cm}/R)(0.98)}$$

$$\Omega = \frac{(9.8 \text{ m/s}^2)(0.17)}{(5 \text{ m/s})(0.98)} = \underline{0.34 \text{ rad/s}}$$

The angular velocity of the wheel is $\omega = (5 \text{ m/s})/0.35 \text{ m} = 14.3 \text{ rad/s}$ and $\omega/\Omega = 42$. Thus $\omega \gg \Omega$ and we are justified in treating the wheel as a "fast top".

PROBLEMS

1. A flywheel in the shape of a uniform disk of mass 70 kg and radius 10 cm accelerates from 0 to 2500 rpm in 8 s. What power must the motor driving the flywheel provide?

2. An engine in a boat delivers 75 hp to a propeller turning at 1000 rpm. What is the torque exerted on the propeller?

3. A modern power plant generates 10^8 watts of electrical power. If the steam turbine rotates at 10,000 rpm, what must be the moment of inertia of the turbine to generate this much power. (Assume the rotational kinetic energy of the turbine is equal to the electrical energy delivered each second)

4. A solid sphere rolls down an inclined plane of angle with the horizontal. What is the minimum value that the coefficient of static friction must have in order that it roll without slipping?

5. A flywheel of moment of inertia 50 kg-m^2 rotating with an angular velocity of 1200 rpm has a torque of 200 N-m applied perpendicularly to its spin axis. What is the resulting precessional angular velocity?

6. A symmetric top of mass 60 g and moment of inertia 0.02 kg-m^2 spins at 200 rpm. The spin axis of the top makes an angle of 20° with the vertical and the center of mass of the top is 2 cm from its base. What is the top's precessional angular velocity?

Chapter 14
Oscillatory Motion

PREVIEW

Periodic behavior is one of the most common and important phenomena in nature. Many physical and biological systems exhibit this type of behavior and this chapter introduces one of the fundamental processes used in the analysis of periodic systems. First we introduce the basic parameters which are used to describe periodic systems and define the conditions under which simple mechanical systems will exhibit oscillatory behavior. We further specialize our consideration to a special type of oscillatory motion, that which is simple harmonic. We present discussions of free simple harmonic motion for undamped as well as damped oscillations. The topic of forced oscillations and resonance in a damped simple harmonic oscillator is also considered.

SUMMARY

14.1 SIMPLE HARMONIC MOTION

Definition. Periodic behavior: a system which changes in such a way that the configuration of the system is exactly reproduced with a constant time interval between each recurrence is said to exhibit periodic behavior.

Definition. The period: the constant time interval between recurrences of the configuration of a periodic system.

The period is usually designated by the symbol T. It is the time required for the system to complete one full cycle of its behavior. A related and useful quantity is the frequency of recurrence. It is simply the inverse of the period. We employ the symbol ν to denote frequency.
$$\nu = 1/T \tag{14.1}$$

Definition. Oscillatory motion: periodic motion in which the system's configuration varies about an equilibrium configuration through which the system passes during the periodic motion.

For a mechanical system to exhibit oscillatory behavior we must have two conditions fulfilled. 1. The net force or torque must act to restore the system to its equilibrium position regardless of the configuration. 2. The system must possess inertia.

The importance of oscillatory behavior can be seen if one thinks of the general form of the potential energy curve of a system which exhibits stable equilibria. Since stable equilibrium points correspond to minima of the potential energy curve, in the neighborhood of such points the system will exhibit oscillatory behavior. If the displacement of the system about the equilibrium position is small, and if the potential energy function is "smooth", the oscillatory motion has special characteristics. The nature of such small oscillations is identical to that of a special class of systems which we now define.

Definition. Simple harmonic motion (SHM): the oscillatory motion exhibited by a system whose acceleration is proportional to the negative of its displacement from its equilibrium position.

Because of Newton's second law the above criterion is equivalent to stating that the net force acting on the system is a restoring force proportional to the displacement of the system from equilibrium. Systems which execute SHM can be characterized by a single natural frequency of oscillation. This frequency, as we shall see, is a characteristic of the physical properties of the system. Apart from its applicability to systems which rigorously satisfy the above definition and its use in describing small oscillations, many periodic systems can be studied using simple harmonic oscillators (SHO) as models.

Linear springs and torsional springs which obey the linear or rotational forms of Hooke's law ($F = -kx$ or $\tau = -\beta\theta$) as well as a simple pendulum which undergoes small oscillations execute SHM. We use the linear spring as a model for the discussion of SHM but the equations can be cast in a general form which holds for any SHO.

Equation of motion of a simple harmonic oscillator. Employing Newton's second law and the fact that the force law obeyed by a SHO is proportional to the negative of the displacement from equilibrium the general equation of motion obeyed by such systems is:
$$d^2x/dt^2 = -\omega^2 x \tag{14.4}$$
Where is a constant related to the physical properties of the specific system.

For a mass attached to a linear spring $\omega^2 = k/m$ (Eq. 14.5), for a simple pendulum $\omega^2 = g/l$ (Eq. 14.6), and for a torsional pendulum $\omega^2 = \beta/I$.

Restoring forces which give rise to SHM are conservative and the total mechanical energy of such a system can be written in the form:
$$E = (1/2)mv^2 + (1/2)kx^2 \tag{14.7}$$
The turning points of the motion of a SHO are the points of its maximum excursion from equilibrium. The magnitude of the turning points is called the amplitude A of the motion. The total mechanical energy can be expressed in terms of the amplitude as $E = (1/2)kA^2$. The maximum velocity of the system occurs when it passes through the equilibrium point and $v_{max} = \sqrt{2E/m} = A\sqrt{k/m} = A\omega$ (Eq. 14.10). A great deal about the nature of the motion of a SHO can be learned by studying the energy relationship alone. A full description of the motion requires solution of Eq. 14.4.

14.2 SOLUTION OF THE HARMONIC OSCILLATOR EQUATION

The harmonic oscillator equation (Eq. 14.4) is typical of the type of differential equation which must be solved in many physical situations. It is officially called a homogeneous linear second order ordinary differential equation with constant coefficients. (The term homogeneous refers to the fact that the right hand side of the equation equals zero.) It is easier to solve it than say it!

Although many of you will not have had experience with such equations, they are the simplest type to solve and you will soon be comfortable with them. Since the equation is of second order in the derivative, solving the equation will generate two "integration constants". In a given physical situation you derive a solution to the problem at hand by specifying what these constants must be from the initial conditions the system must obey.

The most general solution to the SHO equation is:
$$x = A\cos(\omega t + \phi) \tag{14.11}$$

A is the amplitude of the motion since $-1 < \cos(\omega t + \phi) < +1$. The angle ϕ is called the initial phase angle and specifies where in the cycle of motion the time has been chosen to be zero. The velocity and acceleration of the SHO at any time are:
$$dx/dt = v(t) = -\omega A \sin(\omega t + \phi)$$
$$d^2x/dt^2 = a(t) = -\omega^2 A \cos(\omega t + \phi) \tag{14.12}$$

One could choose to write the solution in terms of a sine function rather than a cosine. The only difference is the introduction of a phase angle of $\pi/2$. Since $\omega = \sqrt{k/m}$ already has the units of $1/s$, and since $\omega t + \phi$ must be an angle it is easy to show (and not surprising) that $\omega = 2\pi/T = 2\pi\nu$ (Eq. 14.13). From Eqs. 14.11 and 14.12 we see that the maximum values of x, v, and a are $x_{max} = A$, $v_{max} = \omega A$, $a_{max} = \omega^2 A$. (Eqs. 14.18, 14.19, 14.20)

14.3 APPLICATIONS OF SIMPLE HARMONIC MOTION

This section presents a discussion of a number of applications of simple harmonic motion. Since these are essentially examples we will not review this section specifically.

14.4 UNIFORM CIRCULAR MOTION AND SIMPLE HARMONIC MOTION

The general solution of the SHO equation, $x = A\cos(\omega t)$ has as its angular coordinate t. Thinking back to our discussion of uniform circular motion as well as our definition of angular kinematical quantities it should not be too difficult to see that a particle undergoing uniform circular motion with constant tangential speed has an angular displacement which can be expressed as $\theta = \omega t$. The similarity between the angular coordinates in SHO and uniform circular motion is more than coincidental. If we choose a reference frame with its origin at the center of the circular path of the particle, the x and y coordinates of the particle can be expressed in <u>polar form</u> as:
$$x = r\cos\theta \quad ; \quad y = r\sin\theta$$
where we have chosen $\theta = 0$ when the particle is at the point (r,0). Substituting the expression for θ as a function of t we find:
$$x = r\cos\omega t \quad ; \quad y = r\sin\omega t$$
Differentiating these gives the x and y components of the velocity and acceleration.
$$v_x = -\omega r \sin\omega t \quad ; \quad v_y = \omega r \cos\omega t$$
$$a_x = -\omega^2 r \cos\omega t \quad ; \quad a_y = -\omega^2 r \sin\omega t$$

The x and y coordinates of the particle separately satisfy the definition for SHM. Thus the projection of uniform circular motion onto a diameter of the circular path is simple harmonic. In this instance the amplitude is r, the maximum velocity is $r\omega$, and the maximum acceleration is $\omega^2 r$.

14.5 DAMPED HARMONIC MOTION

So far we have treated SHM in terms of a purely conservative restoring force. All real physical systems experience dissipative forces which decrease their mechanical energy over time. A simple model for friction is to assume that the force is proportional to the negative of the system's velocity. This assures that the frictional force always opposes the motion. Although some other forms for the frictional force are more realistic this type of model exhibits the essential features we wish to discuss and maintains the linearity of the equation of motion of the harmonic oscillator. The force law for the motion becomes:

$$F = -kx - \gamma v$$

where γ is a frictional coefficient. For a mass attached to a linear spring the equation of motion is

$$d^2x/dt^2 + (\gamma/m)dx/dt + (kx/m) = 0 \qquad (14.37)$$

This equation can be written in a generally applicable form by noting that $\omega^2 = k/m$ and defining a new parameter $\tau = 2m/\gamma$.

$$d^2x/dt^2 + (2/\tau)dx/dt + \omega^2 x = 0 \qquad (14.40)$$

The reason for this choice for τ will become evident when we consider the solutions of this equation. The properties of the solution to this equation depend on the relative value of γ and ω. Using the general techniques for solving this type of equation it can be shown that if $\omega > (\gamma/2m)$ the solution will be oscillatory but damped:

$$x = Ae^{-(t/\tau)} \cos \omega_1 t \qquad (14.41)$$

$$\omega_1 = (1 - 1/\omega^2 \tau^2)^{1/2}$$

This solution is called the underdamped solution since the oscillator still oscillates with both positive and negative displacement but with a slowly decreasing amplitude. A graph of such an underdamped oscillator is shown in FIG. 14.19 of the text.

The parameter τ, which has units of seconds, determines how rapidly the system's amplitude and therefore its energy decays. If we set $t = \tau$ in Eq. 14.41, we see that the amplitude of the oscillations will have decreased to 1/e of its original value. This type of exponential decay is very common in physics and the time τ is called the mean time, mean lifetime, or simply the lifetime of the decay process.

If $\omega < (\gamma/2m)$ then the solutions to Eq. (14.41) are not oscillatory at all but are merely decaying exponential.

$$x = 1/(\alpha_2 - \alpha_1)(\alpha_2 e^{-\alpha_1 t} - \alpha_1 e^{-\alpha_2 t}) \qquad (14.43)$$

where α_1 and α_2 are the roots of the quadratic equation $\alpha^2 - (2/\tau)\alpha + \omega^2 = 0$. Solutions for a number of cases are exhibited in FIG. 14.21 of the text.

If $\omega = (\gamma/2m)$ then the solution becomes:

$$x = (A + Bt)e^{-t/\tau}$$

This case is called critical damping. This corresponds to the most rapid nonoscillatory decay of the system back to its equilibrium value. This is the type of damping which is most often designed into mechanical systems. An example of such a system is the automobile. The shock absorbers are tuned to each vehicle so that critical damping is achieved.

Since a damped system includes dissipative forces, the mechanical energy of the system is <u>not</u> conserved but decreases at a rate:

$$dE/dt = -\gamma v^2 \qquad (14.46)$$

14.6 FORCED HARMONIC MOTION: RESONANCE

A somewhat more complex situation arises when a damped harmonic oscillator is forced to oscillate by an imposed external periodic force. This situation arises frequently enough in real life that it is worthwhile to be familiar with the effects that occur. We suppose that in addition to the linear restoring force and the velocity dependent frictional force an oscillating force of the form $F = F_0\cos\omega't$ is applied. The equation of motion can then be written as:

$$d^2x/dt^2 + (2/\tau)dx/dt + \omega^2 x = a_0\cos\omega't \qquad (14.48)$$

with $a_0 = F_0/m$. The general solution to this nonhomogeneous equation can be shown to consist of two parts. The first part is simply the solution to the original homogeneous equation. Since these terms are exponentially damped regardless of the type of damping (under-, over-, or critically damped), if $t \gg \tau$ the contribution of this part to the solution will be negligible. This portion is therefore termed the transient part of the solution.

The oscillator is being forced to vibrate. Eventually it will settle down into a steady state oscillation with a frequency (or angular frequency) equal to that of the impressed force and of the form $x = A\cos(\omega't + \phi)$ (Eq. 14.49). In this case the amplitude and phase angle, while constant, depend on the relative magnitude of the impressed angular frequency and the natural frequency as well as the damping parameters. Although the general steady state solution can be determined (see A. P. French, <u>Vibrations and Waves</u>, W. W. Norton & Co., Publisher, for a thorough but still fairly elementary discussion of forced vibrations and resonance) most of the interesting information can be obtained by considering three limiting cases.

1. $\omega' = \omega$. For this case $A = \tau a_0/2\omega'$ and $\phi = -\pi/2$. The amplitude is primarily determined by the damping coefficient γ through τ. If there is no damping then $\tau \to \infty$ and and the amplitude becomes infinite. The case $\omega' = \omega$ is called <u>resonance</u> and the amplitude of the forced vibration can become very large in this instance. We also see that the driving force reaches its maximum earlier than the displacement by a phase angle of $\pi/2$.

2. $\omega' \ll \omega$. In this case $A = a_0/\omega^2$ and $\phi = 0$. The amplitude becomes constant and the oscillation is slow enough that the displacement "follows" the force exactly.

3. $\omega' \gg \omega$. In this instance $A = a_0/(\omega')^2$ and $\phi = \pi$. The amplitude decreases as ω' increases and approaches zero as $\omega' \to \infty$. The displacement and the driving force are completely out of phase. When the driving force reaches its maximum value the displacement reaches its minimum ($x = -A$) and vice versa.

The response of a forced damped oscillator with respect to phase and amplitude are shown in FIGS. 14.22 and 14.23 of the text.

EXAMPLE PROBLEMS
Example 1

A mass of 10 kg resting on a frictionless surface is attached to a spring with k = 300 N/m.. The mass is pulled 20 cm to the right of the equilibrium position and then released. a) Find the angular frequency, period, and frequency of the motion.
b) What is the maximum velocity of the mass?
c) What is the system's total energy? d) How much
time is required for the mass to move from its initial position to the equilibrium point?

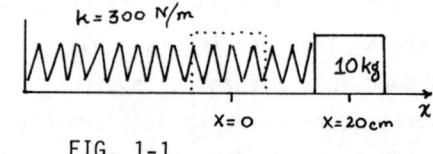

FIG. 1-1

Solution
Given: $k = 300$ N/m
          $m = 10$ kg
          $A = 20$ cm $= 0.2$ m
Required: a) ω, T, ν b) v_{max} c) E_{total} d) time from $x = 0.2$ to $x = 0$

a) The period, frequency and angular frequency are functions of the physical properties of the system only.

$\omega = \sqrt{k/m} = \sqrt{(3200 \text{ N/m})/10 \text{ kg}} = \underline{5.48 \text{ rad/s}}$; $T = 2\pi/\omega = \underline{1.15 \text{ s}}$; $\nu = 1/T = \underline{0.87 \text{ Hz}}$

b) Since the mass is released at 0.2 m, this corresponds to the amplitude of the motion. The maximum velocity is (Eq. 14.19) $v_{max} = A\omega$. Substituting the given quantities

$v_{max} = (0.2 \text{ m})(5.48 \text{ rad/s}) = \underline{1.09 \text{ m/s}}$

c) Using the principle of conservation of mechanical energy

$E_{total} = (1/2)kA^2 = (1/2)(300 \text{ N/m})(0.2 \text{ m})^2 = \underline{6 \text{ J}}$

d) The time required for the system to travel from its point of release to the equilibrium point $x = 0$ is 1/4 of the period

$(1/4)T = (1/4)(1.15 \text{ s}) = \underline{0.29 \text{ s}}$

Example 2

A mass m is attached to a linear spring of force constant k. Determine the amplitude and phase angle ϕ in the general solution $x = A\cos(\omega t + \phi)$ in the following situations. a) The spring is displaced a distance +0.1 m from its equilibrium position and released from rest at $t = 0$.
b) The mass is given an initial velocity of -2 m/s at the equilibrium position at $t = 0$.
c) The mass has velocity +2 m/s at $x = 0.1$ m, $t = 0.5$ s.

Solution
Given: m, k
          a) $x = 0.1$ m, $v_0 = 0$, $t = 0$
          b) $x = 0$, $v_0 = -2$ m/s, $t = 0$
          c) $x = 0.1$ m, $V_0 = 2$ m/s, $t = 0.5$ s
Required: A and ϕ for each case

a) This exercise is to help you understand how the initial conditions in a given situation are used to provide the specific solution from the general solution to the SHO equation. In this case, since the particle is released from rest at $x = 0.1$ m and since the force will be $-k(0.1 \text{ m})$, this value of x is equal to the amplitude, $\underline{A = 0.1 \text{ m}}$. To determine the phase angle, we substitute for A, x, and t in the general solution:

$0.1 \text{ m} = 0.1 \text{ m} \cos(\omega \cdot 0 + \phi) \rightarrow \cos\phi = 1 \rightarrow \underline{\phi = 0}$

(Note 2π and 0 are considered the same angle due to the periodicity of the cosine function.

b) This situation is a bit more difficult, but not much. Since the particle begins its motion at the equilibrium position, the speed at this point must be $v_{max} = A\omega$ (Eq. 14.19). Thus the amplitude is

$\frac{2 \text{ m/s}}{\omega} = A = \underline{2\sqrt{m/k} \text{ m}}$

To determine the phase angle we first consider the general solution with $x = 0$, $t = 0$

$0 = A\cos\phi$

$\cos\phi$ is zero at two angles within a given period, $\pi/2$ and $3\pi/2$. We must use the information given about the velocity to decide which of these is correct. Since

$v = -A\omega \sin(\omega t + \phi)$ at $t = 0$ we have $v = -A \sin\theta$

The initial velocity is negative, therefore $\sin\theta$ must be positive and we see that $\phi = \pi/2$

c) This set of initial conditions presents the largest challenge. Since we know x and v at a given point, we may use conservation of mechanical energy to find the amplitude.

$$1/2\ kx^2 + 1/2\ mv^2 = \text{constant} = (1/2)kA^2$$

Substituting

$$1/2\ k(0.1\ m)^2 + 1/2\ m\ (2\ m/s)^2 = (1/2)kA^2$$

$$.01\ k + 4\ m = kA^2 \quad \text{(suppressing units)}$$

$$A = \sqrt{.01 + 4\ m/k}\ m$$

In order to determine ϕ we now substitute into the general solution for x (we will use A for amplitude)

$$0.1 = A \cos(0.5\omega + \phi)$$

$$(0.1/A) = \cos(0.5\ \omega + \phi)$$

$$\phi = \cos^{-1}(0.1/A) - 0.5\omega\ \text{rad}.$$

If there is any ambiguity in ϕ, this can be resolved as above by recourse to the equation for v.

Example 3

A massless spring of length 0.5 m when it is horizontal and k = 250 N/m, is hung from the ceiling. A block of mass 1 kg is attached to the spring. The block is pulled down 5 cm below the equilibrium position and released from rest. a) What is the angular frequency of the motion, b) What is y(t)? c) What is the total mechanical energy of the spring-mass system?

Solution

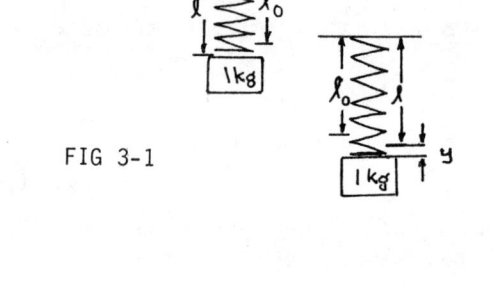

FIG 3-1

Given: k = 250 N/m

ℓ_o = 0.5 m

m = 1 kg

A = y_i = 0.05 m

Required: a) ω, b) y(t), c) E_{total}

a) The spring (since it is massless) has a relaxed length ℓ_o = 0.5 m. When the mass is attached, the spring is stretched to a new equilibrium position ℓ. This equilibrium position is such that $(\ell - \ell_o)k = mg$ (we choose down as positive), therefore ℓ = 0.54 m. The force exerted on the block when it is pulled down a distance y below the <u>new</u> equilibrium point is:

$$F = mg - k[(\ell - \ell_o) + y] = -ky$$

Thus the angular frequency is simply

$$\omega = \sqrt{k/m} = \sqrt{(200\ N/m)/(1\ kg)} = 14.14\ rad/s$$

b) At t = 0 we have y = 0.05 m = A and v_o = 0

$$y = A\cos(\omega t + \phi) = 0.05 \cos(\omega t + 0) = 0.05 \cos(14.14\ rad/s)t$$

c) When the block is attached to the spring and the spring stretches to its new equilibrium length, there will be an increase in the potential energy of

$$U_o = 1/2\, k\, (\ell - \ell_o)^2$$

When the spring and block are further stretched to a position y below, the total energy stored in the spring is:

$$U_s = 1/2\, k\, [(\ell - \ell_o) + y]^2$$

There is a decrease in the gravitational potential energy

$$U_g(y) = -mgy = -ky(\ell - \ell_o)$$

Thus the total mechanical energy is

$$E_{total}(y) = U_g + U_s = (1/2)k\,(\ell - \ell_o)^2 + (1/2)ky^2$$

$$E_{total} = U_o + (1/2)ky^2 \cong 0.2\,J + 0.3\,J = 0.5\,J$$

Example 4

For a damped harmonic oscillator the case of critical damping corresponds to the most rapid approach the system can make to equilibrium without overshooting the equilibrium position. Show this explicitly employing the solutions to Eq. (14.40) for overdamped and critically damped behavior.

Solution

Given:
$$x_{crit}(t) = (a + bt)e^{-t/\tau}$$

$$x_{OD}(t) = \frac{1}{\alpha_2 - \alpha_1}(\alpha_2 e^{-\alpha_1 t} - \alpha_1 e^{-\alpha_2 t})$$

Required: To show $x_{crit} \to 0$ faster than x_{OD}

Before we demonstrate that the required condition holds, it is advantageous to rewrite $x_{OD}(t)$ in terms of the explicit solution to the characteristic equation $\alpha^2 - (2/\tau)\alpha + \omega^2 = 0$ which determines α_1 and α_2. Employing the quadratic formula we find

$$\alpha = \frac{2/\tau \pm \sqrt{(2/\tau)^2 - 4\omega^2}}{2} = (1/\tau) \pm \sqrt{1/\tau^2 - \omega^2}$$

The solution for $x_{OD}(t)$ can now be written in the form

$$x_{OD}(t) = A e^{-t/\tau}(e^{\pm t\sqrt{1/\tau^2 - \omega^2}})$$

Now let us consider the ratio $x_{crit}(t)/x_{OD}(t)$

$$\frac{x_{crit}(t)}{x_{OD}(t)} = \frac{e^{-t/\tau}(a + bt)}{A e^{-t/\tau}(e^{-t\sqrt{1/\tau^2 - \omega^2}} + e^{+t\sqrt{1/\tau^2 - \omega^2}})}$$

as t becomes large the negative exponential terms become small and this expression is dominated by the linear term in the numerator and the positive exponential in the denominator, since $\tau > \tau'$

$$\lim_{t \to \infty} \frac{t}{e^t} = 0$$

Thus the numerator does not increase as fast as the denominator. We can thus conclude that the numerator $x_{crit}(t)$ must approach zero more rapidly than does $x_{OD}(t)$.

Example 5

A resonating organ pipe is equivalent to a damped linear spring harmonic oscillator with $m = 10^{-5}$ kg, $k = 25$ N/m and $\gamma = 1.3 \times 10^{-5}$ kg/s. a) What is the pipe's resonant frequency? b) If the pipe is driven by a force equivalent to 1 N what would be the equivalent amplitude of oscillation? c) How much energy could be stored in the organ pipe?

Solution

Given: $m = 10^{-5}$ kg

$k = 25$ N/m

$\gamma = 1.3 \times 10^{-5}$ kg/s

Required: a) $\omega_{resonant}$ b) A c) E_{total}

a) The resonant frequency is the normal free oscillation frequency of the harmonic oscillator

$\omega = \sqrt{k/m} = \sqrt{(25 \text{ N/m})/(10^{-5} \text{ kg})} = \underline{1581 \text{ rad/s}}$

$\gamma = \omega/2\pi = \underline{252 \text{ Hz}}$

b) The amplitude at resonance is given by Eq. (14.51)

$A = \dfrac{\tau a_0}{2\omega} = \dfrac{(2m/\ )(F_0/m)}{2\omega} = \dfrac{F_0}{\gamma \omega} = \dfrac{1 \text{ N}}{(1.3 \times 10^{-5} \text{kg/s})(1581 \text{ rad/s})} = \underline{48.7 \text{ m}}$

c) The total mechanical energy can be determined from the amplitude

$E = 1/2\ kA^2 = (1/2)(25 \text{ N/m})(48.7)^2 = \underline{29646 \text{ J}}$

PROBLEMS

1. A mass of 10 g hangs on a light spring. When pulled down 15 cm below the equilibrium position and released, it oscillates with a period of 2 s. a) What is its force constant? b) What is its maximum speed?

2. A mass of 5 kg executes SHM of frequency 200 Hz and amplitude 0.5 m. What is the magnitude of the force exerted on the mass when it is at one of its turning points?

3. A spear gun has an elastic band which requires a force of 60 N to stretch 1 cm. A spear of mass 0.2 kg is inserted and the bands are stretched to a length of 20 cm. a) What is the potential energy of the spear? b) With what speed will the spear leave the gun?

4. A mass m is attached to a spring with force constant k. a) What is x(t) if the mass is displaced to a position -A relative to the equilibrium and released? b) What is x(t) if instead it is given a velocity +v at the equilibrium point?

5. A linear spring of force constant k = 75 N/m is attached to a mass of 2 kg. At a certain instant the mass is observed to be at a distance of 1 m and have a velocity of -5.6 m/s. What is x(t)?

6. A mass m resting on a horizontal frictionless surface is connected between two identical springs. The springs are also horizontal and are attached to two vertical supports. (FIG. 6-1) If the mass is pulled to one side and released, what will be the frequency of its motion?

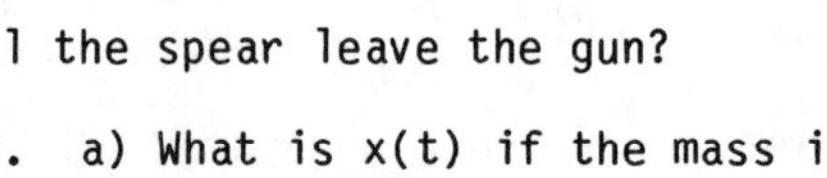

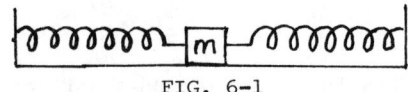

FIG. 6-1

7. A car has a mass of 2000 kg. Its four springs each have a force constant of 10^4 N/m. What damping constant must the shock absorbers have to insure critical damping?

8. A swing of length 2.5 m has a mass of 1.8 kg. When set in motion it requires 45 s for its amplitude to decrease to 1/e of its original value. What is its damping constant?

9. It is desired to make a SHO with mass 0.5 kg, k = 3.2 N/m, and = 0.15 kg/s oscillate at resonance with an amplitude of 0.6 m. What must the magnitude of the oscillating driving force?

Chapter 15
The Mechanical Properties of Matter

PREVIEW

This chapter deals with the mechanical properties of matter, primarily solids, in macroscopic quantities. In previous chapters, the effects of a force on the <u>motion</u> of an object have been described. Here, we find that application of a force to an object will also affect its size and/or shape. A material object subject to a force will be distorted. The concepts of stress and strain are introduced to describe these effects and Hooke's law, an empirical relationship between stress and strain, is formulated.

SUMMARY

15.1 STATES OF MATTER

The properties of matter are generally divided into two classes, macroscopic and microscopic.

<u>Definition</u>. <u>Macroscopic property</u>: a property which depends upon the combined effects of a very large number of atoms or molecules. These properties can be described in terms of a continuum model of matter without invoking the idea of a structure composed of discrete atomic sized particles. The mass of an object is an example of a macroscopic property which describes the amount of matter in the object.

<u>Definition</u>. <u>Microscopic property</u>: a property which depends upon the action of the individual atoms or molecules which make up the structure of the matter. The number of molecules in a sample of matter is an example of a microscopic property of the sample which describes the amount of matter in the sample.

The physical states of matter in the bulk, or macroscopic, form may be divided into two broad classes, solids and fluids.

<u>Definition</u>. <u>A solid</u>: a sample of matter which has a definite size and shape when free of external forces. Large external forces are required to change the size or shape of a solid object.

<u>Definition</u>. <u>A fluid</u>: a sample of matter which has no definite shape. When free of external forces, a fluid will assume a shape determined by the internal forces present in the sample and its shape can be changed by the application of small external forces.

Fluids may be further subdivided into gases and liquids.

<u>Definition</u>. <u>A gas</u>: a fluid which has no definite size. In the absence of external forces, a gas will be uniformly distributed throughout the container in which it is confined, thereby assuming the size of the container.

Definition. A liquid: a fluid which has a definite size (volume) but assumes the shape
of the container in which it is confined. If unconfined, a liquid assumes a shape
determined by the internal and external forces to which it is subject.

Since we deal here with matter in general rather than with specific objects, it is more convenient to describe inertial and gravitational effects in terms of the density rather than the mass.

Definition. Density: the mass per unit volume of a sample of matter. Density is
symbolized by ρ and has units of kilograms per cubic meter.
$$\rho = M/V \tag{15.1}$$

In some cases, the mass is not distributed uniformly throughout the volume of the sample and Eq. 15.1 must be interpreted as the <u>average</u> density.

15.2 STRESS AND STRAIN

To describe the distortion of a sample of matter as a result of the application of forces, we may first assume that the sample is in equilibrium so that the forces are balanced and the situation is not complicated by acceleration. For example, if a rod or wire is distorted by stretching, we suppose that equal and oppositely directed forces act upon the two ends.

Three general types of distortion are recognized and a real situation may be analyzed in terms of one or some combination of these three. For each of these types, a corresponding strain is defined as a measure of the distortion.

A distortion in which a single dimension of a solid object changes is referred to as a tensile or compressive strain depending on whether the change is an increase or a decrease.

Definition. Tensile or compressive strain: the fractional change in length of a solid
object as a result of the application of a stretching or compressive force.
$$\text{Tensile strain} = \Delta \ell / \ell \tag{15.3}$$

A distortion in which the volume of the sample changes as a result of forces more or less uniformly distributed over its surface is referred to as volume strain.

Definition. Volume strain: the fractional change in volume as a result of forces
distributed over the surface of the sample. This concept is particularly well suited to
the study of fluids, but may be applied to solids as well.
$$\text{Volume strain} = \Delta V/V$$

A distortion which results in a change in the shape rather than the size of a solid object is referred to as a shear strain. In this case, the distortion is an angular change in the direction of a dimension of the object perpendicular to the direction of the distorting forces. Shear strain is a result of opposing non-concurrent forces.

Definition. Shear strain: the ratio of the displacement of the surface parallel to the
applied force to the length of the dimension perpendicular to the force. So long as the
strain is small, the angular (in radians) displacement of the dimension perpendicular to
the force is an equivalent definition (See Fig. 15.3 in the textbook). Fluids will not
support a shear strain.

118   Chapter 15

$$\text{Shear strain} = \Delta x/y \simeq \theta \tag{15.6}$$

Note that in each case, strain is a dimensionless quantity.

For each of the three types of distortion, the distorting force may be described by a corresponding <u>stress</u>.

<u>Definition</u>. <u>Tensile or compressive stress</u>: the ratio of the stretching force to the cross sectional area of the member being stretched or compressed.

<u>Definition</u>. <u>Volume stress</u> (also called pressure): the ratio of the total force distributed over and perpendicular to the surface of an object to the surface area.

<u>Definition</u>. <u>Shear stress</u>: the ratio of the force distributed over and parallel to the surface of an object to the surface area.

$$\text{Stress (tensile, volume, shear)} = F/A \tag{15.2,4,5}$$

In each of the three cases, the unit of stress is force per unit area which, in the SI system is called the Pascal (Pa) and is equivalent to the newton per square meter.

15.3 HOOKE'S LAW AND YOUNG'S MODULUS

For each of these three types of distortion, it is found experimentally that over a certain (usually small) range of forces, the stress is directly proportional to the strain. This relationship is known as Hooke's law for tensile on compressive distortion and Hooke's name is often applied to the other two types as well. The proportionality constant used to describe this relationship is called an elastic modulus.

$$\text{Elastic modulus} = \frac{\text{stress}}{\text{strain}} \tag{15.7}$$

Since strain is dimensionless, the units of the elastic modulus must be the same as those for stress.

<u>Warning</u>: Hooke's law is an empirical relationship, not a fundamental law. It is valid only over a limited range of force (stress) and for stresses exceeding this range, the stress and strain are not proportional.

In the case of tensile or compressive distortion, the elastic modulus is called Young's modulus and is defined as

<u>Definition</u>. <u>Young's modulus</u>: the ratio of the tensile stress to the tensile strain.

$$Y = \frac{F/A}{\Delta \ell / \ell} \tag{15.8}$$

15.4 ELASTIC ENERGY AND THE SPRING CONSTANT

When a force applied to the end of a wire causes it to elongate, i.e. causes the end of the wire to move in the direction of the force, work is done on the wire. If the elastic limit of the wire has not been exceeded, this work can be recovered as the wire relaxes back to its original length; thus, the wire in its stretched condition has elastic potential energy.

From Chapter 8, the elastic potential energy of a stretched spring is

$$U_E = 1/2 \, kx^2 \,,$$

where x is the amount by which the spring has been stretched. If we compare the stretched spring of Chapter 8 to the stretched wire of the present chapter, we see that k is analagous to (YA/). Thus, for a stretched wire,

$$U_E = 1/2 \frac{YA}{\ell} \Delta\ell^2 \qquad (15.15)$$

15.5 BULK MODULUS AND COMPRESSIBILITY

When a force normal to the surface of a sample is distributed uniformly over the surface, the result is a <u>decrease</u> in the volume of the sample.

<u>Definition</u>. <u>Bulk modulus</u>: the negative of the ratio of the volume stress to the volume strain. The negative sign is included in the definition so that the bulk modulus will always be a positive number. The bulk modulus is symbolized by the letter B.

$$B = -\frac{P}{\Delta V/V} \qquad (15.21)$$

Note that in most cases, particularly in the case of fluids, the sample is subject to some pressure in its "undistorted" condition. Thus, the P in Eq. 15.21 must be interpreted as the <u>additional</u> pressure required to distort the sample.

In some applications, it is more convenient to use the reciprocal of the bulk modulus, the compressibility.

$$K = 1/B = \frac{\Delta V/V}{P} \qquad (15.22)$$

An important distinction between a gas and a liquid is that the bulk modulus for a liquid is very high while that for a gas is very low.

15.6 SHEAR MODULUS

If the distorting force is parallel to the surface over which it is distributed, the result is a shear strain, a change in the shape of the object without a significant change in its size.

<u>Definition</u>. <u>Shear modulus</u>: the ratio of the shear stress to the shear strain.

$$\mu = \frac{F/A}{\theta} \qquad (15.26)$$

The concepts of shear stress and strain can be extended to such applications as the twisting of a pipe or the bending of a girder.

EXAMPLE PROBLEMS
Example 1

A disabled automobile is being towed by another automobile at a constant speed of 50 mi/hr on a level highway. The tow vehicle requires 25 hp in addition to that necessary to maintain its own speed without the impediment of towing the second car. The two vehicles are connected via a steel cable one quarter inch in diameter, ten feet long, and having a Young's modulus of 15×10^9 Pa. By how much is the cable stretched as a result of the force required to tow the disabled car?

Solution

Given: $v = 50$ mi/hr $= 80.47$ km/hr $= 22.35$ M/s
$P = 25$ Hp $= 18650$ watts
$d = .25$ in $\longrightarrow$ $r = .003175$ M

120   Chapter 15

$$\ell = 10 \text{ ft.} = 3.048 \text{ M}$$
$$Y = 15 \times 10^9 \text{ Pa}$$

Required:

From Eq. 7.17, the force which the cable must exert on the towed vehicle to maintain a constant speed is

$$F = P/v$$

and from Hooke's law,

$$\Delta\ell = \frac{F}{\pi r^2 Y} = \frac{P\ell}{\pi r^2 v Y} \quad .0054 \text{ M}$$

2. Some automobiles use torsion rods rather than springs for the front wheel suspension. These steel rods are 2.6 cm in diameter and 1.5 m long with one end clamped rigidly to the frame. The front wheel is attached to the other end via a 25 cm long lever so that as the wheel moves up and down, the rod twists. If the car has a mass of 2000 Kgm, by how much is the torsion rod twisted when the car is at rest on a level surface. Assume that the weight of the car is distributed evenly on each of the four wheels. The relationship between the applied torque and the angle of twist for a solid cylindrical rod suggested by Fig. 15.7 in the textbook is

$$\tau = \frac{\pi\mu R^4}{\ell}\theta \quad 2$$

Solution

Given: M = 2000 Kgm
         d = 2.6 cm = .026 M
         ℓ = 1.5 M
         μ = 16 × 10^{10} Pa (from Table 15.1)
         Lever arm = A = 25 cm = .25 M

Required: θ

Since each wheel supports one fourth of the weight of the car, the torque applied to the end of the torsion rod is

$$\tau = (1/4)MgA$$

Substituting this expression for the torque into the equation above and solving for θ, we obtain

$$\theta = \frac{MgA}{\pi\mu R^4} = .512 \text{ rad} = 29.3°$$

Example 3.

How much work is required to produce the shear distortion shown in Fig. 15.3 in the textbook?

Solution

In producing the distortion, the applied force, F, moves through a distance, Δx and thus, does an amount of work,

$$W = \int F \cdot dx \ .$$

By Eq. 15.26,

$$F = (\mu A/y)x$$

so that

$$W = (\mu A/y)\int x \, dx = \frac{\mu A}{2y}x^2$$

Example 4

The block pictured in FIG. 15.26 is a cubical block of jello 5 cm on each side and having a mass of 150 gm. The jello is given a shear distortion and released so that it starts to oscillate (shimmy) at a frequency of 3 Hz. What is the shear modulus of the jello?

Solution

Given: $m = 150$ gm $= .150$ Kgm
   $\ell = 5$ cm $= .05$ M
   $\nu = 3$ Hz

Required: μ

When the block of jello is distorted and then released, the restoring force is given by Eq. 15.26

$$F = \mu A \theta$$

which may be translated to a torque about the point of oscillation, the bottom of the cube,

$$\tau = \mu A \ell \theta$$

The cube of jello is an inverted physical pendulum with moment of inertia

$$I = 1/3\, m\ell^2$$

for which the equation of motion is

$$\alpha = -\frac{\mu A \ell}{I} \theta$$

from which we may calculate the frequency of oscillation by

$$\nu = 1/2\pi \sqrt{\frac{3\mu A}{m\ell}}$$

From this equation, we may calculate

$$\mu = 3.55 \times 10^5 \text{ Pa},$$

a rather small value as we should expect for jello.

PROBLEMS

1. A table of elastic constants gives the values listed below for the Young's modulus and yield point for aluminum. How far may an aluminum wire 1 m long be stretched before the yield point is exceeded resulting in a permanent distortion?

2. A steel wire 1 mm in diameter is stretched taut between two rigid supports. Assuming no sag in the unloaded wire, by what angle does the wire sag when a 50 Kgm weight is suspended from its center? (Hint: Use the first two terms of the series expansions for the tangent and sine functions.)

3. A certain modern sculpture consists of a rectangular block of marble with a mass of 50 metric tons resting upon four cylindrical aluminum posts each 3 m long and having a diameter of 10 cm. Using the value of Young's modulus from Table 15.1 in the textbook, calculate the amount by which each of the posts is compressed by the weight of the marble block.

4. A rectangular block of material having a mass of 75 Kgm and a base 2 m long by 1 m wide is pulled across a horizontal floor at a constant speed by a horizontal rope attached to its top edge. If the coefficient of friction between the block and the floor is 0.3, what is the shear stress applied to the block?

5. An analysis similar to that suggested by Fig. 15.7 in the textbook reveals that the torque required to twist one end of a thin-walled pipe through an angle, θ, while holding the other end fixed is given by

$$\tau = \frac{2\pi\mu R^3 t}{\ell} \theta ,$$

where μ is the shear modulus of the material of which the pipe is made, R is the radius, t is the wall thickness, and ℓ is the length. θ is measured in radians. Derive the expression for the elastic energy of the pipe when it is twisted through an angle, .

Chapter 16
Fluid Mechanics

PREVIEW

In this chapter, the principles of mechanics are used to describe the macroscopic behavior of fluids at rest, or hydrostatics, and fluids in motion, known as hydrodynamics. Archimedes' Principle and the pressure-depth relation form a part of hydrostatics, while hydrodynamics includes Bernoulli's equation and the equation of continuity. The chapter also presents an introduction to viscosity and turbulence.

SUMMARY

16.1 FLUID STATICS: ARCHIMEDES' PRINCIPLE

Fluids can flow and change shapes and include both gases and liquids.

> Definition. Fluid: a substance that cannot support a shearing stress.

We will find out later in the chapter, however, that fluids can resist shearing motions to the extent that they possess a property known as viscosity.

> Definition. Hydrostatic equilibrium: a condition wherein each fluid particle is at rest, i.e., has zero velocity and acceleration.

By fluid particle we mean a volume of fluid that is small macroscopically but nonetheless contains many molecules of fluid.

> Definition. Buoyant force: The net effect of the surface forces exerted by a fluid on a body immersed in the fluid, of magnitude equal to the weight of the displaced fluid and directed vertically upward through the center of gravity of the displaced fluid.

The magnitude of the buoyant force can be expressed as $B = \rho_{fl} V_s g$ (Eq. 16.2), where ρ_{fl} is the fluid density and V_s is the volume submerged. For an object that is completely submerged in a fluid, the volume of the displaced fluid is equal to the volume of the object. For a floating object, the volume of the displaced fluid is the volume of the object below the surface of the fluid.

> Definition. Apparent weight: the true weight of a body immersed in a fluid decreased by the buoyant force.

For a floating object, the buoyant force is equal to the true weight, and the apparent weight is zero. For a suspended object, the apparent weight is the tension in the supporting string.

16.2 PRESSURE-DEPTH RELATION

<u>Definition</u>. <u>Pressure</u>: the ratio of the magnitude of a force applied perpendicular to a surface and the area over which the force is exerted.

The unit of pressure is the pascal (Pa). Other common units are pounds per square inch (psi) and atmospheres (atm), with 1 atm = 1.013×10^5 Pa = 14.70 psi.

<u>Definition</u>. <u>Atmospheric pressure</u>: the weight of the atmosphere above a square meter of surface area, approximately 10^5 Pa at the earth's surface.

Pressures measured relative to atmospheric pressure are referred to as gauge pressures, while those referenced to zero pressure are referred to as absolute pressures. These two pressures are related by: Gauge pressure = (Absolute pressure) - (Atmospheric pressure)

<u>Definition</u>. <u>Pressure-depth relation</u>: the pressure exerted on a body immersed in a fluid increases as the depth in the fluid increases according to $P = P_0 + \rho g z$ (Eq. 16.6) where P is the pressure at depth z, P_0 is the pressure at the surface, and ρ is the fluid density, assumed to be constant.

According to the pressure-depth relation, the difference in pressure between two points separated vertically by a distance z is given by $\rho g z$ regardless of their horizontal separation, provided that the two points can be connected by a path that lies within the fluid. This means that a pressure change at one point must result in an equal change at every other point, since the pressure difference between two points depends only on ρ and z and remains unchanged. This conclusion based on the pressure-depth relation is known as Pascal's principle.

<u>Pascal's Principle</u>: pressure applied to an enclosed fluid is transmitted undiminished to every portion of the fluid and the walls of the containing vessel.

In applications of Pascal's principle, it is the pressure and not the associated force that is transmitted undiminished throughout the fluid. Equal pressures at two points can lead to immense force differences when the areas involved are different. Pascal's principle involves only fluids that are contained and would not apply, for example, to the water in a tank with a hole below the water line.

When the fluid density cannot be assumed constant, the differential form of the pressure-depth relation must be used: $dP = \rho g\, dz$ (Eq. 16.5).

16.3 FLUID DYNAMICS

In considering fluid dynamics, the study of fluids in motion, we will deal primarily with so-called ideal fluids in irrotational, steady flow. The only departure from this approach will be in the last two sections of the chapter, where we will discuss viscosity and turbulence. An ideal fluid is incompressible and nonviscous.

<u>Definition</u>. <u>Incompressible fluid</u>: a fluid for which the mass density is constant.

<u>Definition</u>. <u>Nonviscous fluid</u>: a fluid that has no viscosity, a fluid form of internal friction.

Definition. **Irrotational flow**: fluid flow in which no fluid particle possesses rotational motion about its center of mass.

Definition. **Steady flow**: fluid flow in which the fluid velocity at each point of space is constant, although the velocity may change from point to point.

Definition. **Streamline**: a curve whose tangent at any point is in the direction of the fluid velocity at that point.

Definition. **Tube of flow**: the region bounded by the surface outlined by the bundle of streamlines passing through the periphery of a fluid volume.

Since streamlines can never cross, fluid particles cannot enter or leave a tube of flow except through the ends.

16.4 THE EQUATION OF CONTINUITY

Definition. **Discharge rate**: the volume of fluid that crosses a given area per unit time, defined by $Q = Au$ (Eq. 16.9), where A is the cross sectional area and u is the fluid velocity at A.

Equation of Continuity: for steady, incompressible flow, the discharge rate is constant along any tube of flow.

The equation of continuity is a statement of the conservation of mass within a tube of flow. It is most easily applied in the form: $A_1 u_1 = A_2 u_2$ (Eq. 16.8), where 1 and 2 refer to two points along a tube of flow. This form assumes steady, incompressible flow but makes no assumptions about viscosity.

16.5 BERNOULLI'S EQUATION

Definition. **Available energy**: the sum of the kinetic, gravitational potential, and flow energy associated with the macroscopic motions of a fluid.

Bernoulli's Equation: the available energy per unit volume of fluid remains constant along any streamline for steady, incompressible, and nonviscous flow.

Bernoulli's equation is most easily applied in the form:

$$P_1 + \rho g z_1 + \rho u_1^2/2 = P_2 + \rho g z_2 + \rho u_2^2/2 \tag{16.17}$$

where 1 and 2 refer to two points along a streamline.

In using Bernoulli's equation, the two positions in question are arbitrary as long as they lie along the same streamline. The data to be used in the equation routinely comes in a variety of units and must be expressed in a single system of units, thus conversion becomes an important aspect of Bernoulli's equation. It is useful to keep in mind that each term has the dimensions of energy per unit volume. An obvious but often misunderstood point is that if the fluid at a selected position is open to the atmosphere, such as at a hole in a container of fluid, the pressure to be used is atmospheric pressure.

Bernoulli's equation is the result of applying the work-energy principle to the motions of fluid particles undergoing steady, incompressible, and nonviscous flow and is a statement of the conservation of energy within a tube of flow. The three forms of available energy can be transformed into one another, but the total energy remains constant.

Bernoulli's equation contains the pressure-depth relation (Eq. 16.6) as a special case, known as the hydrostatic limit, when the velocities are zero.

> Torricelli's Theorem: a result derived from Bernoulli's equation:
> $u_2 = \sqrt{2gh}$ (Eq. 16.20), where u_2 is the fluid velocity at a point located a distance h below another point where the pressure is the same but the fluid velocity is negligible compared to u_2.

In applying Toricelli's theorem, the equation of continuity can be used to determine if the fluid velocity is negligible compared to u_2: $u_1 = u_2(A_2/A_1)$. If $A_1 \gg A_2$, then $u_1 \ll u_2$.

16.6 VISCOSITY

> Definition. Dynamic viscosity: the property of a fluid that measures its resistance to shearing motions, defined as the ratio of shear stress and rate of shear strain:
> $$\eta = (F/A)/(du/dy) \tag{16.21}$$
> where du is the relative velocity of two layers of fluid separated by a distance dy and F is the viscous friction force exerted on area A.

The dynamic viscosity is often referred to as simply viscosity. There is no name for the SI unit of viscosity; the cgs unit is called the poise.

> Definition. Kinematic viscosity: the ratio of the dynamic viscosity and the mass density, defined by $\nu = \eta/\rho$ (Eq. 16.23).

The cgs unit of kinematic viscosity is the stoke.

16.7 TURBULENCE

> Definition. Laminar flow: fluid flow in which adjacent layers of fluid slide smoothly past each other and a stable pattern of streamlines is established.

> Definition. Turbulent flow: unstable fluid flow which produces swirling eddies and random and irregular fluid motion.

> Definition. Reynold's number: a dimensionless parameter formed from quantities which together determine whether fluid flow is laminar or turbulent, defined as $R = \rho u L/\eta = uL/\nu$ (Eq. 16.24), where ρ is the fluid density, u is the fluid velocity, η and ν are the dynamic and kinematic viscosities, and L is some characteristic length associated with the flow.

For flow past a sphere, L is the diameter of the sphere, while for flow through a pipe, L is the diameter of the pipe.

Fluid flow changes from laminar to turbulent when the Reynolds number goes above the so-called critical Reynold's number, which may be anywhere between one thousand and ten thousand depending on the geometry of the system.

EXAMPLE PROBLEMS

Example 1

A block of unknown material in the shape of a cube of edge 0.2 m has mass 5 kg and floats in water with a portion of its volume submerged. Find the depth of the block below the water line.

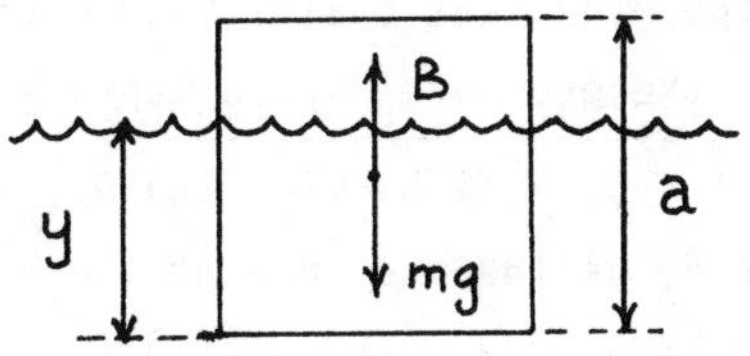

Solution

Given: $a = 0.2$ m
$m = 5$ kg
$\rho_{fl} = 1000$ kg/m^3

Required: y

FIG. 1-1

As shown in FIG. 1-1, for the block to float in equilibrium, the upward buoyant force B must equal the weight mg of the block:

$$B = mg \qquad (1)$$

According to Archimedes' principle, B is given by the weight of the displaced fluid (Eq. 16.2):

$$B = W_{displaced} = \rho_{fl} V_s g \qquad (2)$$

where V_s is the submerged volume of the block and ρ_{fl} is the density of the displaced fluid, here water of density 1000 kg/m^3. Archimedes' principle involves the weight, not just the mass or volume of the displaced fluid, hence the appearance of "g" on the right-hand side of Eq. (2).

The submerged volume V_s is given by the product of the bottom surface area of the block a^2 and the submerged depth y:

$$V_s = a^2 y \qquad (3)$$

Substituting the submerged volume V_s from Eq. (3) into Eq. (2) and the buoyant force B from Eq. (2) into Eq. (1):

$$B = \rho_{fl} a^2 y g = mg$$

Solving this equation for the submerged depth y and substituting for the density of water ρ_{fl} the cube edge a, and the mass of the block m:

$$y = mg/\rho_{fl} a^2 g = (5 \text{ kg})/(1000 \text{ kg/m}^3)(0.2 \text{ m})^2 = \underline{0.125 \text{ m}}$$

Example 2

A container is filled with water to a depth of 0.6 m, and a 0.2-m layer of oil of density 600 kg/m^3 is floating on the water. The upper surface of the oil is exposed to an atmospheric pressure of 10^5 Pa. At the bottom of the container, find a) the absolute pressure and b) the gauge pressure.

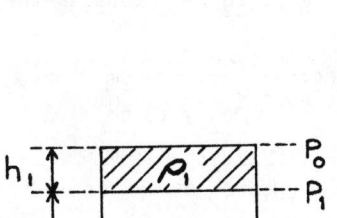

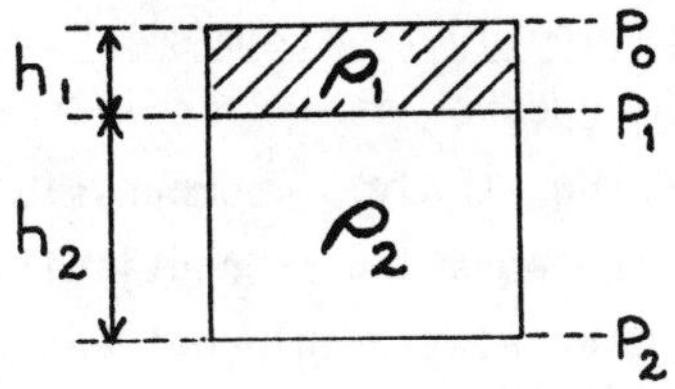

FIG. 2-1

Solution

Given: $h_1 = 0.2$ m
$h_2 = 0.6$ m
$\rho_1 = 600$ kg/m^3
$\rho_2 = 1000$ kg/m^3
$P_o = 1.01 \times 10^5$ Pa

Required: (a) P_2, (b) P(gauge)

(a) In FIG. 2-1, the pressure at the surface of the oil is atmospheric pressure P_o, the pressure at the oil-water interface is P_1, and the pressure at the bottom of the container is P_2.

Assuming that the fluids involved are incompressible, the pressure-depth relation (Eq. 16.6) gives the pressure P at a depth z in a fluid:

$$P = P_o + \rho g z \qquad (1)$$

where P_o is the pressure at the surface of the fluid. Applying Eq. (1) successively to the oil and water:

$$P_1 = P_o + \rho_1 g h_1 \qquad (2)$$
$$P_2 = P_1 + \rho_2 g h_2 \qquad (3)$$

Substituting P_1 from Eq. (2) into Eq. (3):

$$P_2 = P_o + \rho_1 g h_1 + \rho_2 g h_2 \qquad (4)$$

Substituting in Eq. (4) for the densities of oil ρ_1 and water ρ_2, the heights h_1 and h_2, and atmospheric pressure P_o:

$$P_2 = 1.01 \times 10^5 \text{ Pa} + (600 \text{ kg/m}^3)(9.8 \text{ m/s}^2)(0.2\text{m}) + (1000 \text{ kg/m}^3)(9.8 \text{ m/s}^2)(0.6 \text{ m})$$
$$= (1.01 + 0.01176 + 0.0588) \times 10^5 \text{ Pa} = \underline{1.08 \times 10^5 \text{ Pa}}. \qquad (5)$$

This pressure is an absolute pressure since the atmospheric pressure P_o is referenced to zero.

(b) The gauge pressure is pressure above atmospheric pressure, which here is $P_o = 1.01 \times 10^5$ Pa. Gauge pressure can thus be obtained by solving Eq. (5) for $P_2 - 1.01 \times 10^5$ Pa:

$$P(\text{gauge}) = P_2 - 1.01 \times 10^5 \text{ Pa} = (1.08 - 1.01) \times 10^5 \text{ Pa} = \underline{0.07 \times 10^5 \text{ Pa}}$$

Example 3

A horizontal pipeline along which water is flowing at a speed of 0.2 m/s narrows in diameter from 30 cm to 10 cm. What is the flow speed of the water in the narrower section of the pipeline?

Solution

Given: $d_1 = 30$ cm
$d_2 = 10$ cm
$u_1 = 0.2$ m/s

Required: u_2

Assuming steady, incompressible flow, we can use the equation of continuity (Eq. 16.8) to give the relationship between the velocity and cross-sectional area along the tube of flow within the pipeline, as shown in FIG. 3-1.

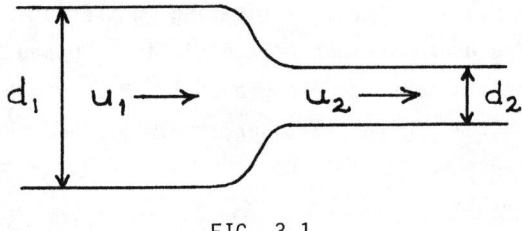

FIG. 3-1

$$u_1 A_1 = u_2 A_2 \tag{1}$$

Solving Eq. (1) for u_2 and expressing the area as $\pi(d/2)^2$:

$$u_2 = u_1 (A_1/A_2) = u_1 (\pi d_1^2/4)/(\pi d_2^2/4) = u_1 (d_1/d_2)^2 \tag{2}$$

Substituting in Eq. (2) for the diameter ratio d_1/d_2 and speed u_1:

$$u_2 = (0.2 \text{ m/s})(30 \text{ cm})/10 \text{ cm})^2 = \underline{1.8 \text{ m/s}}$$

This result can be easily verified. Flow speed is inversely proportional to area, and area is proportional to the square of the diameter. Therefore, if the diameter decreases by a factor of three, the speed should increase by a factor of three squared, or nine, as shown by the calculation.

Note that the discharge rate, $Q = uA$, can be calculated using the values of velocity and cross-sectional area at any point, for example, in the wider section of the pipeline:

$$Q = uA = (0.2 \text{ m/s})\pi(0.30 \text{ m}/2)^2 = \underline{.0141 \text{ m}^3\text{s}}$$

Example 4

The water surface inside a closed water tank is 20 m above ground level, and the air pressure above the water surface is 3 atm. The water flows downward from the tank through a pipe and emerges at a height of 2 m above ground. Find the speed of the water emerging from the pipe, assuming that the cross-sectional area of the tank is four times that of the pipe.

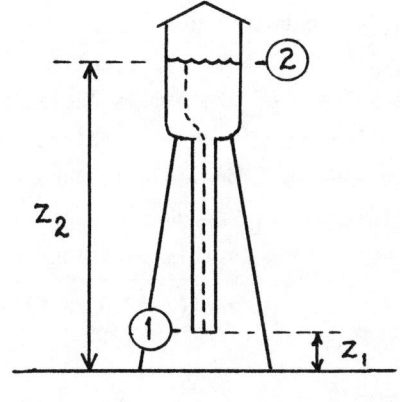

FIG. 4-1

Solution

Given:
$z_1 = 2$ m
$z_2 = 20$ m
$P_2 = 3$ atm
$\rho = 1000$ kg/m^3
$P_0 = 1.01 \times 10^5$ Pa

Required: u_1

Assuming steady, incompressible, and nonviscous flow, we apply Bernoulli's equation (Eq. 16.17) to a streamline that extends from position 2 at the surface of the water in the tank to position 1 at the lower end of the pipe, as shown in FIG. 4-1. Bernoulli's equation equates the available energies per unit volume at positions 1 and 2:

$$P_1 + \rho g z_1 + \rho u_1^2/2 = P_2 + \rho g z_2 + \rho u_2^2/2 \tag{1}$$

The equation of continuity (Eq. 16.8) can be used to give a relationship between the velocities at the two positions:

$$u_2 = u_1 A_1/A_2 \tag{2}$$

Substituting Eq. (2) in Eq. (1) and solving for u_1^2:

$$u_1^2 = [2(P_2 - P_1)/\rho + 2g(z_2 - z_1)]/[1 - (A_1/A_2)^2] \tag{3}$$

Substituting in Eq. (3) for the pressure difference $P_2 - P_1 = 2$ atm, the height difference $z_2 - z_1 = 18$ m, the area ratio $A_1/A_2 = 1/4$, and the density of water ρ:

$$u_1^2 = [2(2)(1.01 \times 10^5 \text{ Pa})/1000 \text{ kg/m}^3 + 2(9.8 \text{ m/s}^2)(18 \text{ m})]/[1 - (1/4)^2]$$
$$= 807 \text{ m}^2/\text{s}^2$$
$$u_1 = \underline{28.4 \text{ m/s}}$$

Example 5

Find the speed with which water flows from a small hole in a dam located 2 m below the level of the water behind the dam.

Solution

Given: $h = 2$ m

Required: u_1

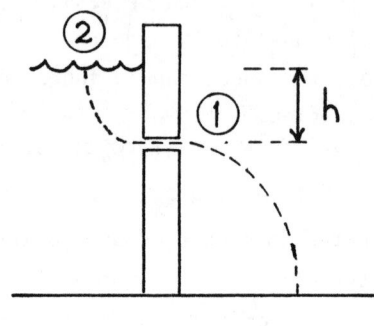

FIG. 5-1

Assuming steady, incompressible, and nonviscous flow, we apply Torricell's theorem (Eq. 16.20) to the streamline that extends from position 2 at the surface of the dam to position 1 at the small hole in the dam, as shown in FIG. 5-1. Torricelli's theorem is applicable in this case because the water velocity at the surface of the dam is negligible compared to that at the small hole, as a result of the large difference in areas, and the pressures at the two positions are the same, as both are open to the atmosphere. Torricelli's theorem gives for the velocity u_1 at the hole:

$$u_1 = \sqrt{2gh}$$

Substituting in this equation for the distance h between positions 1 and 2:

$$u_1 = \sqrt{[(2)(9.8 \text{ m/s}^2)(2 \text{ m})]} = \underline{6.26 \text{ m/s}}$$

Example 6

Find the critical velocity at which the flow of water in a horizontal pipe of diameter 2 cm would change from laminar to turbulent, assuming a critical Reynolds number of 2300.

Solution

Given: $L = D = 2$ cm $= 0.02$ m

$R = 2300$

Required: u

The critical velocity can be obtained from the expression for the Reynolds number (Eq. 16.24):

$$R = \rho u L / \eta \quad (1)$$

Solving Eq. (1) for the critical velocity:

$$u = R\eta/\rho L \quad (2)$$

For a pipe, the characteristic length L is the diameter D of the pipe. The viscosity η of the water is given in Table 16.1. Substituting in Eq. (2) for D and η, the density ρ of water, and the critical Reynolds number R:

$$u = (2300)(1 \times 10^{-3} \text{ N-s/m}^2) / (1000 \text{ kg/m}^3)(0.02 \text{ m}) = \underline{0.115 \text{ m/s}}$$

PROBLEMS

1. A person is standing on a cube of ice of edge 1 m that is floating in water with its top surface level with the water. If the density of the cube is 920 kg/m^3, what is the weight of the person?

2. A beaker contains a 10-cm layer of water floating above a 5-cm layer of mercury. What is the gauge pressure at a) the water-mercury interface and b) the bottom of the beaker?

3. Water is flowing in a hose of radius 3 cm at a speed of 5 m/s. a) What is the flow speed of the water at a nozzle of radius 1.5 cm attached to the hose? b) What is the discharge rate at the nozzle?

4. A large enclosed tank is filled with water to a depth of 4 m, and the pressure above the water level is 2 atm. Find the speed of the water emerging from a small hole in the bottom of the tank.

5. Water is flowing downward through a tapered pipe. If the pressure is 1 atm and the speed is 40 cm/s at a point where the diameter is 2 cm, what is the pressure at a point that is 1 m lower, where the diameter narrows to 0.5 cm?

6. Air of density 1.21 kg/m^3 at 20°C is flowing at a speed of 0.5 m/s through a pipe of diameter 40 cm.
   a) Find the Reynolds number and b) explain whether the flow is more likely to be laminar or turbulent.

Chapter 17
Wave Kinematics

PREVIEW

This chapter gives an introduction to wave motions, leaving the details of specific types of waves to later chapters. Waves are described in terms of kinematic properties, with emphasis on those waves that possess sinusoidal waveforms. This chapter presents the principle of superposition for waves and discusses wave interference and the phenomena of beats and the Doppler effect.

SUMMARY

17.1 WAVE CHARACTERISTICS

Definition. Wave: a disturbance that travels at a definite speed, either in a material medium or through empty space.

In a mechanical wave the disturbance is the displacement of the medium, and in an electromagnetic wave it is the variation of the electric and magnetic fields. Mechanical waves thus require a medium to support them while electromagnetic waves do not. When a mechanical wave passes through or over the surface of a medium, the medium experiences an oscillatory displacement from equilibrium, but there is no permanent displacement of the medium as a whole.

Definition. Transverse (Longitudinal) wave: a wave in which the disturbance is perpendicular (parallel) to the direction of wave travel.

In transverse mechanical waves, the displacement of the medium is perpendicular to the direction of wave propagation. Electromagnetic waves are transverse waves, and it is the variations in the electric and magnetic fields that are perpendicular to the direction of wave travel. Longitudinal mechanical waves are commonly referred to as sound waves. A wave need not be purely transverse or longitudinal. For example, waves on the surface of a body of water have both longitudinal and transverse components.

Definition. Polarization: the direction of the wave disturbance, either a field variation or a medium displacement.

The polarization of a longitudinal wave is along the direction of wave travel and is of little importance, since it cannot be altered. The polarization of a transverse wave is perpendicular to the direction of propagation and can be controlled or varied.

Definition. Waveform: the wave disturbance expressed as a function of spatial and time coordinates.

For example, if the displacement of the medium from equilibrium in a one-dimensional mechanical wave is represented by ψ, the waveform is $\psi(x,t)$. For all waves, the unique relationship between the variables x and t is that they must occur in the waveform $\psi(x,t)$ only as the combination (x ± vt), where v is the speed of propagation of the wave disturbance. The minus sign is for a wave traveling in the positive x-direction at speed v, and the plus sign refers to a wave traveling in the negative x-direction at speed v.

17.2 SINUSOIDAL WAVES

We begin our study of wave kinematics by considering the kinematical aspects of sinusoidal waveforms given by

$$\psi(x,t) = A \sin[2\pi(x - vt)/\lambda] \qquad (17.4)$$

$$\psi(x,t) = B \cos[2\pi(x - vt)/\lambda] \qquad (17.5)$$

Definition. Amplitude: the maximum value of the waveform.

Since the harmonic functions sine and cosine have maximum values of unity and carry no units, the quantities A and B are the amplitudes of the waveforms $\psi(x,t)$ in Eq. (17.4) and Eq. (17.5) and have the units of $\psi(x,t)$. The maximum disturbance is often referred to as a wave crest.

Definition. Wavelength: the distance over which the waveform repeats itself.

Definition. Period: the time required for a wave to advance through a distance equal to the wavelength.

The wavelength is often denoted by λ, and the period is often denoted by T. From their definitions, the basic kinematic relation among the quantities v, λ, and T is given by $\lambda = vT$ (Eq. 17.6).

If a periodic waveform is plotted as a function of position at a specific time, the wave crests will be separated by a distance of one wavelength λ. If the waveform is instead plotted for a given position as a function of time, the crests will be separated by a time of one period T.

Definition. Frequency: the number of crests that pass an observer per second.

The frequency is often denoted by ν, and its unit is the cycle/s, or hertz (Hz). Since frequency is the number of crests per unit time and the period is the time for the passage of one wavelength, the frequency is reciprocally related to the period by $\nu = 1/T$ (Eq. 17.7). A second form of the kinematic relation in terms of frequency is given by $\nu\lambda = v$ (Eq. 17.8).

Introduction of the wave number k as $k = 2\pi/\lambda$ (Eq. 17.10) and the radian or angular frequency as $\omega = 2\pi\nu$, allows an alternate expression of the sinusoidal waveform as
$\psi(x,t) = A \sin(kx - \omega t)$ (Eq. 17.12). The wave speed v is given in terms of ω and k as
$v = \omega/k$ (Eq. 17.13).

It should be emphasized that the wave speed is different from the speed of the medium, which undergoes an oscillatory vibration in a direction perpendicular to the direction of travel of the wave. The wave speed is a constant, given by v, while the speed of the medium at a given point varies with time and is given by $\partial\psi/\partial t$.

17.3 PHASE AND PHASE DIFFERENCE

Definition. Phase angle: the argument of the sine or cosine function that describes the space and time variation of a sinusoidal waveform, often referred to as phase.

The phase angle is measured in radians and is often denoted by ϕ. For a waveform $\psi(x,t) = A \sin(kx - \omega t)$, the phase angle is given by $\phi = kx - \omega t$ (Eq. 17.15).

Definition. Phase difference: the difference in phase for two points along a waveform, defined by $\Delta\phi = 2\pi(\Delta x/\lambda)$ (Eq. 17.17).

The phase difference, like the phase angle, is measured in radians. Since a displacement of one wavelength or a phase difference of 2π leaves the waveform unchanged, any phase angle or phase difference can be increased or reduced by an integral multiple of 2π in order to give an equivalent phase to be used in calculations and results.

17.4 THE PRINCIPLE OF SUPERPOSITION FOR WAVES

Principle of Superposition: a meeting of two or more waves produces a waveform that is the algebraic sum of the waveforms produced by each wave acting separately.

For two waveforms ψ_1 and ψ_2, the resultant waveform is given by $\psi = \psi_1 + \psi_2$. It has been found that the smaller the amplitudes of the waves involved, the more valid is the principle of superposition.

Definition. Interference: the effect of the superposition of waves that are present at the same point in space at the same time.

At points where superposition causes the waves to oppose one another destructive interference is said to occur, and constructive interference occurs at points where superposition causes the waves to reinforce one another. Interference will be considered in more detail in Chapter 40.

Definition. Standing waves: the result of the superposition of two waves having equal amplitudes, wavelengths, and frequencies but traveling in opposite directions.

A standing wave is characterized by no motion of the disturbance and by stationary positions of zero displacement called nodes. Positions of maximum displacement are called antinodes.

17.5 BEATS

Definition. Beats: periodic variations in loudness that occur at a point as the result of the superposition of two waves of nearly the same frequency, with a beat frequency given by the difference in the frequencies of the interfering waves.

17.6 DOPPLER EFFECT

Definition. Doppler effect: the change in the observed frequency (and wavelength) of a wave as a result of relative motion between a wave source and an observer.

The Doppler-shifted frequency is given by $\nu = \nu_0/[1 \pm (u/v)]$ (Eqs. 17.31 and 17.32), where ν is the observed frequency, ν_0 the emitted frequency, u the relative speed of source and observer, and v the wave speed. When the source and observer are receding from each other, the plus sign is used in the Doppler-shift formula and the observed frequency is less than the emitted frequency. When the source and observer are approaching one another, the minus sign is used, and the observed frequency is increased over the emitted frequency.

EXAMPLE PROBLEMS
Example 1
A sinusoidal waveform is given by $\psi(x,t) = 0.5 \sin(0.393x + 1.57t)$, where x and ψ are in meters and t in seconds. Find a) the amplitude, wave number, and angular frequency of the wave, b) the frequency and wavelength, and c) the speed and direction of travel of the wave.
Solution
    Given: $\psi(x,t) = 0.5 \sin(0.393x + 1.57t)$
    Required: a) A, k, ω b) ν, λ,   c) v, direction of travel

a) From the form of a sinusoidal waveform, $\psi(x,t) = A \sin(kx - \omega t)$ (Eq. 17.12), we can immediately identify the amplitude A, wave number k, and angular frequency ω:
$$A = \underline{0.5 \text{ m}}$$
$$k = \underline{0.393 \text{ m}^{-1}}$$
$$\omega = \underline{1.57 \text{ rad/s}}$$
In identifying k and ω, it should be noted that k multiplies x and ω multiplies t, even though the order of kx and ωt in the argument of the sinusoidal function might be reversed. A check of the dimensions or units of the quantities involved will insure that this identification is made properly.

b) The frequency ν and the wavelength λ can be obtained from the angular frequency and wavenumber by solving $\omega = 2\pi\nu$ (Eq. 17.11) and $k = 2\pi/\lambda$ (Eq. 17.10) for the required quantities:
$$\nu = \omega/2\pi = 1.57/2\pi = \underline{0.250 \text{ Hz}}$$
$$\lambda = 2\pi/k = 2\pi/0.393 \text{ m}^{-1} = \underline{16.0 \text{ m}}$$
Again, a dimension or unit check will insure that the quantities involved in these identities are in the correct order.

c) The velocity v can be obtained by substituting in either $v = \nu\lambda$ (Eq. 17.8) or $v = \omega/k$ (Eq. 17.13). Previous calculations of frequency and wavelength can be verified if the velocity is calculated from both expressions:
$$v = \nu\lambda = (0.250 \text{ Hz})(16.0 \text{ m}) = \underline{4 \text{ m/s}}$$
$$v = \omega/k = (1.57 \text{ rad/s})/(0.393 \text{ m}^{-1}) = 4 \text{ m/s}$$
To find the direction of travel of the wave, the given waveform must be compared with the standard expressions for a sinusoidal waveform. The combination (kx - ωt) indicates travel in the positive x-direction, and (kx + ωt) indicates travel in the negative x-direction. The given waveform is therefore traveling in the <u>negative x-direction</u>.

Example 2

A transverse sinusoidal wave of amplitude 0.1 m, wavelength 0.5 m, and angular frequency 25.1 rad/s travels in the positive x-direction along a horizontal rope, such that the height of the wave is zero at x = 0 and t = 0. Find a) the wave speed, b) the waveform $\psi(x,t)$, c) the transverse displacement of the rope at x = 1.75 m and t = 1 s, and d) the transverse velocity of the rope at x = 1.75 m and t = 1 s.

Solution

Given: A = 0.1 m
         λ = 0.5 m
         ω = 25.1 rad/s
         $\psi(0,0) = 0$

Required: a) v
          b) $\psi(x,t)$
          c) $\psi(1.75\text{ m}, 1\text{ s})$
          d) $v_{rope}(1.75\text{ m}, 1\text{ s})$

a) The wave speed v can be obtained from the kinematic relation (Eq. 17.8) by substituting for from $\omega = 2\pi\nu$ (Eq. 17.11):

$$v = \lambda\nu = \lambda(\omega/2\pi) = (0.5\text{ m})(25.1\text{ rad/s})/2\pi = \underline{2\text{ m/s}}$$

b) The waveform $\psi(x,t)$ that is zero at x = 0 and t = 0 is given by Eq. (17.4):

$$\psi(x,t) = A\sin[2\pi(x - vt)/\lambda]$$

Substituting in this equation for the amplitude A, the wavelength λ, and the speed v gives the required waveform:

$$\psi(x,t) = 0.1\sin[2\pi(x - 2t)/0.5]\text{ m} = \underline{0.1\sin[4\pi(x - 2t)]\text{ m}}$$

c) To find the transverse displacement, the waveform determined in b) must be evaluated at the given position x = 1.75 m and time t = 1 s:

$$\psi(1.75\text{ m}, 1\text{ s}) = 0.1\sin[4\pi(1.75 - 2)] = 0.1\sin(-\pi) = \underline{0\text{ m}}$$

d) The transverse velocity of the rope is given by the time derivative of $\psi(x,t)$ with x=constant, as expressed in the equation following Eq. (17.14):

$$v_{rope} = (-2\pi v/\lambda)\,A\cos[2\pi(x - vt)/\lambda]$$
$$= (-8\pi)(0.1)\cos[4\pi(1.75 - 2)] = \underline{2.51\text{ m/s}}$$

The results of parts a) and d) show that the wave and the rope do indeed have different speeds. The wave is moving in the positive x-direction with a constant speed of 2 m/s. The rope, at x = 1.75 m and t = 1 s, is moving perpendicular to the direction of propagation of the wave in the positive or upward direction at a speed of 2.51 m/s.

Example 3

Two speakers are driven in phase at a frequency of 500 Hz, and a listener is situated 4 m from one of the speakers and 5 m from the other. a) What is the phase difference of the sounds from the speakers at the listener's position? b) What is the equivalent phase difference in a)? The speed of sound in air is 340 m/s.

Solution

Given: $\nu = 500$ Hz
$x_1 = 4$ m
$x_2 = 5$ m
$v = 340$ m/s

Required: a) $\Delta\phi$,   b) $\Delta\phi_{equivalent}$

a) The relationship between phase difference $\Delta\phi$ and spatial separation Δx for sounds having the same frequency that are emitted in phase is given by Eq. 17.17:

$$\Delta\phi = 2\pi(\Delta x/\lambda)$$

Substituting in this equation for wavelength λ from the kinematic relation $\nu\lambda = v$ (Eq. 17.8):

$$\Delta\phi = 2\pi(\Delta x\ \nu/v)$$

Substituting here for the frequency ν, velocity v, and difference in path length Δx yields the required phase difference:

$$\Delta\phi = 2\pi[(5\text{ m} - 4\text{ m})(500\text{ Hz})]/340\text{ m/s} = \underline{9.24\text{ rad}}$$

b) The equivalent phase difference is given by subtracting integral multiples of 2π from the phase difference until the result lies between 0 and 2π. Here, subtraction of 2π gives

$$\Delta\phi_{equivalent} = 9.24 - 2\pi = \underline{2.96\text{ rad}}$$

Example 4

Consider the superposition of two waves having waveforms given by:

$$\psi_1(x,t) = 4\sin(6x - 1608t)$$
$$\psi_2(x,t) = 4\sin(6x - 1608t - \pi/2)$$

where ψ and x are in meters and t is in seconds. Find a) the amplitude and b) the frequency of the resultant wave.

Solution

Required: a) A,   b) ν

a) According to the superposition principle, the resultant waveform is the algebraic sum of ψ_1 and ψ_2, which is the sum of the two given sine functions. According to the Trigonometric Formulas section of Appendix 4, such a sum is given by:

$$\sin\alpha + \sin\beta = 2\sin[(\alpha+\beta)/2]\cos[(\alpha-\beta)/2]$$

Identifying α and β in the given waveforms, we can use this identity to obtain the superposition of ψ_1 and ψ_2:

$$\psi = \psi_1 + \psi_2 = (4)(2)\cos(\pi/4)\sin(6x - 1608t - \pi/4) = 5.66\sin(6x - 1608t - \pi/4)$$

The resultant waveform is of the form $\psi = A\sin(kx - \omega t - \phi)$, therefore, the required amplitude A is given by

$$\underline{A = 5.66\text{ m}}$$

b) Comparison of the resultant waveform with the standard form also indicates that the angular frequency $\omega = 1608$ rad/s. The frequency of the resultant wave can be obtained from $\nu = \omega/2\pi$ (Eq. 17.11) and is given by

$$\nu = (1608\text{ rad/s})/2\pi = \underline{256\text{ Hz}}$$

138 Chapter 17

Example 5

Two sounds produce beats at a beat frequency of 4 Hz. One sound has an unknown frequency, and the other has a frequency of 512 Hz. If the known frequency is increased slightly, the beat frequency is observed to decrease to 2 Hz. What is the unknown frequency?

Solution

Given: $\nu_1 = 512$ Hz
$(\nu_{beat})_1 = 4$ Hz
$(\nu_{beat})_2 = 2$ Hz

Required: ν_2

According to Eq. 17.26, the beat frequency is given by the difference between the two superposed frequencies. Since the beat frequency is 4 Hz, the unknown frequency is either 4 Hz above the known frequency of $\nu_1 = 512$ Hz or 4 Hz below 512 Hz, that is, either 508 Hz or 516 Hz.

If the unknown frequency is 508 Hz, increasing the known frequency above 512 Hz would cause the beat frequency to increase above 4 Hz. On the other hand, if the unknown frequency is 516 Hz, increasing the known frequency from 512 Hz towards 516 Hz would cause the beat frequency to decrease, eventually reaching the given value of 2 Hz. Therefore, the unknown frequency must be given by

$$\nu_2 = \underline{516 \text{ Hz}}$$

Example 6

The siren on a police car has a frequency of 500 Hz. What frequency is heard by a stationary pedestrian if the car is approaching at a speed of 30 m/s? The speed of sound in air is 330 m/s.

Solution

Given: $\nu_0 = 500$ Hz
$v = 330$ m/s
$u = 30$ m/s

Required: ν

The Doppler-shifted frequency ν for the case of a source moving toward an observer is given by Eq. 17.32:

$$\nu = \nu_0/[1-(u/v)]$$

Substituting in this equation for the frequency emitted by the source ν_0, the wave speed v, and the speed of the source u:

$$\nu_2 = 500/(1-30/330) \text{ Hz} = \underline{550 \text{ Hz}}$$

The observed frequency is increased from the emitted frequency, as expected when source and observer are moving towards one another.

PROBLEMS

1. A sinusoidal waveform has a displacement of zero at x = 0 and t = 0 and a maximum displacement of 0.8 m. The speed of the wave is 3 m/s in the negative x-direction, and its angular frequency is 4.71 rad/s. a) Find the wavelength, frequency, and wave number of the wave. b) Find an expression for the waveform $\psi(x,t)$.

2. A transverse sinusoidal wave on a string is given by $\psi(x,t) = 2 \sin(3x - 4t)$ m. a) What is the speed and direction of travel of the wave? b) What is the maximum transverse velocity of the string?

3. The waveform for a standing wave is given by $\psi(x,t) = 4 \sin(0.8x) \cos(100t)$ m. Find the amplitude, wavelength, frequency, and speed of the superposed waves.

4. Two speakers radiating in phase at 200 Hz are separated by 5 m. Starting at one speaker and moving toward the other, locate the first three distances from the first speaker at which the phase difference or equivalent phase difference is π radians? The speed of sound in air is 340 m/s.

5. A beat frequency of 4 Hz is heard when a tuning fork is sounded together with a 256-Hz sound. When the tuning fork is loaded with a piece of wax that changes its frequency slightly (< 4 Hz), the beat frequency increases to 6 Hz. What is the original frequency of the tuning fork?

6. A stationary observer hears a frequency of 290 Hz from the whistle of an approaching train. If the observer hears a frequency of 229 Hz after the train passes, what is the speed of the train? The speed of sound in air is 340 m/s.

Chapter 18
Mechanical Waves

PREVIEW

This chapter continues the presentation of waves begun in the previous chapter and specializes to consideration of mechanical waves, or waves that involve the displacement of a material medium. The general dependence of the wave velocity on the properties of the medium and the wave equation governing the propagation of these waves are discussed. The chapter describes the features of several types of mechanical waves, including sound waves and water waves, and introduces the concepts of intensity as a measure of the energy flow in a wave and intensity level as a measure of the corresponding physiological sensation of loudness. Standing waves on a string are used to illustrate the restrictions that boundary conditions impose on the frequencies that can be sustained by a system.

SUMMARY

18.1 WAVES ON A STRING

The sinusoidal waveform $\psi(x,t) = A \sin(kx - \omega t)$ introduced in the previous chapter to represent the disturbance of a wave satisfies a differential equation given by

$$\partial^2\psi/\partial t^2 = v^2 \, (\partial^2\psi/\partial x^2) \tag{18.12}$$

where v is the wave speed.

Eq. 18.12 is the standard form of the wave equation governing the propagation of many types of small-amplitude mechanical waves. The physical significance of the wave disturbance $\psi(x,t)$ and the specific dependence of the wave speed on the properties of the medium differ for the various types of mechanical waves. In general, the speed of a mechanical wave in a medium is proportional to the square root of the ratio of elastic and inertial properties of the medium.

For transverse waves on a string, $\psi(x,t)$ is the height of the wave, and the wave speed is given by

$$v = \sqrt{T/(M/L)} \tag{18.11}$$

where the elastic property is T, the tension in the string, and the inertial property is M/L, the mass per unit length of the string.

18.3 MECHANICAL WAVES: A SAMPLING

The waves on the surface of a body of water have both transverse and longitudinal components. The wave speed for shallow-water waves is given by $v = \sqrt{gh}$ (Eq. 18.20), where h is the depth of the water. For deep-water waves, the wave speed is given by

$$v = \sqrt{(g\lambda/2\pi) + (2\pi S/\rho\lambda)} \tag{18.21}$$

where S is the surface tension and ρ the density of the water.

The wave speeds of longitudinal and transverse waves in solids are given by

$$v_{long} = \sqrt{(B + 4\mu/3)/\rho} \tag{18.24}$$

$$v_{trans} = \sqrt{\mu/\rho} \tag{18.25}$$

where ρ is the mass density, B the bulk modulus, and μ the shear modulus of the solid.

These equations indicate that the wave speed of longitudinal waves is greater than that of transverse waves. Transverse waves cannot propagate in an ideal fluid, defined in Chapter 16 to be incompressible and nonviscous, because the shear modulus is zero for an ideal fluid. The viscosity of real fluids allows the propagation of transverse waves and causes the reduction of the wave amplitude, an effect known as attenuation.

Longitudinal mechanical waves in any medium are commonly referred to as sound waves and categorized by frequency according to the definitions given below.

> Definition Audible waves: sound waves that lie within the range of sensitivity of the human ear, normally 20 Hz to 20,000 Hz.

> Definition. Ultrasonic (Infrasonic waves): sound waves with frequencies higher (lower) than audible waves.

In a sound wave, the wave disturbance $\psi(x,t)$ can be considered either as a pressure variation or as the displacement of the medium. The speed of sound in air is approximately 340 m/s, and the speed of sound in water is nearly five times greater. The speed of sound in many common metals is several times that in water (Table 18.1).

18.4 ENERGY FLOW AND WAVE INTENSITY

> Definition. Intensity: the time average rate at which energy is transported by a wave across a unit area perpendicular to the direction of wave travel, also, the power transported per unit area, defined by $I = E/tA$ (Eq. 18.29), where E is the energy flowing across area A in a time t.

The unit of intensity is the watt per square meter (W/m^2). Since energy flow per unit time is power, the intensity can be written as $I = P/A$. Intensity can also be expressed as $I = uv$ (Eq. 18.30), where v is the wave speed and u is the wave energy per unit volume.

The intensity of a sinusoidal wave can be expressed as $I = \rho\omega^2 D^2 v/2$ (Eq. 18.34), where ρ is the mass density of the medium, and ω, D, and v are the angular frequency, amplitude, and speed of the wave. This expression illustrates the general dependence of intensity on the square of the wave amplitude.

> Definition. Wave front: a surface over which the phase of a wave is constant.

The wave fronts of plane waves are planes perpendicular to the direction of propagation, and the intensity of the wave is constant. The wave fronts of spherical waves from a point source are concentric spherical surfaces centered on the source. Since the area of a spherical surface increases as the wave travels outward, the intensity or power per unit area is inversely proportional to the square of the distance from the source and the amplitude is inversely proportional to the distance from the source.

> Definition. Intensity level: a comparison of the intensity of a sound with the intensity of a reference sound as the logarithm of the intensities, according to

142   Chapter 18

$\beta = 10 \log(I/I_0)$, where β is the intensity level in decibels (dB) and I is the sound intensity. The intensity of the reference sound is $I_0 = 10^{-12}$ W/m^2, which is near the human threshold of hearing (at 1000 Hz).

The difference in the intensity levels of two sounds, I_1 and I_2, is thus given by
$$\beta_2 - \beta_1 = 10 \log(I_2/I_0) - 10 \log(I_1/I_0) = 10 \log(I_2/I_1) \text{ dB}$$
For example, the difference in the intensity levels of two sounds that differ in intensity by a factor of two is ten times the logarithm of 2, or 3 dB. For most humans, the physiological sensation of loudness seems to double if the intensity increases by a factor of 10. Thus, a sound that seems twice as loud has an increase in intensity level of approximately 10 dB, and a sound that increases by 20 dB seems 4 times louder.

18.5 STANDING WAVES AND BOUNDARY CONDITIONS

> **Definition.** <u>Boundary conditions</u>: physical constraints on a medium that restrict the frequencies (and wavelengths) that can be sustained by the medium.

For example, in the case of a uniform string fixed at both ends, the boundary conditions are that the displacement of the string equals zero at the ends. As a result, acceptable sinusoidal standing waves on the string have wavelengths for which the length of the string equals an integral number of half wavelengths, $L = n\lambda/2$ (Eq. 18.41) and frequencies that are an integral multiple of some fundamental, or lowest, frequency:
$$\nu = v/\lambda = n(v/2L) = n\, \nu_{fund} \tag{18.45}$$
The fundamental frequency for any oscillatory system corresponds to the maximum wavelength according to
$$\nu_{fund} = v/\lambda_{max} \tag{18.48}$$
and can be estimated by taking λ_{max} to be the largest dimension of the system.

EXAMPLE PROBLEMS
Example 1
A wire of diameter of 1 mm is made of copper ($\rho = 8960$ kg/m^3) and is under a tension of 220 N. What is the speed of transverse waves along the wire?
Solution
Given: T = 220 N
          $\rho = 8960$ kg/m^3
          d = 1 mm = 10^{-3} m

Required: v

The wave speed for transverse waves on a string under tension is given by Eq. 18.11:
$$v = \sqrt{T/(M/L)} \tag{1}$$
The length of the wire that appears in Eq. (1) will not be required, as the mass per unit length M/L can be obtained by considering the wire as a cylinder of radius r, length L, and volume $\pi r^2 L$:
$$M/L = \rho(\pi r^2 L)/L = \pi r^2 \rho \tag{2}$$
Substituting in Eq. (1) for the tension T and the mass per unit length M/L from Eq. (2):
$$v = \sqrt{T/(\pi r^2 \rho)} = \sqrt{220 \text{ N}/\pi\, (10^{-3} \text{ m}/2)^2 (8960 \text{ kg/m}^3)} = \underline{177 \text{ m/s}}$$

Example 2

Calculate the speed of sound in bone, assuming for the bone medium that $B = 1.2 \times 10^{10}$ N/m^2, $\mu = 0.6 \times 10^{10}$ N/m^2, and $\rho = 1.9 \times 10^3$ kg/m^3.

Solution

Given: $B = 1.2 \times 10^{10}$ N/m^2
 $\mu = 0.6 \times 10^{10}$ N/m^2
 $\rho = 1.9 \times 10^3$ N/m^2

Required: v

To find the speed of sound in bone, we use Eq. 18.24 for the speed of longitudinal waves in a solid medium:

$$v = \sqrt{(B + 4\mu/3)/\rho}$$

Substituting in this equation for the bulk modulus B, the shear modulus μ, and the mass density ρ:

$$v = \sqrt{[1.2 \times 10^{10} \text{ N/m}^2 + (4/3)(0.6) \times 10^{10} \text{ N/m}^2)]/1.9 \times 10^3 \text{ kg/m}^3}$$
$$= \underline{3244 \text{ m/s}}$$

Example 3

Find the tension required for a violin string to sound its fundamental frequency of 440 Hz. Assume that the string has a length of 0.3 m and a mass of 0.002 kg.

Solution

Given: $\nu_{fund} = 440$ Hz
 $L = 0.3$ m
 $M = 0.002$ kg

Required: T

The tension can be obtained from the expression for the velocity of waves on a string under tension (Eq. 18.11):

$$v = \sqrt{T/(M/L)} \qquad (1)$$

Since the violin string is fixed at both ends, the boundary conditions restrict the wavelengths to those given in Eq. 18.41:

$$\lambda = 2L/n, \quad n = 1,2,3... \qquad (2)$$

The fundamental frequency is the minimum frequency and therefore corresponds to the maximum wavelength (Eq. 18.48):

$$\nu_{fund} = v/\lambda_{max} \qquad (3)$$

Solving Eq. (3) for the wave speed v and substituting for the maximum wavelength λ_{max} from Eq. (2) with n = 1:

$$v = \nu_{fund} \lambda_{max} = \nu_{fund} (2L) \qquad (4)$$

Solving Eq. (1) for the tension T and substituting for the wave speed v from Eq. (4):

$$T = (M/L)v^2 = (M/L)[\nu_{fund}(2L)]^2 \qquad (5)$$

Substituting in Eq. (5) for the mass M, length L, and fundamental frequency ν_{fund} of the string:

$$T = (0.002 \text{ kg}/0.3 \text{ m})[(440 \text{ Hz})(2)(0.3 \text{ m})]^2 = \underline{465 \text{ N}}$$

144   Chapter 18

Example 4

How much acoustical power enters one of the ears of an airport worker when the intensity level is 100 dB, if the effective area of an ear is taken to be 20 cm^2?

Solution

Given: β = 100 dB
         A = 20 cm^2 = 0.002 m^2

Required: P

The power can be obtained from the definition of intensity (Eq. 18.29) by replacing the energy flow per unit time with the power:

$$I = E/tA = P/A \qquad (1)$$

We obtain the intensity from the definition of intensity level (Eq. 18.37) by substituting for the intensity level β and the standard of intensity I_o:

$$\beta = 10\ \log(I/I_o)\ dB = 10\ \log(I/10^{-12}) = 100\ dB$$

$$\log(I/10^{-12}) = 100/10 = 10$$

$$I/10^{-12} = 10^{10}$$

$$I = (10^{-12}\ W/m^2)(10^{10}) = 10^{-2}\ W/m^2 \qquad (2)$$

Solving Eq. (1) for the power P and substituting for the intensity I from Eq. (2) and the area A:

$$P = IA = (10^{-2}\ W/m^2)(0.002\ m^2) = 20 \times 10^{-6}\ W = \underline{20\ \mu W}$$

Example 5

How much higher is the intensity level of a trombone quartet than that of a solo trombone, assuming that all instruments have the same intensity?

Solution

Required: $\Delta\beta = \beta_{quartet} - \beta_{solo}$

Since there is no fixed phase relationship among the sounds from the four trombones, there is no interference and the intensity of the quartet is four times that of the solo instrument:

$$I_{quartet} = 4I_{solo} \qquad (1)$$

The intensity level difference is given by:

$$\Delta\beta = 10\ \log(I_{quartet}/I_o) - 10\ \log(I_{solo}/I_o)\ dB$$

$$= 10\ \log(I_{quartet}/I_{solo})\ dB$$

Substituting for the intensity ratio $I_{quartet}/I_{solo}$ from Eq. (1):

$$\Delta\beta = 10\ \log(4) = 10(0.602)\ dB = \underline{6.02\ dB}$$

Recalling that a 10 dB increase is necessary to sense a factor of two in loudness, we see that replacing one trombone with four does not even cause a doubling of the perceived loudness.

Example 6

Standing waves are established on a rope fixed at both ends having a length of 2 m and a fundamental frequency of 2 Hz. If the rope vibrates in three distinct segments separated by nodes, what is a) the wavelength and b) the frequency corresponding to this standing waveform?

Given: $L = 2$ m

$\nu_{fund} = 2$ Hz

\# segments = 3

Required: a) λ,   b) ν

a) The maximum wavelength corresponding to the fundamental frequency is given by Eq. 18.41 with $n = 1$:

$$\lambda_{max} = 2L/n = 2L \qquad (1)$$

The velocity is determined from λ_{max} and the fundamental frequency ν_{fund} using Eq. 18.48:

$$v = \lambda_{max} \nu_{fund} = (2L) \nu_{fund} \qquad (2)$$

The wavelength of the standing wave pattern with 3 segments is given by Eq. 18.41 with $n = 3$:

$$\lambda = 2L/n = 2L/3 = (2)(2 \text{ m})/3 = \underline{1.33 \text{ m}}$$

b) The corresponding frequency is obtained from $\nu\lambda = v$ (Eq. 17.8) by substituting the wavelength λ from a) and the wave speed v from Eq. (2):

$$\nu = v/\lambda = (2L) \nu_{fund}/(2L/3) = 3\nu_{fund} = 3(2 \text{ Hz}) = \underline{6 \text{ Hz}}$$

PROBLEMS

1. What tension is required to produce transverse waves with a wave speed of 500 m/s in a stretched wire having a mass per unit length of 0.003 kg/m?

2. Find the speed of a) longitudinal and b) transverse elastic waves in uranium.

3. How much acoustic power falls on a student's desk of area 0.1 m^2 during a one-hour lecture when the intensity level is 60 dB?

4. The intensity level at a distance of 1 m from a speaker is 130 dB. What is the intensity level at a distance of 100 m?

5. A rope of mass 0.11 kg and length 3 m is under a tension of 65 N. If the rope is fixed at both ends and vibrates in a standing wave pattern that has two distinct segments, find
a) the wavelength and b) the corresponding frequency.

6. If a 100-voice choir has an intensity level of 80 dB, what is the intensity level of a single voice in the choir?

Chapter 19
Special Relativity

PREVIEW

In this chapter we begin an excursion into modern, or post 1900 physics. The special theory of relativity proposed in 1905 by Albert Einstein addresses the mechanics of systems whose velocities approach the speed of light. We introduce in this chapter the effects that arise in such systems. These effects are accounted for by the new perspective that one must develop regarding the nature of events as seen from reference frames that have constant velocities but move very rapidly with respect to one another. The Lorentz transformation equations which replace those of Galilean relativity are developed from Einstein's original postulates. Some of the kinematical consequences of these new transformation equations are considered.

SUMMARY

19.1 EINSTEIN'S POSTULATES OF RELATIVITY

Thus far we have concentrated on the development of Newtonian mechanics in various guises. For over 200 years this theory of nature ruled supreme. However, as man's knowledge of the physical universe expanded and his sophistication grew problems began to arise which could not easily be explained by Newton's synthesis. These problems along with others that arose in conjunction with the classical theory of electricity and magnetism led to a revolution in scientific thinking which began at the turn of the century. Most scientists date the beginning of modern physics and the profoundly deeper understanding of nature which it has given us at or around 1900. Actually, the classical theory of electricity and magnetism as formulated by James Clerk Maxwell in 1864 sowed the seeds which blossomed into what we now term modern physics. The major difference between modern physics and classical physics is the type of systems to which these theories apply. Modern physics deals with systems which have extremely high velocities and very large sizes and with systems which are very, very small. Since Newton's theories were derived mainly from observations of systems of "normal" proportions which occur in our everyday lives, it should not be surprising that the systems of modern physics do not obey the laws of Newtonian mechanics.

Two major problems that existed in the early 1900's were: 1.) Regardless of how hard physicists tried, they could not measure a velocity of light which differed from the constant value of (approximately) 3×10^8 m/s in vacuum. This is a bit surprising since, as we have seen, the earth moves in its orbit with a speed which should make a detectable difference in the velocity of light measured parallel or anti-parallel to the direction of the earth's motion. 2.) The equations of electricity and magnetism are not invariant under the same relativity transformations as Newton's laws of motion.

It was Einstein who had the vision, genius, and intestinal fortitude to realize that it was Newton's laws and Galilean relativity which were incorrect, and that electricity and magnetism was correct. In 1905 he formulated the special theory of relativity (special because it deals with the special case of inertial reference frames) which is based on two very simple postulates which, nonetheless, have profound consequences.

Einstein's postulates of relativity:
1. The speed of light in vacuum is a constant for an observer in any inertial reference frame. In other words, the speed of light in vacuum does not depend on the state of motion of the observer as long as the observer is moving with constant velocity.
2. The laws of nature must have the same form in all inertial reference frames.

Newton's laws of motion are invariant under the Galilean transformation between inertial reference frames, electricity and magnetism is not. The Galilean transformation equations violate the postulate of the constancy of the speed of light. If Einstein's postulates are correct, and experiment confirms that they are, we must abandon Galilean relativity and therefore Newton's laws of motion. Well, not quite. It would be foolish to think that a theory which works as well as Newton's does in most cases has no validity. We shall see that in "normal" situations the special theory of relativity and Newtonian mechanics are the same. It is only when we deal with speeds close to or at the velocity of light that things get weird.

One point which should be stressed is that the relativity postulate demands that no inertial reference frame can be any different from any other. Since the laws of physics are identical in all such frames of reference, there can be no way to distinguish within a given interial frame whether that frame is at rest or moving with constant velocity. Any replacement for the Galilean relativity transformations must be such that they are symmetric with respect to the relative velocity of two observers in different reference frames. This will assure that neither observer can claim his frame of reference is really at rest and the other is really the moving frame. All motion is relative. There is no special fixed inertial reference frame in nature.

19.2 THE RELATIVITY OF TIME

We have always explicitly or implicitly assumed that time is immutable. The time which you or I measure between two events is the same regardless of whether we are moving relative to one another or not. As soon as one admits that the speed of light is a finite, constant quantity in any inertial frame this assumption must be abandoned. Time intervals measured by observers moving relative to one another are not the same but depend on the relative velocity of the observers. A number of thought experiments such as Einstein's train as well as real phenomena demonstrate explicitly that this is the case. This means that when we seek a replacement for the Galilean transformations, we must include the time as one of the variables which will be transformed.

19.3 THE LORENTZ TRANSFORMATIONS

The relativistically correct coordinate transformations between inertial frames of reference can be derived in a number of ways. Either of the relativity postulates can be employed to generate the transformations. The text presents an algebraic derivation based on the assumption that the speed of light is a constant, and accepting the fact that time must be included in the transformation equations as a possible varying quantity.

These transformations were originally developed by H. A. Lorentz while studying the question of the invariance properties of the equations of electricity and magnetism between inertial reference frames. (For a derivation using this approach see H. Yilmaz, <u>Introduction to the Theory of Relativity and the Principles of Modern Physics</u>, Blaisdell Pub. Co, New York).

We consider an inertial reference frame at rest with respect to an observer. This "rest" frame will always be designated with unprimed coordinates in our equations. We consider another reference frame (designated by primed coordinates) moving <u>relative</u> to the first such that the "moving" frames x', y', and z' axes are parallel to the x, y, and z axes of the rest frame and it is moving with constant velocity v in the +x direction. We also choose to measure time such that when the origin of the two reference frames coincide t = t'. This may seem unduly restrictive, but we know that the choice of a coordinate system in a given reference frame is arbitrary and this choice makes the interpretation of the transformation equations much simpler.

The relativistically correct transformation, the Lorentz transformation, between two such reference frames is:

$$x' = (x - vt)/(1 - v^2/c^2)^{1/2} \tag{19.11}$$
$$y' = y$$
$$z' = z$$
$$t' = (t - vx/c^2)/(1 - v^2/c^2)^{1/2} \tag{19.13}$$

The inverse transformations can be simply generated by replacing v by -v and interchanging the primes (Eqs. 19.12 and 19.13). Because of the immensity of the velocity of light squared, it is easy to see that at small relative velocities, $v/c < 1$, the term $v^2/c^2 \ll 1$ and these transformations reduce to the Galilean ones.

We now know that any theory of physical phenomena must be form invariant under a Lorentz transformation. This is the first criteria of its validity as a description of nature.

19.4 THE EINSTEIN VELOCITY ADDITION FORMULA

The Lorentz transformations are linear in the coordinates and displacements and time intervals transform identically to coordinates.

$$\Delta x = (\Delta x' + v \Delta t)/(1 - v^2/c^2)^{1/2} \tag{19.19}$$
$$\Delta t = (\Delta t' + [v/c^2] \Delta x')/(1 - v^2/c^2)^{1/2} \tag{19.20}$$

If a particle moves with instantaneous velocity u in the "rest" frame parallel to the x axis its velocity in the "moving" frame is also parallel to the x' axis. Division of Eq. 19.19 by 19.20 and taking the limit as $\Delta t \to 0$ gives:

$$u = (u' + v)/(1 + u'v/c^2) \tag{19.22}$$

Although not presented in the text, you will find it instructive to consider what happens to the components of a velocity vector u which is not parallel to the x and x' axes.

The obvious and desired result of this new transformation of velocities is that it is impossible for an object to have a velocity greater than the speed of light relative to any single observer.

19.5 LORENTZ-FITZGERALD CONTRACTION

Another unusual consequence of the Lorentz transformation is that when an observer measures the length of a moving object it appears shorter than when it is at rest relative to the observer. This can be seen immediately from the inverse to Eq. 19.19 <u>if we specify that the length measurement made in the "rest" frame is made so that the distances x_1 and x_1 are measured simultaneously.</u>

$$x'_2 - x'_1 = (x_2 - x_1)/(1 - v^2/c^2)^{1/2}$$

$$L = L'(1 - v^2/c^2)^{1/2} \tag{19.25}$$

Although simultaneity cannot be guaranteed from one reference frame to another as shown by the discussion of Einstein's train, there is no difficulty in specifying simultaneous events in a given reference frame. Thus, the measurement referred to above is possible.

19.6 TIME DILATION

Perhaps the strangest and most counterintuitive result of the validity of the Lorentz transformations is that the time interval between two events which happen at the same place in one reference frame is not the same as the time interval measured for the same two events in a frame moving relative to the first. This is the result we discovered from the Einstein train experiment.

If we examine the time interval between two events which occur at the same place in the primed coordinate system ($x' = 0$) we have from Eq. 19.20:

$$\Delta t = \Delta t'/(1 - v^2/c^2)^{1/2} \tag{19.28}$$

The time interval measured in the frame at rest relative to an event is smaller than the time interval measured for the same event when it is viewed from a frame where the event appears to be moving relative to the observer. Moving clocks appear to run slower than clocks which are at rest relative to you. There are a number of confirmations of this effect in elementary particle physics where one deals with particles whose lifetimes are very short when at rest relative to us, but which appear much longer when the particle is moving at velocities near the speed of light. The lifetime of the particle relative to an observer at rest with respect to it remains the same, but since it appears to us that the clocks of this observer count seconds much slower than ours, we measure a much longer lifetime. This effect is referred to as time dilation.

19.7 THE TWIN EFFECT

As required by the relativity postulate, the effects we have discussed, the Lorentz-FitzGerald contraction, time dilation, and relativistic velocity addition, are completely symmetric effects. In particular the clocks in a frame of reference moving with respect to you will appear to run slower than yours. From the point of view of an observer in that reference frame your clocks will appear to run slow. This must be true if, as the principle of relativity demands, there can be no way to distinguish which of the reference frames is "really" at rest. Sometimes this seems a bit paradoxical. The reason this reciprocity works is that both observers are in inertial frames of reference. You may measure a time interval on a single clock, say your wristwatch, but there is no way that you can compare this to your companion observer's wristwatch in the other frame of reference. You can at best see his watch at only one instant of time; when you pass each other. When you compare time intervals you must use two different clocks in his frame of reference which are necessarily at different locations. The same is of course true for him.

One of the most interesting applications of the theory of special relativity deals with time dilation. It is called the twin effect or more often the twin paradox. There really is no paradox. This phenomena demonstrates how careful one must be in applying the theory of relativity.

Suppose two identical twins exist on earth, one of whom is an astronaut. The astronaut twin enters a space ship which leaves the earth and journeys to a star a few light years distance at a velocity close to the speed of light. On her return to earth after some years of travel she goes to visit her sister who is now ensconced in a home for the aged. She is still a relatively young

150 Chapter 19

woman (remember all things are relative). How can this be? By the principle of relativity the time dilation effect must be reciprocal for them.

We have purposefully stated the problem in a way which makes it appear paradoxical. The fact is that the astronaut twin is not an inertial observer. The resolution of this paradox however can be achieved without recourse to the general theory of relativity which deals with accelerated frames of reference. The solution requires a careful consideration of the problem of the synchronization of clocks in the reference frames of the astronaut and the earthbound twin. (For an extensive, elementary discussion of this problem see S. K. Kim, Physics: The Fabric of Reality, MacMillan Pub. Co., Inc., New York.) The fact is that there is no paradox, the astronaut twin will age less than the earthbound twin.

EXAMPLE PROBLEMS

Example 1

Two rocket ships (A and B) each traveling at 0.8 c are approaching one another to engage in battle. A fires a missile with a speed of 0.6 c at B. a) What is the speed of the missile as measured by an observer on a distant planet at "rest" with respect to these two ships? b) What velocity does B measure for the missile? c) What relative velocity does the distant observer measure for the two ships?

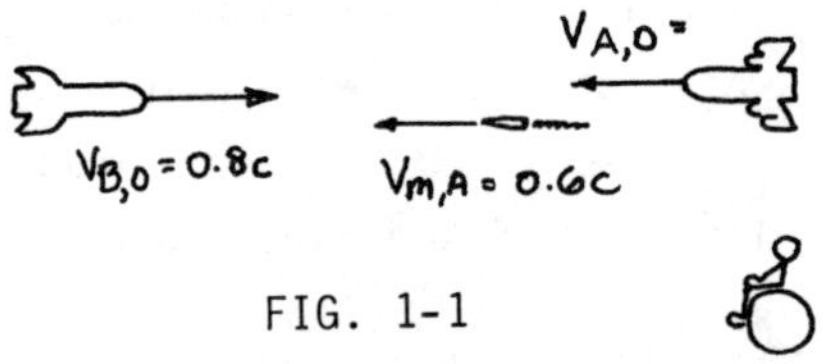

Solution

Given: $v_{A,0} = 0.8$ c
         $v_{B,0} = -0.8$ c
         $v_{m,A} = 0.6$
         $v_0 = 0$ (B,0 means B relative to observer etc)

Required: a) $v_{m,0}$, b) $v_{m,B}$, c) $v_{AB,0}$

FIG. 1-1

a) To the observer on the distant planet spaceship A constitutes a reference frame moving with velocity $v_{A,0} = 0.8$ c relative to him. The velocity of the missile in A's frame of reference is $v_{m,A} = 0.6$ c. We can apply the velocity addition formula directly with $v_{A,0} = v$ and $v_{m,A} = u'$

$$v_{mp} = \frac{u' + v}{1 + u'v/c^2} = \frac{v_{m,A} + v_{A,0}}{1 + v_{m,A}v_{A,0}/c^2} = \frac{0.6\,c + 0.8\,c}{1 + (0.6\,c)(0.8\,c)/c^2} = \underline{0.946\,c}$$

b) This problem requires a bit of thought. Since both rocket ships are moving at 0.8 c we must first decide what B measures as the velocity of ship A. It cannot be 1.6 c since no inertial observer can measure the velocity of another object to be as great as c. We consider B to be at rest. B then sees A moving in a frame of reference approaching B at 0.8 c. The velocity of A in this frame of reference is 0.8 c. Thus v = 0.8 c and u' = 0.8 c in the velocity addition formula

$$v_{AB} = \frac{0.8\,c + 0.8\,c}{1 + (0.8\,c)(0.8\,c)/c^2} = \frac{1.6\,c}{1.64} = \underline{0.975\,c}$$

Now that v_{AB} is determined, B sees the missile, which has velocity 0.6 c relative to v_{AB}, the observed velocity of A by B. Again applying Eq. 19.22 we have

$$v_{m,B} = \frac{0.6\,c + 0.975\,c}{1 + (0.6\,c)(0.975\,c)/c^2} = \frac{1.575\,c}{1.585} = \underline{0.994\,c}$$

c) The relative velocity of A to B as observed by the individual on the distance planet is $V_{AB,0} = 1.6\,c$. This <u>does not</u> contradict the postulates of relativity. The relative velocity of approach of two objects as viewed by a third observer can be almost as large as $2\,c$. The third observer still does not measure the velocity of any object relative to him to be greater than c.

Example 2

Two observers are in relative motion with a constant relative velocity. How fast must they be going relative to one another in order that each measures a meter stick in the others reference frame to be 90 cm in length?

Solution

Given: $L = 90$ cm

            $L' = 1$ m

Required: $v_{relative}$

It does not matter who we choose to be at rest or who is in motion, the principle of relativity guarantees that both observers see the same thing. Applying Eq. 19.25 gives

$$L = L'(1 - v^2/c^2)^{1/2}$$
$$90 \text{ cm} = 100 \text{ cm}\,(1 - v^2/c^2)^{1/2} \rightarrow .9 = (1 - v^2/c^2)^{1/2} \rightarrow .81 = 1 - v^2/c^2$$
$$v^2/c^2 = 0.19 \rightarrow v = \underline{.436\,c} = 1.3 \times 10^8 \text{ m/s}$$

Example 3

π-mesons (lifetime = 2×10^{-8} s) can be created in the laboratory by means of very high energy elementary particle collisions. Suppose when created these particles have a velocity of 2.9×10^8 m/s. a) What would be their lifetime as calculated by a laboratory observer? b) How far does such a particle travel in the laboratory before it decays? c) How far does the π-meson travel from its point of view through the laboratory?

Solution

Given: $v = 2.9 \times 10^{+8}$ m/s

            $\tau_{1/2} = 2 \times 10^{-8}$ s

Required: a) $\tau_{1/2}$ in laboratory frame of reference, b) path length of π's in the laboratory.

a) This is a straight forward application of time dilation, Eq. 19.28. The decay time of the π-meson, as are all decay times, is that measured at rest relative to the π. (Quantities measured in a frame at rest relative to them are called "proper"). The velocity of the π is the velocity of a reference frame attached to it relative to the laboratory, the lifetime measured in the lab is:

$$\Delta t = \Delta t'/(1 - v^2/c^2)^{1/2} = (2 \times 10^{-8})/\sqrt{1 - \left(\frac{2.9 \times 10^8}{3 \times 10^8}\right)^2} = \underline{7.8 \times 10^{-8}} \text{ s}$$

b) The distance traveled in the laboratory is simply

$$x_{lab} = v_{lab}\,\tau_{lab} = (2.9 \times 10^8 \text{ m/s})(7.8 \times 10^{-8} \text{ s}) = \underline{22.62 \text{ m}}$$

c) An observer traveling with the π-meson sees laboratory dimensions contracted due to the Lorentz FitzGerald contraction. The length, 22.62 m, measured in the now moving frame appears to the π-meson as

$$L = L'(1 - v^2/c^2)^{1/2} = L'(1 - 0.934)^{1/2} = (22.62 \text{ m})(0.256) = \underline{5.79 \text{ m}}$$

PROBLEMS

1. How fast must two observers travel relative to one another to see length contractions of
   a) 1/4, b) 1/3, c) 1/2?

2. Show by applying the Lorentz transformations that the quantity $(\Delta x)^2 - c^2(\Delta t)^2 = (\Delta x')^2 - c^2(\Delta t')^2$ and is thus a relativistically invariant quantity.

3. The normal lifespan of a human is about 70 years. Suppose an astronaut 25 years old wishes to travel to Alpha Centauri 4 light years away and return before he dies. How fas must his rocket ship travel? (assume constant velocity)

4. Some elementary particles have lifetimes as short as 10^{-15} s. In order to see such a particle in a detection device it must travel at least 1 mm through the device. What is the minimum velocity it must have to be detectable?

5. A satellite is placed in earth synchronous orbit. After remaining there for one year it is brought back to earth. What will be the difference in time between a clock which was on the satellite and one on the earth if they were originally synchronized before the satellite was put into orbit.

Chapter 20
Relativistic Mechanics

PREVIEW

The previous chapter introduced the relativistic transformations and their kinematical consequences. In this chapter we turn to the dynamical relationships which pertain to relativistic mechanics. The relativistic momentum is defined and Newton's 2nd law in the impulse-momentum form is employed to derive the relativistic expressions for motion under the influence of a constant force. The relativistic equivalent of the work energy principle is developed. Finally relativistic collisions are discussed.

SUMMARY

20.2 RELATIVISTIC LINEAR MOMENTUM

Definition. <u>Relativistic linear moment</u>. The correct expression for the spatial components of the linear momentum of a particle is:
$$p = mv/(1 - v^2/c^2)^{1/2}$$

This expression can be derived from considerations of the properties which systems in relativistic motion must have. Although these conditions are analogous to those kinematical properties we used to derive the Lorentz transformation, they are more complex. We accept the above definition as a starting point for developing relativistic dynamics.

A very important point concerning Eq. 20.1 is that the mass m used in this equation is the invariant <u>rest mass</u> of the system employed. The mass used in all the equations in the text is the rest mass, i.e. the mass of the system measured in a frame of reference at rest with respect to the system. In many texts this mass is denoted by m_o and the quantity $m = m_o/(1 - v^2/c^2)^{1/2}$ is called the relativistic mass of the system. This is not the case in this text for the reasons indicated on page 379. You should take careful note of this difference when reading other texts.

We may use this definition of linear momentum to derive a number of results. The first observation which we make is that this definition of momentum gives a more complex velocity dependence than the nonrelativistic case. We can still apply Newton's 2nd law if we employ the impulse-momentum form. We consider the case of a constant force.
$$dp/dt = F$$
This leads directly upon integration (for a constant force) to:
$$p = Ft$$
Substitution of the relativistic definition of momentum and some algebra yields:
$$v/c = (Ft/mc)/[1 + (Ft/mc)^2]^{1/2} \qquad (20.3)$$
This expression can be integrated again to find the position of the particle as a function of

153

time.
$$x = (mc^2/F)\{[1 + (Ft/mc)^2]^{1/2} - 1\} \tag{20.4}$$

These relativistic expressions have a number of consequences. They both reduce to the classical expressions $v = at$ and $x = (1/2)at^2$ at velocities $v \ll c$. In the extreme relativistic limit they give the expected results also, $v/c = 1$ and $x = ct$. A constant force does not result in a constant acceleration, but a continuously decreasing one until the velocity approaches the constant value of c. The velocity of an object with rest mass can never reach c since the expression for **p** becomes infinite at c. An infinite force would be required to accelerate a system to this velocity.

20.3 RELATIVISTIC ENERGY

The expression for the kinetic energy of a relativistic object can be derived employing the work-energy principle with the appropriate relativistic quantities substituted. The kinetic energy of a particle with initial velocity 0 is:

$$K = \int_0^x F\,dx \tag{20.5}$$

Integration of this equation using $dp/dt = F$ and the relativistic momentum results in the relativistic expression for kinetic energy.

$$K = mc^2/(1 - v^2/c^2)^{1/2} - mc^2 \tag{20.10}$$

As do all relativistic expressions, this reduces to the classical value of $K = (1/2)mv^2$ when $v \ll c$.

The final term in Eq. 20.10 is the equivalent energy stored in the rest mass of the particle. This is one of the most famous results of the special theory of relativity and gives rise to what is probably the best known equation of physics:

$$E_0 = mc^2 \tag{20.11}$$

This expression only provides an equivalence between mass and energy. It is possible in certain physical processes for the interconversion of mass and energy to occur.

The form of the kinetic energy relation leads us to write:

$$K + mc^2 = mc^2/(1 - v^2/c^2)^{1/2}$$

and to define the total relativistic energy of a particle as:

$$E = mc^2/(1 - v^2/c^2)^{1/2} = K + E_0 \tag{20.12}$$

As is the case with the relativistic momentum the total relativistic energy of a particle becomes infinite as $v \to c$.

This definition of the total energy can be combined with the definition of the relativistic momentum to yield an expression relating energy and momentum which is analogous to $p^2/2m = K$ in Newtonian mechanics. We square Eq. 20.1 and multiply it by c^2 and square Eq. 20.12 which gives:

$$E^2 = m^2c^4/(1 - v^2/c^2) \quad ; \quad (pc)^2 = (mvc)^2/(1 - v^2/c^2)$$

Subtracting the momentum expression from the energy expression gives:

$$E^2 - p^2c^2 = (m^2c^4 - m^2c^2v^2)/(1 - v^2/c^2)$$

The expression on the right is merely $(mc^2)^2$.

$$E^2 - p^2c^2 = m^2c^4 \tag{20.13}$$

This expression is important because it tells us that the quantity on the left hand side of this equation is relativistically invariant since the rest mass and c are constants. We may also use this expression to show that particles which travel at the speed of light must have no rest mass, and vice versa, particles with no rest mass must move at the speed of light. There are three known particles which fall into this category, the photon, the graviton, and the neutrino. Eq. 20.13 also tells us that such particles do possess momentum which is related to their total energy by:

$$p = E/c \quad \text{(Valid for zero rest mass particles only)}$$

20.4 CONSERVATION OF ENERGY AND MOMENTUM IN PARTICLE COLLISIONS

Macroscopic objects such as cars, trucks, balls, and even present day rocketships move with velocities much less than c. In all these instances Newtonian mechanics provides a very adequate description of the dynamics of these systems, with relativity providing small corrections. The only major exception to this is in the case of cosmological problems dealing with gravity where Einstein's general theory of relativity must be employed.

On the atomic and elementary particle scale of nature relativistic effects are very important and very evident. It is in this regime that the predictions of mass-energy equivalence, relativistic energies and momenta, etc are commonly verified and employed. The most awesome of these effects is the conversion of matter to energy in stars and in nuclear weapons. The tremendous quantities of energy liberated by the processes involved is testament to the size of c^2. In considering the interaction of elementary particles which have individually very small masses the normal units of energy are not useful. Instead we employ an "atomic" size unit of energy called the electron volt (eV).

$$1 \text{ eV} = 1.60219 \times 10^{-19} \text{ J}$$

The collisions of elementary particles at high energies (i.e. with velocities close to c) must be treated relativistically. In these cases the conservation of relativistic mass-energy and relativistic momentum must be employed. Interactions at this level of nature allow both the conversion of energy into mass by the production of new elementary particles and the conversion of mass into energy.

One example of this which is treated in the text is the case of pair production in particle collisions. In the relativistic case it is much easier to work in the center-of-momentum frame of reference. If the collision of two equal mass particles, one of which is at rest, is treated by Newtonian mechanics, the total kinetic energy available in the center-of-momentum frame of reference is $(1/2)K_i$. K_i is the kinetic energy of the incident particle measured in the laboratory frame. Because the relativistic energy and momentum behave so differently with increasing speed, the results in a relativistic interaction are much different. In the extreme relativistic limit the energy available in the center of mass system increases as the square root of the incident energy as measured in the laboratory frame rather than linearly as in the case of Newtonian mechanics. The collision examples in the text illustrate these effects.

In the final chapter of the text nuclear fission and fusion are considered. These are two of the most familiar processes which convert matter to energy. The potential of this conversion is quite staggering. A 20 kiloton nuclear device liberates 3.8×10^9 J of energy. This corresponds to the conversion of only 4.22×10^{-8} kg of mass into energy.

EXAMPLE PROBLEMS

Example 1

A relativistic particle with mass m_1 and velocity v_1 is incident on an identical particle at rest. They undergo a perfectly inelastic collision. Assuming conservation of momentum, find the final velocity of the particles and show that it reduces to the expected expressions for $v \ll c$ and $v \sim c$.

Solution

Given: $m_1 = m = m_2$, v_1, collision is perfectly inelastic.

Required: v_2, velocity of particles after the collision.

The momentum of the system before the collision is just that of the incident particle.

$$p = m_1 v_1 / (1 - v_1^2/c^2)^{1/2}$$

After the collision the momentum must be

$$p = (m_1 + m_2) v_2 / (1 - v_2^2/c^2)^{1/2} = 2 m_1 v_2 / (1 - v_2^2/c^2)^{1/2}$$

Momentum is conserved in this collision and

$$m_1 v_1 / (1 - v_1^2/c^2)^{1/2} = 2 m_1 v_2 / (1 - v_2^2/c^2)^{1/2}$$

If we square both sides of this equation and perform a lot of algebra we find

$$v_2 = v_1 / (4 - 3 v_1^2/c^2)^{1/2}$$

If $v \ll c$ then this reduces to $v_2 = 1/2\, v_1$, the classical result. If $v \sim c$ then the radical in the denominator approaches 1 and $v_2 \simeq v_1 \sim c$, again as one would suspect.

Example 2

A proposed rocket engine for interstellar travel is an ion propulsion device which produces a constant force of 10 N. The advantage of such a device is that it uses relatively little mass. Suppose we desire to accelerate a space ship of mass 2×10^4 kg to a velocity of 0.5 c employing such an engine. Neglecting the decrease in mass of the rocket, how long would this take?

Solution

Given: $m = 2 \times 10^4$ kg

$F = 10$ N

$v = 0.5\, c$

Required: time to reach this velocity

We employ Eq. 20.3 which we solve algebraically for t

$$t = mv / F (1 - v^2/c^2)^{1/2}$$

substituting we have

$$t = (2 \times 10^4 \text{ kg})(0.5)(3 \times 10^8 \text{ m/s}) / (10 \text{ N})(1 - (0.5)^2)^{1/2}$$

$$t = (3 \times 10^{12}) / (10)(1 - 0.25)^{1/2} = 3.46 \times 10^{11} \text{ s} \simeq 1.1 \times 10^3 \text{ years}$$

This device, like the chemical propulsion device of Example 13 in Chapter 10 of the text doesn't look too promising. It appears interstellar space flight requires the commitment of generations of space travelers or some new physics.

Example 3

An electron and a positron (anti-electron) at rest annihilate each other converting all their mass to energy in the form of photons. a) What energy results? b) What is the minimum number of photons which can be emitted? (Problem 32, page 390)

Solution

Given: $m_{e+} = m_{e-} = 9.109 \times 10^{-31}$ kg

Required: a) Energy released, b) Minimum number of photons.

a) The energy released in this process is $2\,mc^2$ since the conversion of matter to energy is 100% efficient in this case.

$$E = 2\,mc^2 = 2(9.109 \times 10^{-31} \text{ kg})(9 \times 10^{16} \text{ m}^2/\text{s}^2)$$
$$= 1.64 \times 10^{-13} \text{ J} = 1.02 \times 10^6 \text{ eV}$$

b) The electron and positron are at rest. The initial linear momentum of the system is zero, therefore the final momentum after the annihilation must also be zero. Since the photon is a zero rest mass particle, it must move with velocity c and have momentum E/c. There is no way one photon can have zero momentum. Thus, the annihilation results in the emission of two identical photons, each with energy 0.501 MeV and they are emitted in opposite directions.

PROBLEMS

1. At what velocity does the momentum of a particle of mass m differ by 10% from its classical value?

2. An electron has a velocity of 0.9 c. a) What velocity must a proton have in order to have the same momentum? b) The same kinetic energy? c) In either case is the proton traveling at relativistic speed?

3. How much work must be done to accelerate a particle of mass m from rest to 0.9 c? If this much work is performed on the particle again ($v_i = 0.9$ c) what will be its final velocity?

4. Compute the ratio of the kinetic energy of an electron to its rest mass energy for $v = 10^6$ m/s, 10^7 m/s, 10^8 m/s, and 2.9×10^8 m/s.

5. A proton is incident on a proton at rest. a) How much energy must the moving proton have in the laboratory frame of reference in order that the resulting collision produces a Kaon, anti-Kaon pair plus the two protons? b) What velocity does this proton have? c) What would be the corresponding values if one employed a proton-proton colliding beam device where the colliding protons have equal and opposite momenta in the laboratory frame?

Chapter 21
Temperature and Heat

PREVIEW

This chapter serves as an introduction to thermodynamics, the study of bulk matter under conditions when the effects of heat and temperature cannot be ignored, and presents two of the laws of thermodynamics, the zeroth law and the first law. In connection with the zeroth law, the fundamental quantity of thermodynamics, temperature, is defined and temperature scales and thermal equilibrium are discussed. The presentation of the first law, a statement of the conservation of energy, involves discussion of its ingredients: internal energy, thermodynamic work, and heat as a form of energy. The chapter concludes with a number of applications of the first law to various types of thermodynamic processes.

SUMMARY

21.1 TEMPERATURE AND THE ZEROTH LAW OF THERMODYNAMICS

The fundamental quantity of thermodynamics and the fourth fundamental quantity of the SI system that we have encountered is temperature. In a later chapter, we will find that on the microscopic level temperature measures the average kinetic energy of molecular motion. From a macroscopic viewpoint, temperature is a property of an object that indicates when it is in thermal equilibrium with another object. When two objects are in thermal equilibrium, they have the same temperature.

<u>Zeroth Law of Thermodynamics</u>: Two objects, each in thermal equilibrium with a third object, are in thermal equilibrium with each other.

The zeroth law recognizes that the single property temperature is both necessary and sufficient to ensure thermal equilibrium. An operational definition of temperature must include instructions for its measurement, that is, for making a thermometer.

21.2 TEMPERATURE MEASUREMENT

<u>Definition</u>. <u>Thermometer</u>: an instrument used to measure temperature using any system that has a property that changes with temperature.

The property that changes with temperature is known as a thermometric property, and those that have been used in thermometers include color, electrical resistance, the length of a column, and the color of a liquid crystal. A thermometer also requires a fixed point (or points) for which a particular temperature is assigned to a point on the temperature scale defined by the thermometer.

Over the range of common temperatures, the standard thermometer that provides the operational definition of temperature is the constant-volume gas thermometer that uses the ideal gas temperature scale defined by:

$$T = 273.16 \text{ K} \lim_{P_3 \to 0}(P/P_3) \tag{21.3}$$

where P, the pressure of the confined gas, is the thermometric property of the thermometer and P_3 is the pressure at the standard fixed point, chosen to be the triple point of water and assigned the temperature of 273.16 K. This temperature scale is identical with the Kelvin scale (described in Chapter 24), thus the temperature unit is the kelvin (K) and T is referred to as the Kelvin temperature.

Other common temperature scales are the Celsius, Fahrenheit, and Rankine scales. Celsius temperature t_C is related to the Kelvin temperature T by:

$$t_C = (T - 273.15)\,°C \tag{21.4}$$

The Celsius degree and the Kelvin have the same size but the zeros of the two scales differ by 273.15 degrees. The triple point of water is thus 0.01°C.

Temperatures on the Fahrenheit scale are related to those of the Celsius scale by:

$$t_F = (32 + 9t_C/5)\,°F \tag{21.5}$$

The Fahrenheit degree is smaller than the Celsius degree and the Kelvin and is related to the Celsius degree by 1 F° = (5/9) C°. This relation can be used as the conversion between temperature intervals on the two scales, for example, 10 C° = 18 F°.

The Rankine temperature scale has the same zero as the Kelvin scale and the same size degree as the Fahrenheit scale.

The standard practice is to distinguish the units of temperature and temperature interval, for example, a temperature is in degrees Celsius (°C) while a temperature interval is in Celsius degrees (C°).

21.3 ENERGY AND THE FIRST LAW OF THERMODYNAMICS

<u>The First Law of Thermodynamics</u>: energy is conserved in all physical processes and cannot be created or destroyed.

The first law is a statement of the conservation of energy. It allows for the conversion of one form of energy into another and includes heat as one of the forms of energy.

<u>Definition</u>. <u>Perpetual motion machine of the first kind</u>: a device that produces more energy than is absorbed in its operation and therefore violates the first law of thermodynamics.

21.4 HEAT AND SPECIFIC HEAT CAPACITY

<u>Definition</u>. <u>Heat</u>: energy transferred spontaneously from a region of higher temperature to a region of cooler temperature by virtue of a temperature difference.

The unit of heat, since it is a form of energy, is the joule (J). Other heat units include the British thermal unit (Btu) and the kilocalorie (kcal), which is sometimes referred to as the food calorie or Calorie (Cal), with a capital C. The kilocalorie is the amount of heat required to raise the temperature of 1 kg of water from 14.5°C to 15.5°C. The equivalences are

$$1 \text{ kcal} = 4.1858 \text{ kJ} \tag{21.8}$$
$$1 \text{ Btu} = 0.252 \text{ kcal} = 1.05 \text{ kJ}$$
$$1 \text{ kcal} = 1 \text{ Cal}$$

> **Definition** Specific heat capacity: the quantity of heat required to raise a unit of material one degree in temperature in a specified way, defined by $Q = MC\,\Delta T$ (Eq. 21.9), where C is the average specific heat capacity over the temperature range ΔT and Q is the heat absorbed or rejected when a mass m of the material undergoes a temperature change ΔT.

The units of temperature T must be consistent with the units of the specific heat capacity C. Since the size of the degree is the same on the Kelvin and Celsius temperature scales, the temperature change ΔT is the same on both scales and can be calculated using either kelvins or Celsius degrees.

A subscript on the specific heat capacity is used to specify the conditions under which the heat is absorbed or rejected, for example, C_p and C_v. Also, it is assumed that in the process specified no phase or chemical change occurs. The specific heat capacities at constant pressure Cp for a number of solids, liquids, and gases are given in Table 21.1.

> **Definition.** Heat capacity: the quantity of heat required to raise a system one degree in a specified way, defined by heat capacity = mC, where m is the mass of the system and C is the specific heat capacity of the material of the system.

The two quantities, specific heat capacity and heat capacity, have similar names but have different dimensions and units, since heat capacity is the product of specific heat capacity and the mass of the system involved. Specific heat capacity is specific to a given material, while heat capacity refers to a particular system composed of that material.

> **Definition.** Method of mixtures: a technique that can be used for measuring specific heat capacity in which known amounts of materials at different temperatures are combined in a calorimeter and it is assumed that there is no transfer of heat to or from the surroundings.

Since the method of mixtures assumes that there is no gain or loss of heat by the contents of the calorimeter, the heat rejected by the initially warmer parts of the system is equal to the heat absorbed by the initially cooler parts of the system, and the heat transfers for all parts of the system sum to zero: $Q = 0$.

21.5 THERMODYNAMIC WORK

> **Definition.** Reversible process: a quasi-static process that can be reversed by an infinitesimal change in the surroundings.

A quasi-static process is one in which the surroundings and thermodynamic variables of a system change so slowly that the process can be viewed as passing through a succession of equilibrium states. Irreversible processes occur spontaneously within a system that is not in thermodynamic equilibrium. Heat flow and processes involving friction are examples of irreversible processes.

> **Definition.** Thermodynamic work: the work done by a system in a reversible process, so called because it can be expressed entirely in terms of the thermodynamic variables of the system.

For a system consisting of a gas at pressure P, the thermodynamic work dW done by the gas during a reversible change of volume dV is given by

$$dW = P\,dV \tag{21.15}$$

The total work done during a finite change of volume is obtained by integrating this equation between initial and final states. If the pressure P is constant, the total work done can be expressed as $W = P\,\Delta V$, where ΔV is the change in volume of the system.

On a P-V phase diagram, the total work done by the system corresponds to the area beneath the path connecting the initial state and the final state. We say that the work is a path-dependent quantity because it depends not only on the initial and final states but also on the particular path connecting these states. The work can be negative if the volume of the system is decreasing.

Thermodynamic work can take forms other than the mechanical work we have considered for a system consisting of a gas. In fact, work includes all forms of energy transfer except heat.

21.6 INTERNAL ENERGY

> Definition. Adiabatic process: a transformation of a system during which heat is neither absorbed nor rejected.

> Definition. Internal energy: a characteristic property of a system that is equal to the total energy stored within the system, defined in an adiabatic process by $\Delta U + W = 0$ (Eq. 21.17), where ΔU is the change in the internal energy of the system and W is the work done by the system on its surroundings.

The interpretation of this operational definition of internal energy is that if the work done by the system is positive, the change in internal energy is negative and the internal energy of the system decreases in the process. Likewise, if the work done by the system is negative, the change in internal energy is positive and the internal energy increases.

21.7 THE FIRST LAW OF THERMODYNAMICS

The three forms of energy that are related by the first law of thermodynamics have been discussed in the previous sections of this chapter and include heat, thermodynamic work, and internal energy. Summarizing, we have found that internal energy is a property of a system that includes the total energy stored in the system and heat and thermodynamic work are path-dependent mechanisms by which energy is transferred between the system and its surroundings. A body possesses internal energy, but it does not contain heat any more than it contains work.

The first law of thermodynamics can therefore be stated as: heat added to a system appears as a change in the internal energy and/or is used to do work. On the basis of this statement, a quantitative formulation of the first law is

$$Q = U_f - U_i + W \tag{21.18}$$

where Q is the net amount of heat <u>added</u> to the system, $U_f - U_i = \Delta U$ is the change in the internal energy, and W is the net work done <u>by</u> the system.

This definition of the first law implies a particular sign convention for Q and W: Q is positive if heat flows into the system and Q is negative if heat flows out of the system; W is positive if work is done by the system and W is negative if work is done on the system. According to the usual definition of change, ΔU is the final internal energy minus the initial internal energy of the system.

21.8 APPLICATIONS OF THE FIRST LAW

1. Cyclic process. The initial and final states of the system are the same, so that $\Delta U = 0$, and the first law reduces to $Q = W$ (Eq. 21.20). On a P-V phase diagram, a cyclic process is described by a closed path, and the net work done corresponds to the area enclosed by the path. A clockwise path on a P-V diagram corresponds to positive work and a counterclockwise path to negative work according to the sign convention of the first law.

2. Adiabatic process. No heat is absorbed or rejected by the system, so that $Q = 0$, and the first law states that $\Delta U + W = 0$ (Eq. 21.17). As we have seen, this so-called adiabatic form of the first law serves as an operational definition of the change in internal energy.

3. Isobaric process. For this process, the pressure P = constant, $W = P \Delta V$, and the first law does not assume any special form.

4. Isothermal process. For this process, the temperature T = constant, and the first law does not assume any special form. An example of an isothermal process is a phase change, such as the melting of a solid or the freezing of a liquid.

EXAMPLE PROBLEMS
Example 1

A person has a fever of 4 F° above normal body temperature. How much is the fever above normal on the Celsius temperature scale?

Solution

Given: $(T - T_{normal}) = 4$ F°

Required: $(T - T_{normal})$ C°

The actual value of the normal body temperature is of no consequence, as we are asked to convert a temperature interval beginning at this temperature and express it on the Celsius scale. We require the conversion factor relating temperature intervals on the two temperature scales.

The temperature interval between the ice and steam points is divided into 100 C° and 180 F°, therefore, the required conversion factor is given by:

100 C° = 180 F°

1 F° = (5/9) C°

The notation F° and C° is used here because these are temperature intervals and not temperatures.

To convert the given temperature interval of 4 F°, we multiply by the conversion factor:

4 F° = 4 F° [(5/9 C°)/1 F°] = 2.22 C°

$T - T_{normal}$ = <u>2.22 C°</u>

The 4 F° and the 2.22 C° are temperature intervals and represent the temperature change above normal body temperature on the two scales.

This technique of converting temperature intervals above and below well-known fixed points can be used to convert temperatures from one system to another. This requires adding (or subtracting) the calculated interval to (from) the appropriate fixed-point temperature.

Example 2

The method of mixtures is used to measure the specific heat capacity of a 0.4-kg sample of an unknown material. The mass of the aluminum calorimeter can (C = 0.9 kJ/kg-C°) is 0.2 kg, the mass of the water is 0.3 kg, the initial temperature of the calorimeter is 25°C, and the final temperature of the calorimeter and sample is 30°C. If the sample is heated to 77°C before it is placed into the calorimeter, what is the specific heat of the unknown material?

Solution

Given:
m_{sample} = 0.4 kg
m_{can} = 0.2 kg
m_{water} = 0.3 kg
T_{can} = 25°C
T_{water} = 25°C
T_{sample} = 77°C
T_f = 30°C
C_{water} = 4.186 kJ/kg-C° (Table 21.1)
C_{Al} = 0.9 kJ/kg-C° (Table 21.1)

Required: C_{sample}

In the method of mixtures, it is assumed that there is no transfer of heat to or from the surroundings and that the sum of the heat transfers (Q) for all parts of the system is zero. Here, there are three parts: the calorimeter can, the water, and the sample. The heat for each can be calculated by substituting the appropriate values into $Q = mC \Delta T$ (Eq. 21.9).

The temperature units must be consistent with the units of the specific heat capacity, which here include Celsius degrees. The temperature changes are the same on both the Kelvin and Celsius scales, thus either Kelvins or Celsius degrees can be used in their evaluation. The Celsius temperatures given here are consistent with the specific heat capacities and can be used without conversion:

Q_{can} = $m_{can} C_{Al} \Delta T_{can}$
= (0.2 kg)(0.9 kJ/kg-C°)(30 - 25)C° = 0.9 kJ

Q_{water} = $m_{water} C_{water} \Delta T_{water}$
= (0.3 kg)(4.186 kJ/kg-C°)(30 - 25)C° = 6.28 kJ

Q_{sample} = $m_{sample} C_{sample} \Delta T_{sample}$
= (0.4 kg) C_{sample} (30 - 77)C° = -(18.8 kg-C°)C_{sample}

In accordance with the sign convention for Q, the initially warmer part of the system (sample) that loses heat has a negative Q, and the initially cooler parts (can and water) that gain heat have a positive Q.

The sum of the three heat transfers must be zero:

$Q_{can} + Q_{water} + Q_{sample} = 0$

0.9 kJ + 6.28 kJ - (18.8 kg-C°)C_{sample} = 0

Solving this equation for the specific heat capacity of the sample C_{sample}:

C_{sample} = <u>0.382 kJ/kg-C°</u>

The specific heat capacity of the unknown material is nearly equal to that of copper (Table 21.1), thus, we would say that copper is a candidate for the material of the sample.

Because of the variety of units used for energy and temperature, it is important to insure that all quantities are in the same system of units when applying the method of mixtures. Units should be checked and required conversions carried out prior to beginning the heat calculations.

164 Chapter 21

Example 3

a) What quantity of heat is required to raise the temperature of 8 kg of water through an interval of 10 C°? b) How much lead (C = 0.13 kJ/kg-C°) could this quantity of heat take through a similar temperature interval?

Solution

Given: a) m_{water} = 8 kg
            ΔT_{water} = 10 C°
            C_{water} = 4.186 kJ/kg-C° (Table 21.1)
         b) ΔT_{lead} = 10 C°
            C_{lead} = 0.13 kJ/kg-C°

Required: a) Q,   b) m_{lead}

a) The quantity of heat required for the water is given by Eq. 21.9:

$$Q = mC \Delta T$$

The Celsius temperatures given are consistent with the specific heat capacities and can be used without conversion.

Substituting in this equation for the mass m_{water}, the specific heat capacity C_{water}, and the temperature interval ΔT_{water}:

$$Q = (8 \text{ kg})(4.186 \text{ kJ/kg-C°})(10 \text{ C°}) = \underline{335 \text{ kJ}}$$

b) Solving the heat equation for the mass m:

$$m = Q/C \Delta T$$

Substituting in this equation for the quantity of heat Q from a), the specific heat capacity C_{lead}, and the temperature interval ΔT_{lead}:

$$m_{lead} = (335 \text{ kJ})/(0.13 \text{ kJ/kg-C°})(10 \text{ C°}) = \underline{258 \text{ kg}}$$

This result, 8 kg water vs 258 kg lead, reflects the relatively high value of the specific heat capacity of water and is, in fact, the ratio of the specific heat capacities of the two substances.

Example 4

A gas at a pressure of 4×10^5 Pa is confined in a cylinder with a movable piston of radius 2×10^{-2} m. The piston moves inward a distance of 8×10^{-2} m, and the gas is compressed at constant pressure. How much work is done by the gas?

Solution

Given: P = 4×10^5 Pa
         r = 2×10^{-2} m
         Δx = 8×10^{-2} m

Required: W

Evaluating Eq. 21.16 for the work done by a gas during a volume change ΔV from an initial volume V_i to a final volume V_f at a constant pressure P:

$$W = P(V_f - V_i) = P \Delta V \tag{1}$$

Since the piston moves inward, the final volume V_f is the initial volume V_i decreased by the volume moved through by the piston, given by the product of the cross-sectional area and the distance moved:

$$V_f = V_i - A \Delta x$$

The volume change ΔV of the gas is negative and is given by:

$$\Delta V = V_f - V_i = - A \Delta x = - \pi r^2 \Delta x$$

Substituting in Eq. (1) for the volume change ΔV and the pressure P:

$$W = - \pi r^2 P \Delta x = - \pi (2 \times 10^{-2} \text{ m})^2 (4 \times 10^5 \text{ Pa}) (8 \times 10^{-2} \text{ m}) = \underline{40.2 \text{ J}}$$

The work done by the gas during this process is negative as a result of the decreasing volume of the gas. This means that work is done on the gas and is in agreement with the sign convention of the first law of thermodynamics (Eq. 21.18).

Example 5

What is the change in the internal energy of a system in a process in which the net work done by the system is 6 kJ and the net heat added to the system is -8 kJ?

Solution

Given: W = 6 kJ
          Q = -8 kJ

Required: ΔU

The change in the internal energy can be obtained from the first law of thermodynamics (Eq. 21.18):

$$Q = U_f - U_i + W$$

Solving this equation for the change in internal energy $U = U_f - U_i$ and substituting for the net work W done by the system and the net heat Q added to the system:

$$\Delta U = Q - W = -8 \text{kJ} - 6 \text{ kJ} = \underline{-14 \text{ kJ}}$$

The minus sign on Q indicates that heat is flowing out of the system. The negative value for ΔU is the result of both this heat loss and the system doing work.

Example 6

A confined gas is taken through a cycle ABCDA, as shown in FIG. 6-1. Find a) the work done by the gas on each leg of the cycle, b) the net work done by the gas, and c) the net amount of heat added to the gas during the cycle:

Solution

Given: $P_A = P_D = 2 \times 10^5$ Pa
          $P_B = P_C = 6 \times 10^5$ Pa
          $V_A = V_B = 10^{-3}$ m^3
          $V_C = V_D = 7 \times 10^{-3}$ m^3

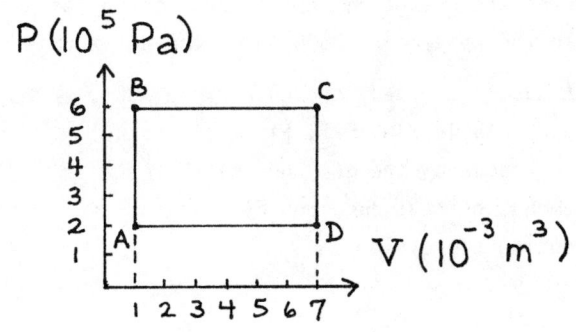

FIG. 6-1

Required: a) W_AB, W_BC, W_CD, W_DA, b) W, c) Q

a) The work done is obtained by evaluating Eq. 21.16 for each of the four paths. For paths AB and CD, there is no volume change, and the work done is zero:

$$W_{AB} = W_{CD} = 0$$

For constant pressure, Eq. 21.16 becomes $W = P \Delta V$, and the work done by the gas on paths BC and DA is given by:

$$W_{BC} = P_B (V_C - V_B) = (6 \times 10^5 \text{ Pa})(7 - 1) \times 10^{-3} \text{ m}^3 = \underline{3600 \text{ J}}$$
$$W_{DA} = P_A (V_A - V_D) = (2 \times 10^5 \text{ Pa})(1 - 7) \times 10^{-3} \text{ m}^3 = \underline{-1200 \text{ J}}$$

b) The net work done by the gas is the sum of the work done on each path:

$$W = W_{AB} + W_{BC} + W_{CD} = W_{DA} = 0 + 3600 \text{ J} + 0 - 1200 \text{ J} = \underline{2400 \text{ J}}$$

The net work done can also be calculated from the area enclosed by the path in a cyclic process. Here, the area enclosed is a rectangle with an area given by:

$$\text{Area} = W = [(6 - 2) \times 10^5 \text{ Pa}][(7 - 1) \times 10^{-3} \text{ Pa}] = 2400 \text{ J}$$

The net work done is positive, as the closed path is traversed in a clockwise direction, in agreement with the calculation above.

c) Since the initial and final states of the gas are the same for a cyclic process, there is no change in internal energy, and the first law of thermodynamics is given by Eq. 21.20:

$$Q = W$$

Substituting in this equation for the net work W from b):

$$Q = \underline{2400 \text{ J}}$$

PROBLEMS

1. The difference between the normal boiling points of nitrogen and oxygen is 12.8 C°. What is this temperature interval on the Fahrenheit temperature scale?

2. A 0.4-kg sample of iron (C = 0.448 kJ/kg-C°) at 20°C is placed into a 0.3-kg aluminum calorimeter can (C = 0.9 kJ/kg-C°) containing 0.2 kg of water at 60°C. What is the final temperature of the calorimeter and sample?

3. A gas is confined at a pressure of 3×10^5 Pa in a cylinder with a movable piston of cross-sectional area 8×10^{-3} m². If the gas expands at constant pressure so that the piston moves outward a distance of 6×10^{-2} m, how much work is done by the gas?

4. Find the change in the internal energy of a system in a process in which the net heat added to the system is -2000 J and the net work done by the system is -5000 J.

5. A gas is carried through a cycle ABCA, as shown in FIG. 5-1. Find a) the net work done by the gas and b) the net amount of heat absorbed by the gas during the cycle.

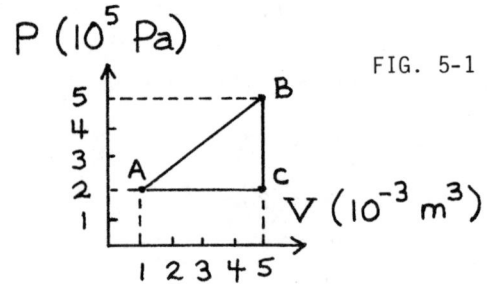

FIG. 5-1

Chapter 22
Thermal Properties of Matter

PREVIEW

This chapter analyzes the thermal expansion of solids and liquids and discusses change of phase and P-V-T surfaces for pure substances. The chapter also introduces the concept of an ideal gas, a model of a gas that incorporates the basic properties common to all gases, and presents the three relations that characterize an ideal gas: the equation of state and expressions for the internal energy and specific heat capacity.

SUMMARY

22.1 THERMAL EXPANSION

Definition. <u>Coefficient of linear expansion</u>: the fractional change in length of a solid for a one-degree change in temperature.

The coefficient of linear expansion is commonly denoted by α and has units of $(C°)^{-1}$. For the most common materials, α is in the range $1\text{-}10 \times 10^{-5}/C°$ (Table 22.1).

The change in length ΔL of a solid of length L and coefficient of linear expansion α for a change in temperature ΔT is thus given by:

$$\Delta L = \alpha L\, \Delta T \qquad (22.1)$$

The coefficient of linear expansion is positive, therefore, this equation describes expansion when the temperature change is positive and contraction when it is negative. The units of temperature must be consistent with the units of α. Since the size of the degree is the same on the Kelvin and Celsius temperature scales, temperature change is the same on both scales and can be calculated using either kelvins or Celsius degrees. The coefficients of expansion for a number of solids are given in Table 22.1.

The expansion of a solid is like a photographic enlargement in that the distance between any two points in the solid, including those at the ends of diameter of a hole in the solid, increases according to Eq. 22.1.

Definition. <u>Coefficient of volume expansion</u>: the fractional change in volume of a substance for a one-degree change in temperature.

The coefficient of volume expansion is commonly denoted by β and has the units of $(C°)^{-1}$. Typical values of β for liquids are near $10^{-3}/C°$ (Table 22.1). For isotropic solids, β is approximately equal to 3α.

The change ΔV in the volume V of a substance for a change in temperature ΔT is thus given by:

$$\Delta V = \beta V\, \Delta T \qquad (22.3)$$

Again, the units of T and β must be consistent, and the temperature interval ΔT can be calculated using either kelvins or Celsius degrees. The coefficients of volume expansion for a number of liquids are given in Table 22.1. Most substances have positive β, but a few exhibit negative values over certain ranges of temperature. For example, water has a negative value of β from 0°C to 3.98°C. This means that water expands upon cooling in this temperature range and has its maximum density at 3.98°C.

22.2 THE IDEAL GAS

Definition. *Ideal gas*: a generalized model of a gas that describes the basic properties of real gases and is based on universal behavior of all gases under conditions of low pressure (below 1 atm) and high temperature (above 200 K).

The ideal gas is characterized by an equation of state and expressions for internal energy and specific heat capacity. The equation of state for the ideal gas is given by

$$PV = nRT \tag{22.9}$$

where P, V, and T are the pressure, volume, and temperature of the gas, n is the number of moles of gas, and R is the universal gas constant, given by $R = 8.31434$ J/K (Eq. 22.10). In this form of the ideal gas equation of state, it is essential that the temperature be on the Kelvin or absolute temperature scale.

The internal energy equation for the ideal gas is given by

$$U = mC_v T \tag{22.18}$$

and the specific heat capacity at constant volume is given by C_v = constant.

The equation of state for the ideal gas is known as the ideal gas law. A convenient method of using the ideal gas law for a given system having a fixed mass (constant n) is to equate the constant nR for two different states of the system:

$$nR = \text{constant} = P_1 V_1 / T_1 = P_2 V_2 / T_2 \tag{22.33}$$

which can be solved for any one of the six quantities representing the initial and final states of the system.

If the ideal gas law is evaluated for both fixed mass of gas (constant n) and constant temperature, the result is known as Boyle's law:

$$PV = \text{constant} \tag{22.6}$$

For fixed mass and pressure, the result is known as the law of Charles and Gay-Lussac:

$$V/T = \text{constant} \tag{22.7}$$

The specific heat capacity is larger at constant pressure than at constant volume because there is no work done at constant volume, while at constant pressure some of the added heat is converted into work and is not available to increase the internal energy. For an ideal gas, the relation between the molar heat capacities at constant volume and pressure, MC_v and MC_p, is given by

$$MC_p - MC_v = R \tag{22.23}$$

where M is the molecular weight, given by $M = m/n$. The molar heat capacities of many real gases are in good agreement with this expression (Table 22.2).

For an adiabatic transformation of the ideal gas, the temperature T and volume V are related by

$$TV^{\gamma-1} = \text{constant} \tag{22.30}$$

where γ is the ratio of specific heat capacities C_p/C_v.

As in the case of the ideal gas law, this relation and two other equivalent relations can be expressed in terms of quantities representing two states of a given system during an adiabatic process:

$$T_1 V_1^{\gamma-1} = T_2 V_2^{\gamma-1} \qquad (22.31)$$
$$P_1 V_1^{\gamma} = P_2 V_2^{\gamma} \qquad (22.34)$$
$$T_2/T_1 = (P_2/P_1)^{(\gamma-1)/\gamma} \qquad (22.36)$$

22.3 P-V-T SURFACES

The equilibrium states of a pure substance can be described by three thermodynamic variables: pressure P, temperature T, and molar volume V/n. The equation of state is a relation among these variables and thus defines a surface, known as a P-V-T surface, in a space in which P, V, and T are the coordinates. Every equilibrium state of a system lies on the P-V-T surface, and non-equilibrium states lie off the surface. A quasi-static process passes through equilibrium states and therefore must have a path that lies entirely on the surface.

There are regions on a P-V-T surface where two or three phases can coexist in thermal equilibrium. The triple line gives states where all three phases are in equilibrium. The high temperature-high pressure point of the liquid-vapor region is called the critical point. In order to be liquified, a gas must have a temperature below the critical temperature, and under these conditions the gas is sometimes referred to as a vapor.

> Definition. Phase diagram: a projection of a P-V-T surface onto a plane in order to indicate the values of the thermodynamic variables at which phase transitions occur.

All the states on a two-dimensional phase diagram are equilibrium points, since they are projections of the equilibrium states on a P-V-T surface. Non-equilibrium states cannot be represented on a phase diagram.

On a P-T phase diagram, the coexistence regions on the P-V-T surface are collapsed into lines. Liquid and vapor coexist along the vaporization curve. Likewise, the sublimation curve separates the solid and vapor regions, and the fusion curve gives states with solid and liquid in equilibrium.

22.4 CHANGE OF PHASE

> Definition. Latent heat: the amount of heat absorbed or given off per kilogram during a change of phase of a substance.

The heats of fusion L_f, vaporization L_v, and sublimation L_s refer, respectively, to the solid-liquid, liquid-vapor, and solid-vapor transitions. The latent heats of fusion and vaporization for a number of substances are given in Table 22.3. For water, L_f = 334 kJ/kg and L_v = 2260 kJ/kg. The heat given off or absorbed in a phase change is given by Q = mL. The sign convention is the same as for the first law of thermodynamics: Q is positive when a system absorbs heat and negative when heat is given off.

> Definition. Heating curve: a curve inversely proportional to specific heat capacity that gives the relation between temperature and amount of heat absorbed during a heating process.

> Definition. Saturated vapor pressure: the vapor pressure at which a liquid and its vapor are in phase equilibrium, dependent only on temperature and given by the vaporization curve.

Definition. Boiling point: the temperature at which atmospheric pressure equals the saturated vapor pressure, given by the vaporization curve.

The positive slope of the vaporization curve on a P-T phase diagram indicates the general increase with temperature of the saturated vapor pressure and therefore also the increase with pressure of the boiling point. For example, the boiling point of water in a pressure cooker at 2 atm is approximately 120°C.

EXAMPLE PROBLEMS

Example 1

A silver wire has a length of 500 m at 40°C. When the temperature falls to -30 C, what is a) the change in length and b) the final length?

Solution

Given:
L_i = 500 m
T_i = 40°C
T_f = -30°C
α_{silver} = 1.89 x 10^{-5}/C° (Table 22.1)

Required: a) ΔL, b) L_f

a) The change in length of a solid is given by Eq. 22.1:

$$\Delta L = L\alpha \Delta T = L\alpha(T_f - T_i)$$

The Celsius temperatures given are consistent with the expansion coefficient and can be used to calculate the temperature change without conversion. Substituting for the coefficient of expansion for silver α, the initial length L_i, and the temperature change $\Delta T = T_f - T_i$:

$$L = (500 \text{ m})(1.89 \times 10^{-5}/\text{C°})(-30 - 40) \text{ C°} = \underline{-0.662 \text{ m}}$$

b) The final length is given by the sum of the initial length and the change in length:

$$L_f = L_i + \Delta L = 500 \text{ m} - 0.662 \text{ m} = \underline{499.3 \text{ m}}$$

The units of the expansion coefficient and the temperature change should be checked and any required conversions carried out before making the length-change calculation in this type of problem. Any length units can be used, as these units are not altered by the calculation and directly become the units of the change in length.

Example 2

A glass bottle of volume 1.5 liters is filled with benzene at 10°C. How much benzene will overflow if the temperature rises to 30°C?

Solution

Given:
V = 1.5 liters
T_i = 10°C
T_f = 30°C
$\beta_{benzene}$ = 3.72 x 10^{-3}/C° (Table 22.1)
α_{glass} = 0.9 x 10^{-5}/C° (Table 22.1)

Required: Overflow = $\Delta V_{benzene} - \Delta V_{glass}$

The change in volume accompanying a temperature change for both solids and liquids is given by Eq. 22.3:

$$\Delta V = \beta V \Delta T$$

Although both the glass bottle and the benzene have the same initial volume, their volume expansions will differ because they have different coefficients of volume expansion.

For a solid, the coefficient of volume expansion β is three times the coefficient of linear expansion α. Substituting in this equation for the coefficient of volume expansion of glass $\beta_{glass} = 3\alpha_{glass}$, the initial volume V, and the temperature change $\Delta T = T_f - T_i$:

$$\Delta V_{glass} = (1.5 \text{ liters})(3)(0.9 \times 10^{-5}/C°)(30 - 10) \text{ C°} = 8.10 \times 10^{-4} \text{ liter}$$

Repeating this procedure for the benzene using the coefficient of volume expansion of benzene $\beta_{benzene}$:

$$\Delta V_{benzene} = (1.5 \text{ liters})(3.72 \times 10^{-3}/C°)(30 - 10) \text{ C°} = 1.12 \times 10^{-1} \text{ liter}$$

The volume expansions of the glass bottle and the benzene differ by two orders of magnitude. The overflow is given by the difference between the two volume changes:

$$\text{Overflow} = \Delta V_{benzene} - \Delta V_{glass}$$
$$= (112 - 0.81) \times 10^{-3} \text{ liter} = 111 \times 10^{-3} \text{ liter}$$
$$= \underline{111 \text{ ml}}$$

Example 3

An ideal gas at a temperature of 300 K is compressed to one-third of its original volume while the pressure increases to four times the original pressure. What is the final temperature?

Solution

Given: $T_1 = 300$ K
$V_2 = V_1/3$, $P_2 = 4P_1$

Required: T_2

Using the method of Eq. 22.33 for a fixed mass of gas (constant n):

$$P_1 V_1/T_1 = P_2 V_2/T_2 \tag{1}$$

Solving Eq. (1) for the final temperature T_2:

$$T_2 = T_1(P_2/P_1)(V_2/V_1) \tag{2}$$

Substituting in Eq. (2) for the initial temperature T_1 and the initial and final pressures and volumes P_1, P_2, V_1, V_2:

$$T_2 = (300 \text{ K})(4P_1/P_1)(V_1/3V_1) = \underline{400 \text{ K}}$$

Example 4

An ideal gas at a temperature of 300 K is compressed adiabatically to one-third of its original volume. What is a) the final temperature and b) the ratio of final and initial pressures? The specific heat ratio for an ideal gas is γ = 1.67.

Solution

Given: $T_1 = 300$ K
$V_2 = V_1/3$
γ = 1.67

Required: a) T_2, b) P_2/P_1

a) We require an expression relating the temperatures and volumes of two states of an adiabatic process (Eq. 22.31):

$$T_1 V_1^{\gamma-1} = T_2 V_2^{\gamma-1} \tag{1}$$

Solving Eq. (1) for the final temperature T_2:

$$T_2 = T_1(V_1/V_2)^{\gamma-1} \tag{2}$$

Substituting in Eq. (2) for the initial temperature T_1 and the initial and final volumes V_1 and V_2, and the specific heat ratio γ:

$$T_2 = (300 \text{ K})[(V_1)/(V_1/3)]^{1.67-1} = (300 \text{ K})(3)^{0.67} = \underline{624 \text{ K}}$$

b) Eq. (22.34) relates the pressures and volumes of two states of an adiabatic process:

$$P_1 V_1^\gamma = P_2 V_2^\gamma \tag{3}$$

Solving Eq. (3) for the ratio of final and initial pressures:

$$P_2/P_1 = (V_1/V_2)^\gamma \tag{4}$$

Substituting in Eq. (4) for the initial and final volumes V_1 and V_2 and the specific heat ratio:

$$P_2/P_1 = [(V_1)/(V_1/3)]^{1.67} = (3)^{1.67} = \underline{6.26}$$

These results can be verified by using the ideal gas law in the form of Eq. 22.33:

$$T_2/T_1 = (P_2/P_1)(V_2/V_1)$$
$$624 \text{ K}/300 \text{ K} = (6.26)(1/3)$$
$$2.08 = 2.08$$

We can also compare these results with those of Example 3, which also involves an ideal gas with an initial temperature of 300 K and a compression ratio of 3 but does not involve an adiabatic process.

Example 5

Three 40-g ice cubes at a temperature of 0°C are placed into an insulated cup containing 400 g of water at 30°C. What is the final temperature of the contents of the cup, assuming that the temperature of the cup does not change?

Solution

Given: $m_{cube} = 40 \text{ g} = 0.040 \text{ kg}$, $T_{cube} = 0°C$
$m_{water} = 400 \text{ g} = 0.4 \text{ kg}$, $T_{water} = 30°C$
$C_{water} = 4.19 \text{ kJ/kg-C°}$ (Table 21.1)
$L_f = 334 \text{ kJ/kg}$ (Table 22.3)

Required: T_f

If we assume that the temperature of the cup does not change and that there is no heat loss to the surroundings, the sum of the heat transfers of the ice cubes and the water must be zero.

The heat for the ice cubes is the heat absorbed during melting:

$$Q_i = mL_f$$

Substituting in this equation for the mass m of the ice cubes and the heat of fusion L_f:

$$Q_i = 3m_{cube}L_f = (3)(0.040 \text{ kg})(334 \text{ kJ/kg}) = 40.1 \text{ kJ}$$

The sign of the heat Q must be determined using the sign convention adopted for the first law of thermodynamics, which is positive for heat absorbed and negative for heat given off. In this case, the heat is absorbed and the sign of Q is positive.

The heat of the water includes both the heat given off by the warmer water as it cools to the final temperature and the heat absorbed by the water that came from the ice cubes as it warms to the final temperature. Both terms are of the form $Q = mC \Delta T$ (Eq. 21.9):

$$Q_w = 3m_{cube}C_{water}(T_f - T_{cube}) + m_{water}C_{water}(T_f - T_{water})$$
$$= (3)(.040 \text{ kg})(4.19 \text{ kJ/kg-C°})(T_f - 0°C) + (0.4 \text{ kg})(4.19 \text{ kJ/kg-C°})(T_f - 30°C)$$

Equating the sum of the heat transfers Q_i and Q_w of the ice and water to zero and solving for the final temperature T_f:

$$T_f = \underline{4.68°C}$$

In problems such as this involving a phase change, the new phase of the substance must be taken into account when calculating additional heat transfers. In this case, the water that came from the ice cubes had to be considered in determining the heat transfers of the water.

Example 6

A tank containing 400-kg of water at 0°C freezes completely during an eight-hour period. Find a) the energy and b) the average power given off by the water to the surroundings during this period. Assume that the tank does not absorb energy.

Solution

Given: m = 400 kg
Δt = 8 hr = (8)(3600) s
L_f = 334 kJ/kg   (Table 22.3)

Required: a) Q, b) P

a) The energy delivered to the surroundings is the heat given off during the freezing of the water, given by:

$$Q = mL_f$$

Substituting in this equation for the mass m and the heat of fusion L_f:

$$Q = (400 \text{ kg})(334 \text{ kJ/kg}) = \underline{1.34 \times 10^5 \text{ kJ}}$$

b) The power delivered is given by the ratio of the heat energy Q and the time interval Δt:

$$P = Q/\Delta t = (1.34 \times 10^5 \text{ kJ})/(8)(3600 \text{ s}) = \underline{4.64 \text{ kW}}$$

During the time that the water is freezing, the heat of fusion is being given off to the surroundings, and the water is functioning as a 4000-W heater.

PROBLEMS

1. An iron beam has a length of 20 m when installed at 30°C. What is the change in its length when the temperature drops to -20°C?

2. A 200-liter steel drum is filled with acetone at 0°C. How much acetone overflows from the drum when the temperature reaches 25°C?

3. An ideal gas at a temperature of 350 K and a pressure of 1 atm is expanded to four times its original volume at a pressure of one-fifth atmosphere. What is the final temperature of the gas?

4. An ideal gas at a temperature of 350 K and a pressure of 1 atm is expanded adiabatically to four times it origial volume. What is a) the final temperature and b) the final pressure? For an ideal gas, γ = 1.67.

5. How much heat is given off if a mixture of 3 kg of steam and 2 kg of water at 100°C is converted entirely to water and cools to 0°C?

6. What is the final temperature if 6 kg of ice at 0°C is mixed with 1 kg of steam at 100°C?

Chapter 23
Heat Transfer

PREVIEW

In this chapter we investigate the three forms of heat transfer: conduction in solids and fluids, convection in fluids, and thermal radiation. Fourier's law is introduced as a quantitative description of convection and is later used in the derivation of Newton's law of cooling for certain types of convective heat transfer. The chapter concludes with a discussion of thermal radiation and the Stefan-Boltzmann law.

SUMMARY

23.1 CONDUCTION

    <u>Definition</u>. <u>Conduction</u>: a transfer of heat in solids and fluids without a flow of matter.

    <u>Fourier's Law of Heat Conduction</u>: the rate of heat flow through a material is proportional to the area perpendicular to the direction of flow and to the negative of the temperature gradient in the direction of flow and is given by $q = -kA(dT/dx)$ (Eq. 23.2), where k is the thermal conductivity and dT/dx is the temperature gradient.

If the temperature gradient is constant, Eq. 23.2 can be expressed as

$$q = -kA(\Delta T/\Delta x) \tag{23.1}$$

where ΔT is the temperature difference across an area A of material of thickness Δx and thermal conductivity k. The negative sign means that if temperature T increases in the direction of increasing x, the gradient is positive and the heat flow is in the negative x-direction, as heat always flows from hot to cold. As in previous heat calculations, the temperature units must be consistent with the units of the thermal conductivity, and the temperature difference can be evaluated on either the Kelvin or Celsius scales.

The value of the thermal conductivity k depends primarily on the composition of the material. Materials with large values of k are described as good thermal conductors, and materials with small values of k are known as good insulators. Metals are good thermal conductors, and gases and various manufactured materials are good insulators (Table 23.1).

For two-dimensional heat flow outward through a pipe, known as cylindrical heat flow, Fourier's law can be integrated to give

$$T(r) - T_i = -(q/2\pi kL) \ln(r/r_i) \tag{23.8}$$

where r_i is the inner radius of the pipe at temperature T_i and $T(r)$ is the temperature at some other radius r.

23.2 CONVECTION

    <u>Definition</u>. <u>Convection</u>: a transfer of heat in fluids involving flow of matter.

In convective heat flow, circulating currents transfer fluid between regions of different temperature. Natural convection results from a temperature gradient in the fluid, while forced convection is produced by an external agent.

In convective heat transfer by a fluid that is in contact with a solid surface, most of the temperature change takes place across a very thin boundary layer by conduction. If the full surface-to-fluid temperature change is assumed to occur across this boundary layer, Fourier's law of conduction can be used to derive an approximate result known as Newton's law of cooling.

> Newton's Law of Cooling: the rate of convective heat transfer between a solid surface and a fluid is proportional to the surface area and temperature difference between the surface and fluid and is given by $q = Ah(T_s - T_f)$ (Eq. 23.10), where T_s and T_f are the temperatures of the surface and the fluid, A is the surface area, and h is the heat transfer coefficient.

The heat transfer coefficient is determined experimentally and depends on the type of fluid, the speed and nature of the fluid flow, and the surface. Typical values are given in Table 23.2.

A relation between the temperature of an object and the time that it has been cooling by convection is given by

$$(T_s - T_f)/(T_0 - T_f) = e^{-(t/t_c)} \tag{23.16}$$

where the cooling time $t_c = mC/Ah$ (Eq. 23.17) is the time required for the temperature difference $(T_s - T_f)$ between the object and the fluid to be divided by a factor of $e = 2.718$. T_0 is the initial temperature of the object, and m, C, and A are the mass, specific heat capacity, and surface area of the object. The cooling time can be taken as a figure of merit for the time required for an object to cool to the temperature of its environment.

23.3 THERMAL RADIATION

> Definition. Thermal radiation: energy given off by an object in the form of electromagnetic waves as the result of its temperature.

We refer to this energy as radiant energy. An object also absorbs radiant energy from its surroundings, and radiative equilibrium exists when the rates of absorption and emission are equal.

> Definition. Blackbody: an idealized system that absorbs all incident thermal radiation and emits radiation having a spectrum dependent only on the absolute temperature.

> Stefan-Boltzmann Law: the rate at which a blackbody emits thermal radiation per unit surface area is proportional to the fourth power of its Kelvin temperature and is given by $F = \sigma T^4$ (Eq. 23.18), where the surface flux F is the energy radiated per second per square meter, T is the Kelvin temperature, and Stefan's constant $\sigma = 5.6703 \times 10^{-8}$ W/m^2-K^4 (Eq. 23.19).

If the object emitting thermal radiation is not a blackbody, its surface is described by a parameter ε, known as the emissivity, and the surface flux is given by

$$F = \varepsilon \sigma T^4 \tag{23.20}$$

where σ is Stefan's constant and ε and T are the emissivity and Kelvin temperature of the object. The temperature must be in kelvins, since two temperatures on the Celsius scale that are positive and negative the same number of degrees would yield the same (incorrect) result for the surface flux.

Definition. <u>Emissivity</u>: the ratio of the surface flux emitted by an object to that emitted by a blackbody at the same temperature.

The emissivity depends on the physical composition of the surface and on its temperature. A blackbody is a perfect emitter and has $\varepsilon = 1$. For all other objects $0 < \varepsilon < 1$.

Definition. <u>Luminosity</u>: the total rate at which an object radiates energy, given by $L = FA = \varepsilon \sigma T^4 A$ (Eq. 23.22).

The net thermal radiation flux emitted by an object which is both radiating energy and absorbing thermal radiation from its environment is given by

$$F_{net} = \varepsilon \sigma (T^4 - T_e^4) \tag{23.21}$$

where T_e is the temperature of the environment. Also, the net rate at which an object emits radiant energy is given by $L_{net} = F_{net} A$.

EXAMPLE PROBLEMS
Example 1

A wooden house door has dimensions 0.9 m x 2.1 m x 0.045 m. What is the rate of conductive heat flow through the door if the inside temperature is 20°C and the outside temperature is 0°C?

Solution

Given: $A = 0.9 \text{ m} \times 2.1 \text{ m}$
                $T_i = 20°C$
                $T_f = 0°C$
                $\Delta x = 0.045 \text{ m}$
                $k = 0.19 \text{ W/m-C°}$ (Table 23.1)

Required: q

Assuming a constant temperature gradient across the door, the rate of heat flow by conduction is given by Eq. 23.1:

$$q = -kA(\Delta T / \Delta x)$$

The given Celsius temperatures are consistent with the units of the thermal conductivity and can be used in the calculation of the temperature change without conversion. If the outward direction is taken to be positive, positive Δx corresponds to negative ΔT since the outside temperature is lower than the inside temperature, and the minus sign on the equation then gives a positive or outward flow of heat.

Substituting in this equation for the thermal conductivity k of wood, the area A of the door, the outside-inside temperature difference ΔT, and the thickness of the door Δx:

$$q = -(0.19 \text{ W/m-C°})(0.9 \text{ m} \times 2.1 \text{ m})(0 - 20) \text{ C°}/0.045 \text{ m} = \underline{160 \text{ W, outward}}$$

Example 2

A 2-kg casserole dish with surface area 0.2 m^2 is removed from an oven at 90°C and placed in a room where the temperature is 20°C. a) If the heat transfer coefficient for air moving past the dish is 10 W/m^2-C°, find the rate of convective heat flow. b) If the specific heat capacity of the dish is 0.8 KJ/kg-C°, find the time for the temperature difference to be reduced by a factor of e = 2.7 from 70 C° to 26 C°.

Solution

Given: $A = 0.2 \text{ m}^2$
$h = 10 \text{ W/m}^2\text{-C°}$
$T_s = 90°C$
$T_f = 20°C$
$m = 2 \text{ kg}$
$C = 0.8 \text{ kJ/kg-C°}$

Required: a) q, b) t_c

a) Assuming that Newton's law of cooling is applicable in this situation, the rate of convective heat flow from the dish is given by Eq. 23.10:

$$q = Ah(T_s - T_f)$$

The temperatures of the solid surface T_s and the fluid T_f can be used, as given, in Celsius degrees. Since the temperature of the surface is higher than that of the fluid, the heat flow will be positive or outward.

Substituting in this equation for the area A of the dish, the heat transfer coefficient h, and the temperatures T_s and T_f of the dish and air:

$$q = (0.2 \text{ m}^2)(10 \text{ W/m}^2\text{-C°})(90 - 20) \text{ C°} = \underline{140 \text{ W, outward}}$$

b) The cooling time, or the time required for the temperature difference between the dish and the air to be divided by e = 2.7 is given by Eq. (23.17):

$$t_c = mC/Ah$$

Substituting in this equation for the mass m, the specific heat capacity C and surface area A of the dish, and the heat transfer coefficient h:

$$t_c = (2 \text{ kg})(0.8 \text{ kJ/kg-C°})/(0.2 \text{ m}^2)(10 \text{ J/m}^2\text{-C°}) = \underline{800 \text{ s}}$$

Example 3

The filament of a light bulb has an emissivity of 0.3 and a surface area of $8 \times 10^{-6} \text{ m}^2$. What is the rate of emission of radiant energy by the filament at a temperature of 3000 K.

Solution

Given: $\varepsilon = 0.3$
$A = 8 \times 10^{-6} \text{ m}^2$
$T = 3000 \text{ K}$
$\sigma = 5.67 \times 10^{-8} \text{ W/m}^2\text{-K}^4$ (Eq. 23.19)

Required: L = FA

The rate of emission of radiant energy is given by the luminosity (Eq. 23.22):

$$L = FA = \varepsilon \sigma T^4 A$$

The temperature must be on the Kelvin scale, and since the temperature of the filament is given in kelvins no conversion is necessary. The flow of radiant energy from filament is positive or outward.

Substituting in this equation for the emissivity ε, Stefan's constant σ, and the area A and temperature T of the filament:

$$L = (0.3)(5.67 \times 10^{-8} \text{ W/m}^2\text{-K}^4)(8 \times 10^{-6} \text{ m}^2)(3000 \text{ K})^4 = \underline{11.0 \text{ W, outward}}$$

Example 4

A person has a skin temperature of 37°C and is sitting in a sauna where the temperature is 60°C. a) What is the net rate of emission of thermal radiation from the person's body, assuming that the emissivity of the body is 0.97 and the surface area is 1.5 m^2? b) Express the result in kcal/hr.

Solution

Given: $\varepsilon = 0.97$
$A = 1.5$ m^2
$T = 37°C = 310$ K
$T_e = 60°C = 333$ K
$\sigma = 5.67 \times 10^{-8}$ W/m^2-K^4 (Eq. 23.19)

Required: a) $L_{net} = F_{net}A$

a) The net thermal radiation flux for an object radiating into an environment is given by Eq. 23.21, and the net rate of radiation of thermal energy, or net luminosity, is given by Eq. 23.22:

$$L_{net} = F_{net}A = \varepsilon\sigma A(T^4 - T_e^4)$$

The temperatures must be on the Kelvin scale, therefore the given Celsius temperatures must be converted to kelvins using Eq. 21.4. Here, the temperature of the environment T_e is larger than the temperature of the body, thus the energy flow is negative or inwards.

Substituting in this equation for the emissivity ε, Stefan's constant σ, the surface area A and the temperature T of the body, and the temperature T_e of the room:

$$L_{net} = (0.97)(5.67 \times 10^{-8} \text{ W/}m^2\text{-}K^4)(1.5 \text{ }m^2)[(310)^4 - (333)^4] \text{ }K^4 = \underline{-253 \text{ W}}$$

The negative sign indicates that the body has a net absorption of radiant energy.

b) To convert the result from W to cal/hr, the energy units must be converted from J to kcal and the time units must be converted from s to hr. There are 4186 J in 1 kcal, and there are 3600 s in 1 hr. Applying these conversion factors to the result of 253 W:

$$253 \text{ W} = 253 \text{ W (kcal/4186 J)(3600 s/hr)} = \underline{217 \text{ kcal/hr}}$$

If the body did not give off energy in some other way, the body temperature of an 80-kg person would increase in an hour by $\Delta T = Q/mC = 217$ kcal/(80 kg)(0.83 kcal/kg-C°) = 3.26 C° due to this absorption of radiant energy.

PROBLEMS

1. A cooler in the form of a cube of edge 40 cm is made of material of thickness 2 cm and thermal conductivity 0.01 W/m-C°. If the temperature inside the cooler is 3°C, and the outside temperature is 25°C, what is the total rate of conductive heat flow through the six walls of the cooler?

2. A 0.9-m x 2.1-m house door has an outer surface temperature of 0°C. If the heat transfer coefficient for air at -15°C blowing past the door is 25 W/m^2-C°, what is the rate of convective heat flow?

3. A 1000-W electric heater has a heating element with an emissivity of 0.5 and a temperature of 1500 K. If thermal radiation accounts for all of the energy emitted by the heater, find the surface area of the heating element.

4. A warming tray with an emissivity of 0.8 and a surface area of 0.1 m^2 has a net rate of emission of radiant energy of 5 W in a room where the temperature is 27°C. What is the surface temperature of the tray?

Chapter 24
The Second Law of Thermodynamics

PREVIEW

This chapter presents some of the many aspects of the second law of thermodynamics. These include the Kelvin and Clausius statements of the second law, heat engines, and the Carnot engine, the most efficient form of cyclic process for converting heat into work. The Carnot engine is used to define the Kelvin temperature scale and quantify the limit on the conversion of heat into work imposed by the second law. The concept of entropy is introduced as a measure of disorder on the microscopic level and used to give a formulation of the second law known as the entropy principle.

SUMMARY

24.1 HEAT ENGINES AND THERMODYNAMIC EFFICIENCY

There is a fundamental difference in the first and second laws of thermodynamics: the first law is an equality stating that energy is conserved, while the second law is an inequality stating that the complete conversion of heat into work is impossible.

> Definition. Heat reservoir: a body of uniform temperature whose mass is large enough to keep its temperature constant when heat is added or removed.

The heat reservoir is really an energy reservoir, as we have seen in a previous chapter that a body can possess energy but not heat (or work).

> Definition. Heat engine: any device, operating in a cycle that absorbs heat from a high-temperature reservoir, converts part of it into work, and ejects the remainder into a low-temperature reservoir.

In terms of heat engines, the second law states that no heat engine can carry out the conversion of heat completely into work.

In order to discuss the energy flow in a heat engine, we make the following definitions:

Q_H = heat absorbed from a high-temperature reservoir
W = net work done by the heat engine
Q_L = heat ejected to a low-temperature reservoir

These quantities are numbers carrying no algebraic signs. The sign is attached at the time of use according to the sign convention adopted for the first law of thermodynamics.

Since a heat engine operates in a cycle, there is no change in its internal energy, and the first law requires that the net work done equal the net heat absorbed:

$$W = Q = Q_H - Q_L \qquad (24.1)$$

The signs are assigned with respect to the working substance of the heat engine. Heat Q_H is absorbed by the substance and is entered into the first law with a plus sign, while heat Q_L is ejected by the substance and is entered with a minus sign.

> Definition. Thermodynamic efficiency: the fraction of the heat absorbed by a heat engine that is converted into work, defined by $\eta = W/Q$, where W is the net work done and Q the heat absorbed in the process.

The thermodynamic efficiency can be expressed entirely in terms of the heat transfers in a heat engine by substituting for the net work from the expression for the first law given in Eq. 24.1:

$$\eta = W/Q = (Q_H - Q_L)/Q_H \tag{24.3}$$

In terms of thermodynamic efficiency, the second law does not permit an efficiency of unity, while the first law allows any efficiency between (and including) zero and unity.

A refrigeration cycle applies to refrigerators and heat pumps and is a heat engine cycle operated in reverse: heat is extracted from a low-temperature reservoir, net work is performed on the device, and heat is ejected to a high-temperature reservoir.

> Definition. Coefficient of performance: for a refrigeration cycle, the ratio of a heat transfer of interest to the net work supplied to the device, defined by $COP = Q/W$, where Q is the heat extracted from the low-temperature reservoir for a refrigerator and the heat ejected to the high-temperature reservoir for a heat pump.

Since the signs of Q_H, Q_L, and W all reverse in going from a heat engine to a refrigeration cycle, Eq. 24.1 remains a valid statement of the first law and allows the coefficient of performance for a refrigerator to be expressed as

$$COP = Q_L/W = Q_L/(Q_H - Q_L) \tag{24.4}$$

24.2 THE CARNOT CYCLE

> Definition. Carnot cycle: a cycle that takes an ideal gas through four reversible processes consisting of two isothermal processes connected by two adiabatic processes.

In the Carnot cycle, all heat transfers take place at specific temperatures, as heat Q_H is absorbed during the isothermal expansion at temperature T_H, and heat Q_L is ejected during the isothermal compression at temperature T_L. There is, of course, no heat transfer during the adiabatic processes. Some of the heat extracted from the high-temperature reservoir is converted into work, therefore, the Carnot cycle is often referred to as a Carnot engine.

24.3 THE SECOND LAW OF THERMODYNAMICS

The Kelvin and Clausius statements of the second law are named after the men by whom they were formulated and have been shown to be equivalent.

> Kelvin Statement of the Second Law of Thermodynamics: It is impossible to devise a process whose only result is to convert heat extracted from a single reservoir into work.

An interpretation of the Kelvin statement focuses on the words only and single: heat can be converted into work in a process using a single reservoir if the process has some other result, or heat can be converted into work in a process involving multiple reservoirs if some heat is ejected to a reservoir at a lower temperature.

<u>Clausius Statement of the Second Law of Thermodynamics</u>: It is impossible to devise a process whose only result is to extract heat from a reservoir and eject it to a reservoir at a higher temperature.

An interpretation of the Clausius statement is that heat flows spontaneously from a high-temperature reservoir to a low-temperature reservoir in an irreversible process and heat cannot flow in the opposite direction unless work is performed to accomplish the transfer.

Definition. <u>Perpetual motion machine of the second kind</u>: a device that extracts heat from a reservoir and converts it completely into other forms of energy and therefore violates the second law of thermodynamics.

<u>Carnot's Theorem</u>: No heat engine operating between two heat reservoirs can be more efficient than a Carnot engine operating between the same two reservoirs.

We see that evaluating the efficiency of a Carnot engine will allow us to obtain a quantitative expression of the previous statements of the second law in terms of heat engines and thermodynamic efficiency. In addition, the fact that the efficiency of a Carnot engine is independent of the working substance allows us to define a thermodynamic temperature scale that is dependent only on the laws of thermodynamics.

24.4 THE KELVIN TEMPERATURE SCALE

A thermodynamic temperature is defined by taking it to be proportional to the heat transferred at that temperature in a Carnot engine:

$$Q_H/T_H = Q_L/T_L \qquad (24.9)$$

Definition. <u>Kelvin temperature scale</u>: the temperature of the high-temperature reservoir of a Carnot engine when the low-temperature reservoir is at the triple point of water, defined to be 273.16 K.

Using Eq. 24.9, the Kelvin temperature is, therefore, given by

$$T = 273.16 \, (Q/Q_3) \text{ K} \qquad (24.10)$$

where Q_3 is the heat ejected to the reservoir of a Carnot engine at the triple point of water and Q is the heat absorbed from the reservoir at the Kelvin temperature T.

A quantitative relation for the fundamental limit imposed by Carnot's theorem on the conversion of heat into work is obtained by expressing the efficiency of a Carnot engine in terms of the Kelvin temperatures of its reservoirs using the definition of thermodynamic temperature (Eq. 24.9):

$$\eta_C = 1 - (Q_L/Q_H) = 1 - (T_L/T_H) \qquad (24.11)$$

Definition. <u>Carnot efficiency</u>: the efficiency of a Carnot engine operating between a high-temperature reservoir at T_H and a low-temperature reservoir at T_L, given by $\eta_C = 1 - (T_L/T_H)$.

According to Carnot's theorem, the Carnot efficiency is the maximum thermodynamic efficiency for any cyclic heat engine operating between reservoirs at two given temperatures. In evaluating the Carnot efficiency, it is essential that the temperatures be on the Kelvin temperature scale.

24.5 ENTROPY

Entropy is probably one of the least well understood physical quantities. Reasons for this may be that thermodynamics does not give a microscopic interpretation of entropy and that entropy, unlike other thermodynamic quantities such as pressure, volume, and temperature, is not directly measureable.

Entropy can be introduced by considering the differential form of the first law (Eq. 21.19): $dU = dQ - dW$. Here, the change in a state variable, the internal energy, is given in terms of two path-dependent variables, heat and work, which themselves can be expressed in terms of other state variables and their changes. We have seen that for a reversible process, the work is given by $dW = P\,dV$ (Eq. 21.15). For a reversible process, the heat can be expressed in a similar form in terms of the temperature and the change in a quantity known as entropy: $dQ = T\,dS$ (Eq. 24.16).

> <u>Definition</u>. <u>Entropy</u>: a property of a system whose change in a reversible differential process is the ratio of the heat absorbed by the system and the absolute temperature of the system, defined by $dS = dQ/T$ (Eq. 24.17), where dQ is the heat absorbed at the temperature T.

Several important aspects of entropy are that the defining equation is for a reversible process, the temperature must be on the Kelvin temperature scale, and entropy is a state variable. Since such a variable depends only on the state of the system, the entropy change in a cyclic process is zero. The units of entropy are joules per kelvin (J/K).

By integrating the defining equation, the total entropy change in a reversible process can be obtained:
$$\Delta S = S_f - S_i = \int_i^f (dQ/T) \tag{24.18}$$
where S_f and S_i are the entropies of the final and initial states.

For an isothermal process carried out at temperature T, the integral can be easily evaluated to give
$$\Delta S = Q/T \tag{24.20}$$
where Q is the total heat absorbed by the system.

The entropy change of a system in a process carried out at constant pressure for which the heat added is given by $dQ = mC_p\,dT$ is given by
$$S = mC_p \ln(T_f/T_i) \tag{24.22}$$
where T_i and T_f are the initial and final temperatures of the system.

This equation illustrates the relationship between entropy and disorder on the molecular level. According to the equation, if the temperature rises, the entropy increases. An increase in temperature also increases the average molecular speed, decreases the molecular interactions, and thus increases the disorder. Hence, increasing entropy corresponds to increasing disorder, or randomness.

The entropy change of an ideal gas in a reversible process in which the gas is carried from state 1 to state 2 can be expressed as
$$\Delta S = mC_v \ln(T_2/T_1) + nR \ln(V_2/V_1) \tag{24.28}$$
where the first term is zero if the process is isothermal and the second term is zero if the process is at constant volume.

These entropy changes have been calculated on the basis of an equation that is valid only for reversible processes. However, the fact that entropy is a state variable allows us to find the entropy change in irreversible processes. The change in entropy depends only on the initial and final states and not on the process itself. Therefore, the entropy change for an irreversible process can be evaluated if a reversible process can be found connecting the same states. An example is the use of Eq. 24.28 to calculate the entropy change for the irreversible free expansion of a gas.

24.6 ENTROPY FORMULATION OF THE SECOND LAW

A fundamental difference between energy and entropy is that entropy is not a conserved quantity. The energy of an isolated system is constant, while its entropy can either remain constant or be increased by irreversible processes. We can say that irreversible processes generate entropy and that the second law is a consequence of irreversibility. The entropy principle is a statement of the second law in terms of entropy.

Entropy Principle: the entropy of an isolated system never decreases.

A quantitative form of the entropy principle is given by

$$\Delta S_{\text{isolated system}} \geq 0 \tag{24.29}$$

where ΔS is the entropy change brought about by any processes that connect equilibrium states of an isolated system. If an isolated system is not in equilibrium, irreversible processes generate entropy as they drive the system toward equilibrium. The equality in the entropy principle applies to reversible processes in which the entropy of an isolated system remains constant.

The entropy principle applies only to isolated systems. The entropy of a system that is not isolated can, in fact, decrease. However, if all parts of a system are included, the entropy principle states that the total entropy of the system increases or remains constant.

EXAMPLE PROBLEMS

Example 1

A heat engine absorbs 175 kJ of heat from a reservoir at 1400°C and ejects 105 kJ of heat to a reservoir at 400°C. Find a) the thermodynamic efficiency and b) the Carnot efficiency of the engine.

Solution

Given: T_H = 1400°C = 1673 K
T_L = 400°C = 673 K
Q_H = 175 kJ
Q_L = 105 kJ

Required: a) η, b) η_C

a) The thermodynamic efficiency is given by Eq. 24.2:

$$\eta = W/Q_H \tag{1}$$

The work W can be obtained from the first law of thermodynamics (Eq. 24.1):

$$W = Q_H - Q_L$$

Substituting in this equation for the heat absorbed Q_H and the heat ejected Q_L:

$$W = 175 \text{ kJ} - 105 \text{ kJ} = 70 \text{ kJ}$$

Substituting this result for the work W and the heat absorbed Q_H into Eq. (1):
$$\eta = 70 \text{ kJ}/175 \text{ kJ} = \underline{0.4}$$

b) The thermodynamic efficiency calculated in a) is the actual efficiency of the heat engine, while the Carnot efficiency is the maximum efficiency for the given reservoir temperatures. The Carnot efficiency is calculated using Eq. 24.13:
$$\eta_C = 1 - (T_L/T_H)$$
Substituting for the Kelvin temperatures T_L and T_H of the low-temperature and high-temperature reservoirs:
$$\eta_C = 1 - (673 \text{ K}/1673 \text{ K}) = 1 - 0.402 = \underline{0.598}$$

The given Celsius temperatures must be converted to Kelvin temperatures using $T = t_C + 273.15$ K (Eq. 21.4) prior to their use in the expression for the Carnot efficiency. In this case, the actual efficiency of $\eta = 0.4$ is approximately thirty percent less than the maximum efficiency for these temperatures.

Example 2

What is the entropy change of 5 kg of nitrogen that is liquified at a temperature of −196°C and a pressure of 1 atm? The heat of vaporization of nitrogen at 1 atm is 201 kJ/kg.

Solution

Given: $m = 5$ kg
$T = -196°C = 77$ K
$L_v = 201$ kJ/kg

Required: ΔS

This phase change to liquid nitrogen takes place at constant temperature, therefore, the entropy change can be calculated using Eq. 24.20:
$$\Delta S = Q/T \tag{1}$$

The nitrogen gives off heat, therefore, according to the sign convention of the first law Q in this equation is negative and is given by
$$Q = -mL_v$$
Substituting into Eq. (1) for the heat Q, the mass m, and the Kelvin temperature T:
$$\Delta S = -mL_v/T = -(5 \text{ kg})(201 \text{ kJ/kg})/77 \text{ K} = \underline{-13.1 \text{ kJ/K}}$$

The given Celsius temperature must be converted to kelvins before use in the expression for the entropy change in an isothermal process. The entropy change of the nitrogen is negative, however, the nitrogen is not an isolated system. In this process, an isolated system would have to include at least the object whose entropy increases as a result of absorbing heat from the nitrogen.

Example 3

If a sample of 2 moles of helium is cooled at constant volume from 300°C to 100°C, what is its change in entropy?

Solution

Given: $n = 2$ moles
$T_i = 300°C = 573$ K
$T_f = 100°C = 373$ K
$MC_v = 12.61$ J/K (Table 22.2)

Required: ΔS

186   Chapter 24

Assuming that the helium behaves as an ideal gas, the entropy change in a constant volume process that takes the gas from temperature T_i to temperature T_f is given by Eq. 24.28:

$$\Delta S = mC_v \ln(T_f/T_i) \tag{1}$$

where m is the mass and C_v the specific heat capacity at constant volume.

The mass can be expressed as the product of the number of moles n and the mass per mole, or molecular weight M: m = nM, so that Eq. (1) becomes:

$$\Delta S = nMC_v \ln(T_f/T_i)$$

where the quantity (MC_v) is the molar heat capacity.

Substituting into this equation for the number of moles n, the molar heat capacity (MC_v), and the initial and final Kelvin temperatures T_i and T_f:

$$\Delta S = (2)(12.61 \text{ J/K}) \ln(373/573) = \underline{-10.8 \text{ J/K}}$$

The entropy of the sample of helium decreases in this cooling process. According to the entropy principle, the entropy of an isolated system cannot decrease, thus, we can conclude that the helium is not an isolated system.

Example 4

Two masses of water, 1 kg at 80°C and 3 kg at 40°C, are mixed in a cup having negligible specific heat capacity, and the mixture comes to a final temperature of 50°C. What is the entropy change of a) the cooler water and   b) the warmer water?   c) What is the entropy change of the system consisting of both masses of water?

Solution

Given: T_w = 80°C = 353 K
          T_c = 40°C = 313 K
          T_f = 50°C = 323 K
          C = 4.19 kJ/kg-C°   (Table 21.1)

Required: a) ΔS_c ,   b) ΔS_w ,   c) ΔS

a) The final temperature of 50°C assumes that the heat lost by the warmer water is gained by the cooler water, therefore, the system consisting of both masses of water is an isolated system.

The entropy change of a liquid in an isobaric process is given by Eq. 24.22:

$$\Delta S = mC_p \ln(T_f/T_i)$$

Substituting into this equation for the specific heat capacity of water C, the final Kelvin temperature T_f, and the mass m and initial Kelvin temperature T_c of the cooler water:

$$\Delta S_c = (3 \text{ kg})(4.19 \text{ kJ/kg-C°}) \ln(323 \text{ K}/313 \text{ K}) = \underline{0.395 \text{ kJ/K}}$$

b) Substituting the corresponding quantities for the warmer water yields:

$$\Delta S_w = (1 \text{ kg})(4.19 \text{ kJ/kg-C°}) \ln(323 \text{ K}/353 \text{ K}) = \underline{-0.372 \text{ kJ/K}}$$

c) The entropy change of the system consisting of both masses of water is obtained by adding the results of a) and b):

$$\Delta S = \Delta S_c + \Delta S_w = (0.395 - 0.372) \text{ kJ/K} = \underline{0.023 \text{ kJ/K}}$$

The increase in entropy of the cooler water is greater than the decrease in entropy of the warmer water, thus, the entropy change of the isolated system consisting of both masses of water is positive. This result is in agreement with the prediction of the entropy principle for this irreversible mixing process.

PROBLEMS

1. Find a) the thermodynamic efficiency and b) the maximum efficiency for a heat engine that absorbs 60 kJ of heat from a reservoir at a temperature of 177°C and ejects 45 kJ of heat to a reservoir at a temperature of 27°C.

2. What is the change in entropy of 10 kg of water that freezes at a pressure of 1 atm?

3. What is the entropy change of one mole of an ideal gas that is heated at constant volume from a temperature of 173°C to a temperature of 273°C? The molar heat capacity of an ideal gas is 3R/2.

4. A mass of 2 kg of water at 38°C absorbs heat at 1 atm from a reservoir at 50°C until its temperature reaches 42°C. Find a) the entropy change of the water, b) the entropy change of the reservoir, and c) the total entropy change for the process.

Chapter 25
Kinetic Theory

PREVIEW

This chapter relates the measurable quantities of thermodynamics, such as temperature and pressure, to the microscopic properties of atoms and molecules, which are not directly measurable. After some discussion of the elements of the atomic model, the chapter develops the ideal gas equation of state and a kinetic interpretation of temperature. The chapter also introduces the concept of a distribution function and discusses the Maxwellian distribution of molecular speeds for a gas in thermal equilibrium.

SUMMARY

25.1 THE ATOMIC MODEL OF MATTER

> Definition. Kinetic theory: a theory based on the atomic model that uses statistical methods to relate the average microscopic properties of atoms and molecules to the behavior of matter on the macroscopic level.

The quantity of a substance is related to the number of atoms or molecules it contains by the mole (mol), the fifth fundamental quantity of the SI system of units we have considered.

> Definition. Mole: the amount of a substance that contains a number of atoms or molecules equal to the number of atoms in 12 gm of carbon 12, or Avogadro's number.

> Definition. Avogadro's number: the number of atoms in 12 gm of carbon 12, given by $N_A = 6.022169 \times 10^{23}$ (Eq. 25.1)

> Definition. Gram molecular weight: the mass (in gm) of one mole of a substance on a scale on which the gram molecular weight of carbon 12 is 12 gm.

The gram molecular weight is denoted by M. For molecules, M is determined by summing the gram molecular weights of the constituent atoms, which are usually given as the average elemental atomic masses in atomic weight tables. For example, using the atomic masses for hydrogen and oxygen of 1.008 and 15.999, we obtain a gram molecular weight for water (H_2O) of 18.015. Many of the average elemental atomic masses are nearly integral, therefore, many gram molecular weights are well approximated by an integral number of grams.

> Definition. Atomic mass: the mass of a neutral atom, usually expressed on a scale on which the carbon 12 atom is assigned a mass of exactly 12 atomic mass units (u).

The definitions of gram molecular weight and atomic mass indicate that the gram molecular weight of monatomic molecules is numerically equal to the atomic weight of the atoms that constitute the molecules and that the absolute mass of an atomic mass unit is given by

1 u = 1.66053 × 10^{-27} kg.

A reliable estimate of atomic size based on the observation that solids and liquids are nearly incompressible can be made by assuming that individual molecules occupy cubic volumes with no empty space between them. The edge of these cubic volumes is given by

$$d = (M/N_A \rho)^{1/3} \tag{25.2}$$

The length d is therefore a characteristic atomic and molecular dimension and is found to lie between 2-5 angstroms (Å) for a wide variety of solids and liquids (Table 25.1).

25.2 MEAN FREE PATH AND CROSS SECTION

> **Definition.** <u>Mean free path</u>: the average distance traveled by the atoms of a gas between randomly occurring collisions.

> **Definition.** <u>Cross section</u>: a circular area associated with each atom of a gas whose radius is a measure of the range of interatomic forces.

The cross section $\sigma = \pi R_0^2$ (Eq. 25.3) is interpreted to be an area presented by each atom of a gas to other atoms, such that a collision occurs if another atom intercepts the area.

A fundamental result of kinetic theory relating the mean free path λ and the cross section σ to the number density n, or the number of atoms per unit volume, is given by $n\sigma\lambda = 1$ (Eq. 25.6). Calculations using this expression show that the atoms in a gas spend most of their time in free flight.

25.3 THE IDEAL GAS: KINETIC INTERPRETATION OF TEMPERATURE

In our previous discussion of the ideal gas model, we considered the ideal gas equation of state: PV = nRT (Eq. 22.9). An alternate form is given by PV = NkT (Eq. 25.11), where $k = R/N_A = 1.38062 \times 10^{-23}$ J/K (Eq. 25.10) is called the Boltzmann constant.

The ideal gas equation of state can be derived in kinetic theory on the basis of an atomic model of an ideal gas. This model assumes that the gas consists of a large number of non-interacting point particles that move about randomly inside a container and collide elastically with the walls. This means that the atoms possess only translational kinetic energy and linear momentum. Since only center-of-mass motion is considered, the derivation is valid for any type of molecule. In this model, collisions of the atoms with the container walls give rise to a force and thus to a pressure. The force is obtained using Newton's second law by calculating the average momentum transferred to a wall per second. The force divided by the area of the wall is the gas pressure.

The result of the kinetic theory calculation can be compared with the ideal gas law:

$$PV = 2N\langle K \rangle/3 = NkT \tag{25.25}$$

where $\langle K \rangle = m\langle v^2 \rangle/2$ is the average kinetic energy per atom.

The conclusion is that the temperature of an ideal gas is proportional to the average kinetic energy of the random motions of its molecules. This interpretation is valid only for temperatures on the Kelvin temperature scale:

$$\langle K \rangle = 3kT/2 \tag{25.26}$$

The square root of $\langle v^2 \rangle$ is called the root-mean-square (rms) speed and is given by

$$v_{rms} = \sqrt{3kT/m} \tag{25.28}$$

This equation indicates that the rms speed of a gas molecule increases with increasing temperature and decreasing mass. Evaluation of the speeds of air molecules at room temperature reveals that they are moving rapidly at speeds comparable with the speed of sound, for example, v_{rms} = 517 m/s for nitrogen molecules at 300 K. In general, the rms speed is somewhat larger than the average speed.

25.4 THE DISTRIBUTION OF MOLECULAR SPEEDS

There is a great variation in the speeds of the molecules in a gas, and the distribution of speeds is described by a quantity called a distribution function. The distribution function $f(v)$ is defined by

$$dN = f(v) \, dv \tag{25.37}$$

where dN is the number of molecules with speeds in the range v to $v + dv$. On a graph of $f(v)$ vs v, dN is the area under the $f(v)$ curve between v and $v + dv$.

The number of molecules with speeds in the range from v_1 to v_2 is the integral of Eq. 25.37 and corresponds to the area under the $f(v)$ curve between v_1 and v_2. The total number of molecules in the gas is the total area:

$$N = \int_0^\infty f(v) \, dv \tag{25.39}$$

Another use of the distribution function is to calculate the average value $\langle O \rangle$ of a quantity $O(v)$ which is a function of v:

$$\langle O \rangle = (1/N) \int_0^\infty O \, f(v) \, dv$$

This expression can be used, for example, to calculate the rms speed with $O = v^2$ and the average speed with $O = v$ if the distribution function is known.

The Maxwellian distribution function is the distribution function that describes a gas of N molecules in thermal equilibrium at temperature T:

$$f(v) = 4\pi N (m/2\pi kt)^{3/2} \, v^2 e^{-(mv^2/2kT)} \tag{25.40}$$

Using the Maxwellian distribution function, we can calculate the three speeds that characterize the distribution of molecular speeds in a gas in thermal equilibrium: the rms speed, the average speed, and the most probable speed, or the speed for which $f(v)$ is a maximum. The most probable speed is given by $v_{mp} = \sqrt{2kT/m}$ and the average speed lies between this value and $v_{rms} = \sqrt{3kT/m}$.

EXAMPLE PROBLEMS
Example 1

What is a) the gram molecular weight and b) the molecular mass of ethyl alcohol (C_2H_5OH), assuming the following atomic weights: hydrogen (1.008), carbon (12.011), oxygen (15.999)?

Solution

Given:

Element	Atomic Weight
H	1.008
C	12.011
O	15.999

$$N_A = 6.022 \times 10^{23} \quad \text{(Eq. 25.1)}$$

Required: a) M, b) m

a) The given atomic weights are average atomic masses such as might be found in an atomic mass table. They are summed for all atoms in a molecule to obtain the gram molecular weight in grams. For the ethyl alcohol molecule (C_2H_5OH):

$$\begin{aligned} 6H &= 6(1.008) = 6.048 \\ 2C &= 2(12.011) = 24.022 \\ O &= 15.999 \\ \hline C_2H_5OH &= 46.069 \end{aligned}$$

The gram molecular weight of ethyl alcohol is M = __46.1 gm__.

b) The gram molecular weight is the mass of one mole, which contains Avogadro's number of molecules. The mass of one molecule is therefore the gram molecular weight divided by Avogadro's number. For ethyl alcohol:

$$m = M/N_A = 46.069 \text{ gm}/6.022 \times 10^{23} = 7.650 \times 10^{-23} \text{ gm}$$
$$m = \underline{7.65 \times 10^{-26} \text{ kg}}$$

Example 2

Find a) the number density and b) the mean free path for oxygen molecules at a temperature of 27°C and a pressure of 1/5 atm, assuming an effective molecular radius of 3 Å.

Solution

Given:
$$T = 27°C = 300 \text{ K}$$
$$P = 1/5 \text{ atm} = (1/5)(1.01 \times 10^5 \text{ Pa})$$
$$R_0 = 3 \text{ Å} = 3 \times 10^{-10} \text{ m}$$
$$k = 1.38 \times 10^{-23} \text{ J/K} \quad \text{(Eq. 25.10)}$$

Required: a) n, b) λ

a) Assuming an ideal gas, we can obtain the number density from the equation of state in the form $PV = NkT$ (Eq. 25.11):

$$n = N/V = P/kT \tag{1}$$

Substituting in Eq. (1) for the pressure P, the temperature T in kelvins, and Boltzmann's constant k:

$$n = (1/5)(1.01 \times 10^5 \text{ Pa})/(1.38 \times 10^{-23})(300 \text{ K}) = \underline{4.88 \times 10^{24}/m^3}$$

b) The mean free path can be obtained from the expression relating it to the number density and the cross section (Eq. 25.6):

$$n\sigma\lambda = 1$$

Solving this equation for the mean free path and substituting for the number density from Eq. (1) and the cross section $\sigma = \pi R_0^2$ (Eq. 25.3):

$$\lambda = 1/n\sigma = kT/\pi R_0^2 P$$

Substituting in this equation for the Kelvin temperature T, Boltzmann's constant k, the effective molecular radius R_0, and the pressure P:

192   Chapter 25

$$= (1.38 \times 10^{-23} \text{ J/K})(300 \text{ K})/\pi(3 \times 10^{-10} \text{ m})^2(1/5)(1.01 \times 10^5 \text{ Pa})$$
$$= \underline{7.23 \times 10^{-7} \text{ m}}$$

Thus, we see that at a pressure of 1/5 atmosphere and temperature of 300 K, the mean free path of a typical air molecule is more than 2400 times as large as the effective molecular radius assumed.

Example 3

Find a) the total kinetic energy of random motion of 6 moles of helium confined to a volume of 0.5 m^3 at a pressure of 1 atm and b) the average kinetic energy of a helium molecule.

Solution

Given: $V = 0.5$ m^3
$P = 1$ atm $= 1.01 \times 10^5$ Pa
$R = 8.314$ J/K   (Eq. 22.10)

Required: a) K,   b) <K>

a) Using Eq. 25.26, we can obtain the total random kinetic energy as

$$K = N<K> = 3NkT/2 \qquad (1)$$

The ideal gas law (Eqs. 25.7 and 25.11) is given by

$$PV = nRT = NkT \qquad (2)$$

and can be used to express Eq. (1) as

$$K = 3nRT/2 \qquad (3)$$

Solving Eq. (2) for the Kelvin temperature

$$T = PV/nR$$

Substituting in this equation for the pressure P, volume V, number of moles n, and gas constant R

$$T = (1.01 \times 10^5 \text{ Pa})(0.5 \text{ m}^3)/(6)(8.314 \text{ J/K}) = 1012 \text{ K}$$

Substituting this result for the Kelvin temperature T, the number of moles n, and the gas constant R into Eq. (3):

$$K = (3)(6)(8.314 \text{ J/K})(1012 \text{ K})/2 = \underline{7.58 \times 10^4 \text{ J}}$$

b) The average kinetic energy per molecule depends only on the Kelvin temperature and is given by Eq. (1) as

$$<K> = 3kT/2$$

Substituting in this equation for the Kelvin temperature T and Boltzmann's constant k:

$$<K> = (3)(1.38 \times 10^{-23} \text{ J/K})(1012 \text{ K})/2 = \underline{2.09 \times 10^{-20} \text{ J}}$$

In an alternate method of solution, we use Eq. (2) to express Eq. (1) as $K = 3PV/2$ and find b) by dividing the result of a) by the total number of molecules in 6 moles, which is $6N_A$.

Example 4

For oxygen (O$_2$) molecules (M = 32 gm) at 27°C, find a) the rms speed, b) the most probable speed, and c) the temperature at which the rms speed is doubled.

Solution

Given: M = 32 gm = 0.032 kg

$T = 27°C = 300$ K
$R = 8.314$ J/K

Required: a) v_{rms}, b) v_{mp}, c) T_2

a) The rms speed is given by Eq. 25.28:

$$v_{rms} = \sqrt{3kT/m}$$

Expressing the mass of a molecule as $m = M/N_A$ and Boltzmann's constant as $k = R/N_A$, this equation becomes

$$v_{rms} = \sqrt{3RT/M}$$

Substituting in this equation for the Kelvin temperature T, the gram molecular weight expressed in kg, and the gas constant R:

$$v_{rms} = \sqrt{(3)(8.314 \text{ J/K})(300 \text{ K})/0.032 \text{ kg}} = \underline{484 \text{ m/s}}$$

b) Assuming thermal equilibrium, we can obtain the ratio of most probable speed to rms speed using Eq. 25.28 and Eq. 25.42

$$v_{mp}/v_{rms} = \sqrt{(2kT/m)/(3kT/m)} = \sqrt{2/3} = 0.816$$
$$v_{mp} = 0.816 \, v_{rms}$$

Substituting in this equation for v_{rms} from a):

$$v_{mp} = (0.816)(484 \text{ m/s}) = \underline{395 \text{ m/s}}$$

c) If T_1 is the temperature corresponding to a given rms speed and T_2 is the temperature at which the speed is multiplied by a factor of X

$$X\sqrt{3kT_1/m} = \sqrt{3kX^2T_1/m} = \sqrt{3kT_2/m}$$
$$T_2 = X^2 T_1$$

Here, $X = 2$, therefore

$$T_2 = (2)^2 T_1 = 4T_1 = 4(300 \text{ K}) = 1200 \text{ K} = \underline{927 \text{ C}}$$

The relationship between the rms speed and temperature is valid only for Kelvin temperatures and indicates that the speed increases as the square root of the temperature. The effects of not using Kelvin temperatures are illustrated by the ratios of the final and initial temperatures in this example: the ratio of the Kelvin temperatures is 4, while the ratio of the Celsius temperatures is 927/27 = 34.

PROBLEMS

1. For carbon dioxide (CO_2), find a) the gram molecular weight and b) the mass of one molecule, assuming that the atomic weights of carbon and oxygen are 12.011 and 15.999, respectively.

2. Find a) the number density and b) the gas pressure for air molecules at a temperature of 200 K, assuming an effective molecular radius of 2.8 Å and a mean free path of 10^5 m.

3. What is a) the total random kinetic energy of one mole of helium at 5000 K and b) the average kinetic energy per molecule of helium? c) At what temperature would the average kinetic energy increase by 50 percent?

4. For neon (Ne) molecules (M = 20.2 gm) at 19°C, what is a) the rms speed and b) the most probable speed? c) At what Celsius temperature will the rms speed be cut in half?

Chapter 26
Electric Charge

PREVIEW

In this chapter, a brief review of the development of mankind's knowledge and understanding of the natural phenomena associated with electricity is presented. The concept of electric charge is described as a fundamental property of matter, and the interaction between charged particles at rest is described quantitatively in the form of Coulomb's Law which is accepted as one of the fundamental laws of physics. Application of Coulomb's Law to extended objects is presented in terms of the Principle of Superposition.

SUMMARY

26.1 DISCOVERY OF ELECTRICITY

The phenomenon of electricity has been known since the earliest times, but it was not until the beginning of the seventeenth century that any scientific, quantitative investigation of electricity was undertaken. This study was actually a "spin-off" of William Gilbert's intensive study of the phenomenon of magnetism which was undertaken to improve the navigational capabilities of the British navy and Gilbert certainly recognized the similarities between the two phenomena, but it was more than two centuries later during the first half of the nineteenth century that the connection between electricity and magnetism was fully elucidated.

26.2 ELECTRIC CHARGE

We presently recognize electric charge as one of the several fundamental properties of matter in much the same context as we recognize mass as a fundamental property. We can no more explain why the elementary particles which constitute bulk matter, such as the electron and the proton, bear an electric charge than we can explain why these particles have mass. Unlike mass, charge is recognized as existing in two forms which we call (using Benjamin Franklin's convention) positive and negative. These names have no intrinsic meaning except to imply an "oppositeness"; they do, however aid in applications in which it is necessary to add or subtract charges.

Definition. Electric Charge: a fundamental property of matter by virtue of which two charged particles or objects interact with each other without physical contact.

Charge is quantized. There exists a certain minimum amount of electric charge and all charges which appear on either elementary particles or extended objects must be integral multiples of this charge. The smallest charge which has been isolated so far is the charge on an electron which is negative and has the magnitude,

$$e = 1.602 \times 10^{-19} \text{ coulombs.}$$

The proton has a positive charge of the same magnitude. It should be noted that theorists have postulated an even smaller charge (having a magnitude 1/3 this value) but it has not been observed experimentally and it is not clear that such a charge could exist on an isolated particle.

>Definition. Quantization: a quantity which is quantized can exist only as an integral multiple of a specific minimum magnitude.

LAW OF CONSERVATION OF CHARGE Charge, like mass, is conserved. Charge may be neither created nor destroyed.

An important difference between charge and mass is that charge can exist in either the positive of negative form, and thus while the total amount of charge in the universe must remain constant, charge can be created or destroyed provided that equal amounts of positive and negative charge are created or destroyed simultaneously. An important example of this aspect of conservation of charge is the creation or annihilation interaction between the negatively charged electron and the positively charged positron. While not directly related to the electrical characteristics of the particles, an absorption or release of energy must accompany these interactions since there is a simultaneous creation or annihilation of mass.

When the Law of Conservation of Charge is applied to extended objects which bear more than a single quantum of charge, it requires that charge be neither created nor destroyed, but does allow charge to be transferred from one object to another or from one position to another on an object. Our concept of the nature of electric charge then implies that it is actually one of the elementary particles, almost always an electron, which is actually being transferred.

26.3 COULOMB'S LAW

The interaction between two charged particles was first measured experimentally by Charles Augustin Coulomb and is given quantitatively by the law that bears his name. If two charged particles are separated from one another by a distance r, each of these charges exerts a force on the other which is given by

$$f_c = k_e \frac{q_1 q_2}{r^2} \tag{26.2}$$

If the two charges are alike in sign, the forces are repulsive, while if the charges are of different sign, the forces are attractive. The value of the proportionality constant is determined experimentally and has been found to be

$$k_e = 8.98755 \times 10^9 \text{N m}^2/\text{C}$$

in the m.k.s. system of units. Note that the constant is not dimensionless and therefore would have a different value in a different system of units.

26.4 SUPERPOSITION

If a certain application requires that one charged particle interact with two or more other such particles, the resultant interaction (force) can be calculated by calculating the interaction with each of the other particles individually and then adding these forces together according to the rules of vector addition.

Since an extended object may be treated as a collection of a very large number of particles of infinitesimal size, the Principle of Superposition may be extended to such objects by substituting the process of vector integration for that of vector addition.

EXAMPLE PROBLEMS

Example 1

Two small spheres (they may be considered particles) are suspended as shown in FIG. 1-1 with the first one hanging from the ceiling by a 5 cm length of fine silk thread and the second one hanging from an additional 2 cm length of thread attached to the first. Each has a mass of 10 gm and bears a positive charge of 0.2 microcoulombs. What is the tension in each section of the thread?

Solution

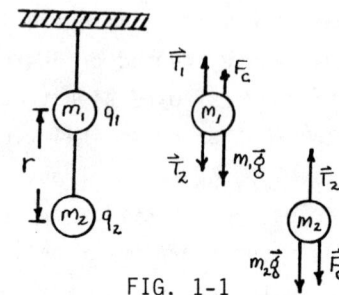

FIG. 1-1

Given: $q_1 = q_2 = +0.2 \times 10^{-6}$ coulombs
$m_1 = m_2 = .01$ Kgm
$r = .02$ m

Required: T_1; T_2

The upper sphere (1) is acted upon by four forces: its weight acting downward, the tension in the upper thread acting upward, the tension in the lower thread acting downward, and the coulomb force exerted on it by the lower sphere (2) acting upward. Thus

$$T_1 - T_2 - m_1 g + F_c = 0 \qquad (1)$$

The lower sphere (2) is acted upon by three forces: its weight acting downward, the tension in the lower thread acting upward, and the coulomb force exerted on it by the upper sphere acting downward. Thus

$$T_2 - m_2 g - F_c = 0 \qquad (2)$$

Adding Eqs. (1) and (2) together yields

$$T_1 = (m_1 + m_2)g$$

which is not surprising. Eq. (2) gives

$$T_2 = m_2 g + k \frac{q_1 q_2}{r^2}$$

Substituting the appropriate numbers into these equations yields a tension of about .2 newtons (.045 lb) for the upper thread and about 1.0 newtons (.22 lb) for the lower thread.

Example 2

How many electrons must be added to or removed from (which?) each of the spheres in Problem 1 to achieve the state of charge specified there?

Solution

Since the electron bears a negative charge, electrons must be removed from the spheres to give them the specified positive charge. The number of electrons may be calculated from

$$\text{Number of electrons} = \frac{\text{Charge on the sphere}}{\text{Charge per electron}}$$

giving a total of 1.25×10^{12} electrons on each sphere.

Example 3

FIG. 3-1 shows three small spheres are arranged at the corners of a right triangle so that a positive charge of 1.0 microcoulomb is at the junction of the two perpendicular sides and negative charges of 0.1 microcoulomb each are at the other two corners. If the two perpendicular sides of the triangle are each 10 cm long, what is the magnitude and direction of the net force on the sphere bearing the 1.0 microcoulomb charge?

Solution

Given: $q_1 = q_2 = 0.1 \times 10^{-6}$ coulombs
$q_3 = 1.0 \times 10^{-6}$ coulombs
$r = 0.10$ m

Required: F

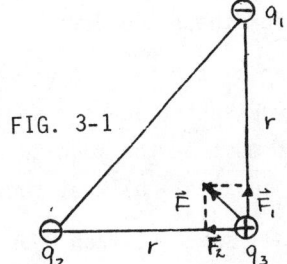

FIG. 3-1

Since the two negative charges are equidistant from the positive charge, the forces they exert on it will be equal in magnitude, attractive, and at right angles to each other. The magnitude of the force on the sphere bearing the positive charge may be calculated by use of the Pythagorean Theorem.

$$F = (F_1^2 + F_2^2)^{1/2}$$

with the two component forces being equal and calculated from Coulomb's Law. The direction of this force will be toward and perpendicular to the line joining the two negatively charged spheres.

$$F_1 = F_2 = k_3 \frac{q_2 q_3}{r^2}$$

Substitution of the appropriate numerical values into Coulomb's Law gives for the net force,

$$F = .13 \text{ N}$$

Example 4

A positive charge of 50 nanocoulombs is uniformly distributed over a thin semicircular ring of radius 15 cm. (See FIG. 4-1) What force will this ring exert on an electron placed at the center of the semicircular arc?

Solution

Given: $Q = 50 \times 10^{-9}$ coulombs
$q = e = -1.602 \times 10^{-19}$ coulombs
$R = .15$ m

Required: F

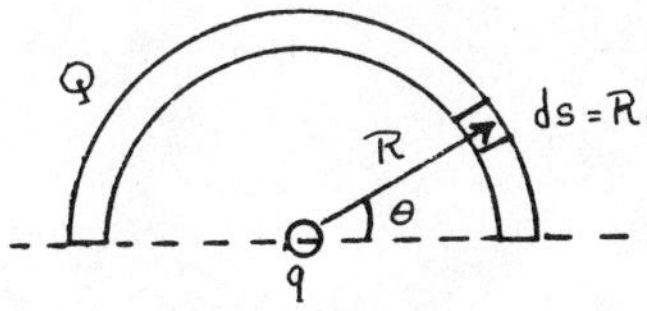

FIG. 4-1

The amount of charge, dQ, on the element of the ring, ds, is

$$dQ = \frac{Q}{\pi R} ds = \frac{Q}{\pi} d\theta$$

from which the force components may be calculated by integrating over the whole half ring.

$$F_x = \frac{k_e qQ}{\pi R^2} \int_0^\pi \cos\theta \, d\theta = 0$$

and

$$F_y = \frac{k_e qQ}{\pi R^2} \int_0^\pi \sin\theta \, d\theta = 0$$

Substituting numerical values into this equation gives
$$F_x = 0; \quad F_y = 20.4 \times 10^{-16} \text{ newtons}$$
Since there is no X component of force, the total force is the same as the Y component and is directed toward the center of the arc.

PROBLEMS

1. Two electrons repel each other because of their identical electric charges; they also exert gravitational attractive forces on each other. How massive would the electrons have to be in order for these two forces to be equal and thus cancel?

2. Consider an arrangement of charged spheres identical to that in Example Problem 1 except that the lower sphere bears a negative unknown charge.
   a) What must be the magnitude of this negative charge in order that the tension in the lower thread be zero? b) What then will be the tension in the upper thread?

3. Two small spheres, each bearing a positive charge Q, are fixed a distance R apart in space. A small negatively charged sphere is constrained to move on the perpendicular bisector of the line joining the two positive charges but is free to move along that bisector. Show that for small displacements from the equilibrium position, the motion of the negative charge is simple harmonic motion.

4. The midpoint of the line joining two like charges such as described in Problem 3 is a position of equilibrium for either a positive or a negative charge placed there. Describe this equilibrium in terms of stability for small displacements both along and perpendicular to the line joining the two charges for both positive and negative charges placed at the equilibrium position.

5. A thin plastic rod 20 cm long and bearing a negative charge of 5 microcoulombs is bent into an arc so that it forms one quadrant of a circle. A small glass sphere bearing a positive charge of 5 nanocoulombs is placed at the center of curvature of the arc. What force (magnitude and direction) is exerted on the glass sphere?

Chapter 27
Electric Field and Gauss' Law

PREVIEW

The concept of <u>Electric Field</u> is introduced as a useful intermediate step in determining the interaction between a charged object and a neighboring distribution of charges. The electric field is often described in terms of <u>electric field lines</u> which are an artifice often useful in visualizing the electric field in a semi-quantitative way. Gauss' Law is derived from Coulomb's Law as a tool for determining the electric field produced by a given distribution of charge, and is found to be particularly useful for charge distributions which have a high degree of symmetry.

SUMMARY

27.1 ELECTRIC FIELD

The electric field is a vector quantity and its magnitude and direction are defined operationally as follows:

<u>Definition</u>. Electric <u>Field</u>: the electric field at a point in space is the electric force per unit charge on a positive test charge placed at this point. The direction of the field is the same as the direction of the force on the test charge.

Symbolically, $E = \dfrac{F}{q_0}$ newtons/coulomb (27.1)

where F is the (vector) force on the test charge of magnitude q_0.

The electric field produced by a single point charge of magnitude Q is easily determined by use of Eq. (27.4). Clearly, the field produced by a positive point charge is radially <u>outward</u> from the point charge while that produced by a negative point charge is radially <u>inward</u> toward the point charge.

$$E = k \dfrac{Q}{r^2} \hat{r} \text{ newtons per coulomb}$$

It will turn out to be convenient later, particularly in applications of Gauss' Law, if we replace the constant of proportionality in this equation with a different set of constants; thus

$$k_e = \dfrac{1}{4\pi\varepsilon_0}$$

where $\varepsilon_0 = 8.85 \times 10^{-12}$ C^2/Nm^2.

27.2 ELECTRIC FIELD LINES

The concept of the electric field line was developed by Michael Faraday to assist in visualizing the electric field in the space surrounding a distribution of charges.

> Definition. Electric Field Lines: an imaginary line in space, often called a line of force, which originates on a positive charge, terminates on a negative charge (or at infinity) and is everywhere along its length tangent to the direction of the electric field.

Comparing this definition with the definition of the electric field given above, we can see that each electric field line must originate on a positive charge and terminate on a negative charge, and that two field lines may never intersect each other.

> DENSITY OF FIELD LINES: We may make the concept quantitative by requiring that the number of lines passing normally through a unit area be equal to the magnitude of the field at that point.

This requirement sometimes causes some confusion when first encountered because of the possibility of a "fractional" field line, but the confusion disappears when one remembers that these lines are not real, but an artifice developed to assist in visualization.

27.3 ELECTRIC FLUX

In applications involving the electric field, it is often necessary to consider the effect of the field over an extended region of space rather than just at a point. For these purposes, we introduce the concept of electric flux.

> Definition. Electric Flux: the electric flux passing through a given surface area is the surface integral of the normal component of the electric field integrated over that area. It may also be interpreted as the number of lines of force passing through that area.

As we shall see below, the concept of electric flux also makes the statement of Gauss' Law much more concise than it might otherwise be. Symbolically, the electric flux may be defined as

$$\Phi_E = \int E \cdot \hat{n} \, dA \tag{27.8}$$

27.4 GAUSS' LAW

Gauss' Law states that the surface integral of the normal component of the electric field integrated over any closed surface is equal to the total charge contained within that surface divided by the permativity of free space. Symbollically, the law may be written as:

$$\int E \cdot \hat{n} \, dA = q/\varepsilon_0 \tag{27.15}$$

27.5 APPLICATIONS OF GAUSS' LAW

Gauss' Law is particularly useful in working problems involving the calculation of electric fields in those cases in which a sufficiently high degree of symmetry exists to allow easy evaluation of the integral. This usefulness will be illustrated in the examples below.

27.6 MOTION OF POINT CHARGES IN A STATIC ELECTRIC FIELD

As we have seen above, an electrically charged particle situated in an electric field experiences a force in (or opposite to) the direction of the field. According to Newton's Second Law, such a force, unless balanced by other forces, will cause the particle to be accelerated in the direction of the force.

$$F = qE = ma \longrightarrow a = E(q/m) \tag{27.19}$$

Note that unlike the case of the gravitational force, this force is not proportional to the mass of the particle and the acceleration is not the same for every particle but rather depends upon both the mass of the particle and its charge. Except for this important difference, problems involving the motion of a charged particle may be solved in exactly the same manner as problems involving any other type of force.

27.7 ELECTRIC DIPOLE IN AN ELECTRIC FIELD

<u>Definition</u>. <u>Electric Dipole</u>: two equal charges of opposite sign separated from one another by a distance ℓ.

<u>Definition</u>. <u>Dipole Moment</u>: the dipole moment of a dipole consisting of two charges of equal magnitude, q, and opposite sign separated by a distance ℓ is a vector of magnitude ql directed from the negative toward the positive charge.

This is an important configuration since many molecules may be treated as dipoles and their motion in an electric field bears directly upon some of the important characteristics of matter composed of such molecules.

Since the net charge on the dipole is zero, it is clear that in a <u>uniform</u> electric field, the dipole will experience no net force. However, in a non-uniform field, the forces acting on the two charges will not be of the same magnitude and thus will not cancel. The dipole situated in a non-uniform field will experience a force which is directly proportional to the dipole moment and to the space rate of change of the field. The direction of this force will be the direction in which the strength of the field is most rapidly increasing and may be calculated by

$$F = q\ell \frac{dE}{dx} \tag{27.26}$$

Although a dipole will experience no net force in a uniform electric field, it will experience a torque in either a uniform or non-uniform field. This torque will be in a direction which will tend to align the dipole moment parallel to the field and may be calculated by

$$\tau = q\ell E \sin\theta \tag{27.24}$$

EXAMPLE PROBLEMS
Example 1

Two small spheres (they may be considered particles) are suspended with the first one hanging from the ceiling by a 50 cm length of fine silk thread and the second one hanging from an additional 20 cm length of thread attached to the first. Each has a mass of 10 gm ; the upper sphere has a positive charge of 20 microcoulombs and the second has a negative charge of the same magnitude. A uniform horizontal electric field of 500 N/C exists in the space occupied by the two spheres. Calculate the angle between each of the thread segments and the vertical.

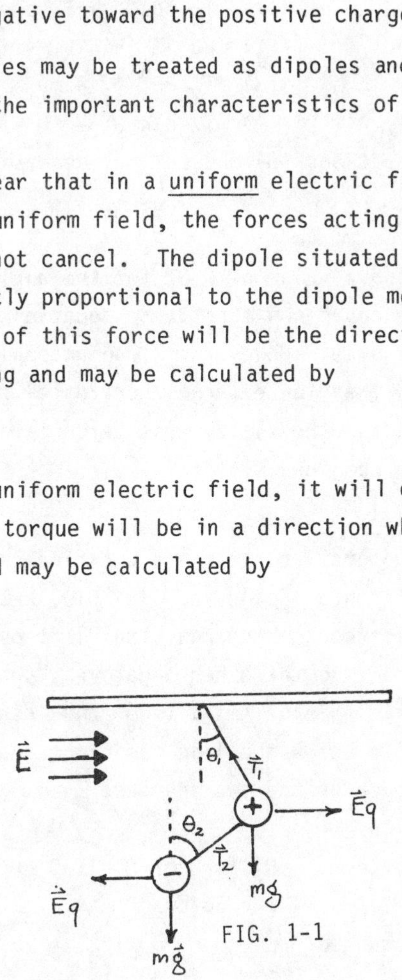

FIG. 1-1

Solution

Given:
$m_1 = m_2 = m = 10$ gm
$q_1 = -q_2 = 20$ C
$E = 500$ N/C

Required: θ_1 ; θ_2

The diagram in FIG. 1-1 shows the physical arrangement of the two spheres as well as the forces acting on each of them. Since the system is clearly in equilibrium, Newton's first law applies (both to the system as a whole and to each sphere individually). Thus, we may begin by writing the (four) equations representing the first condition for equilibrium for each of the two objects.

(1) $T_1 \cos\theta_1 - T_2 \cos\theta_2 - mg = 0$
(2) $Eq - T_1 \sin\theta_1 - T_1 \sin\theta_2 = 0$
(3) $T_2 \cos\theta_2 - mg = 0$
(4) $T_2 \sin\theta_2 - Eq = 0$

Equations (3) and (4) may be easily solved to yield

$$\tan\theta_2 = \frac{Eq}{mg}$$

yielding a value of 5.8° for the angle.

We could now substitute equations (3) and (4) into equations (1) and (2) and solve for the other angle. However, we can accomplish this end by looking at the two spheres as "a system" and noting that the external horizontal forces on the system add to zero and thus, the thread suspending the system must be vertical, and the tension in this thread must be equal to the weight of the combined system.

Example 2

Two small objects are positioned 50 cm apart along a horizontal line as in FIG. 2-1. Each bears a 100 microcoulomb charge, the leftmost being positively charged and the other negatively charged. Calculate the electric field (magnitude and direction) at a point which is below the line joining the charges and 50 cm from each of the two charges.

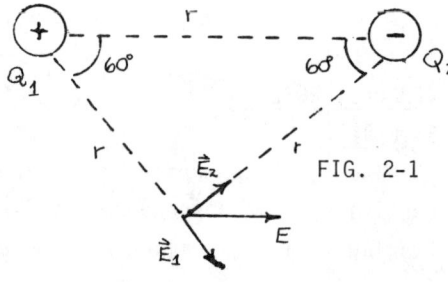

FIG. 2-1

Solution

Given:
$Q_1 = -Q_2 = Q = 100$ μC
$r = 50$ cm

Required: E

At the point of interest, the positive charge produces a field directed away from itself at an angle of 60° from the horizontal while the negative charge produces a field directed toward itself also at an angle of 60° from the horizontal. Using the principle of superposition to combine these two fields, we find that the vertical components of the two fields are equal in magnitude and opposite in direction,

$E_{1y} = -E_{2y}$ from which
$E_y = 0$,

while the horizontal components are also equal in magnitude but in the same direction.

$$E_{1x} = E_{2x} = \frac{1}{4\pi\epsilon_o} \frac{Q}{r^2} \cos 60°$$

Thus, the combined field is horizontal and has a magnitude of

$$E = 6.23 \times 10^6 \text{ N/C}$$

Example 3

A small object bearing a positive charge of 150 microcoulombs is placed 30 centimeters from another small object bearing a negative charge of 300 microcoulombs. Calculate the solid angle which contains all of those field lines which originate on the positively charged object and terminate on the negatively charged object. Sketch the pattern of field lines both near to and far from the two objects.

Solution

In the immediate vicinity of each of the charges (so close that the field lines are unaffected by the presence of the other charge) the field lines are radial and are uniformly distributed throughout the total solid angle surrounding the object. Since the negative charge is twice as large as the positive charge, only half of the field lines which terminate on the negative charge may originate on the positive charge. Thus, all of the lines which originate on the positive charge will be bent around so that they terminate on the negative charge and will approach the negative charge within the 6.28 steradians of solid angle around the line joining the two charges. The remaining half of the field lines terminating on the negative charge must come from infinity and at a distance large compared to the separation of the two charges will have the same appearance as if the two charges were coincident, i.e. as if they were terminating on a single point charge of -150 microcoulombs.

Example 4

A very large thin slab of non-conducting material (See FIG. 4-1) bears a positive charge distributed uniformly throughout its volume with a density of .01 coulombs per cubic meter. Use Gauss' Law to describe the electric field both inside and outside the material at positions not too near the edges.

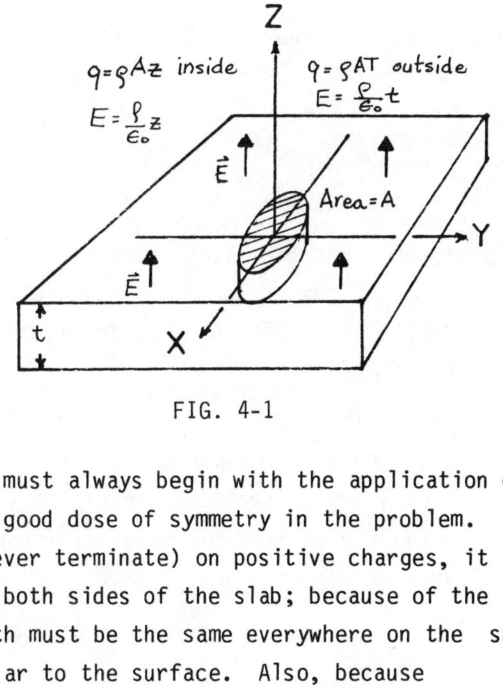

Solution

Given: $\rho = .01$ coulombs/cubic meter
Required: E_1 and E_0

FIG. 4-1

Using Gauss' Law to solve electric field problems must always begin with the application of intuition and is never successful in the absence of a good dose of symmetry in the problem. Because electric field lines must always originate (never terminate) on positive charges, it is clear that the field lines must be directed away from both sides of the slab; because of the symmetry of the charge distribution, the field strength must be the same everywhere on the surface (far from the edges) and the field must be perpendicular to the surface. Also, because of the symmetry, the field must be zero everywhere on the central plane of the slab. Thus, we may draw a gaussian surface with the shape of a cylindrical "pillbox" with its axis perpendicular to the surface of the slab and one of its flat surfaces at the central plane of the slab. The other flat surface should be at the position at which we wish to calculate the electric field, a distance z from the central plane. In using Gauss' Law with such a surface, we need integrate only over the second flat surface, the one not on the central plane.

$EA = q/\varepsilon_0$ and
$q = \rho Az$, yielding
$E_i = \rho z/\varepsilon_0$
$E_0 = E_i$ (at the surface)

PROBLEMS

1. A small particle bearing a positive charge Q is placed a distance 2d away from a second positively charged particle bearing the same charge. Show that for positions on the line joining the two charges and quite near the point midway between them, the electric field is directed toward this midpoint and is directly proportional to the displacement from it.

2. Show that a small positively charged particle placed midway between the two charges in Problem 1 will, if displaced slightly and released, execute simple harmonic motion and derive the expression for the frequency of that motion.

3. A large non-conducting hollow sphere bears a uniformly distributed positive charge, Q. A small solid sphere of the same material placed inside the hollow sphere bears a negative charge of the same magnitude. (a) Use Gauss' Law to calculate the electric field both inside and outside the larger sphere if the centers of the two spheres coincide. (b) Use the principle of superposition to show qualitatively that, if the centers of the two spheres do not coincide, the electric field outside the larger sphere is identical to that of a dipole.

4. A positive charge Q is uniformly distributed over a spherical volume of radius R (see problem 40 in textbook) so that the charge density is

$$\rho = \frac{3Q}{4\pi R^3}$$

A negative point charge of magnitude Q is placed at the center of the sphere. Use Gauss' Law to find the electric field at points both inside and outside the sphere.

5. Assume that the negative point charge is free to move about within the spherical charge distribution of Problem 4 (they exert no mechanical forces on one another). Describe quantitatively the effect of subjecting this charge distribution to an externally applied uniform electric field, E.

6. An electric dipole is placed near (i.e. in the electric field of) a point electric charge. Does the dipole experience a (a) torque (b) force? If the answer to either of these is "yes", does the direction of the torque or force depend upon the sign of the point charge? Explain your answers in sufficient detail to demonstrate an understanding of the principles involved.

Chapter 28
Electric Potential

PREVIEW

Just as the concept of energy provided an alternative, and in some cases easier, method of attacking problems in mechanics, we shall see in this chapter that the concept of ELECTRIC POTENTIAL provides a completely analagous alternative to the electric field for the solution of problems in electromagnetism. Electric potential will be defined as a scalar quantity which is derived from the electric field in the same way that the scalar gravitational potential energy is derived from the gravitational field.

SUMMARY

28.1 POTENTIAL DIFFERENCE

The scalar concept of potential difference may be used as an alternative to the vector concept of electric field in attacking problems in electrostatics. Since it involves the ubiquitous concept of energy, it also provides a convenient bridge between electrostatics and other fields such as mechanics and thermodynamics.

> Definition. The potential difference between point A and point B is equal to the amount of work per unit charge required to move a charge from point B to point A. Since both work and charge are scalars, so potential difference is a scalar. This definition assumes that the only force which must be overcome in moving the charge from one point to another is of an electrical nature, i.e. an electric field. In symbolic form,

$$V_{AB} = \frac{W_{B-A}}{q} \qquad (28.6)$$

Since the work done on the charge (or any other object) is defined as the integrated dot product of the force and the displacement, this equation may be written as

$$V_{AB} = \int \frac{F}{q} \cdot ds$$

and since the force required to overcome the electric field is

$$F = -Eq$$

we finally have

$$V_{AB} = - \int E \cdot ds \qquad (28.9)$$

where the integration is carried out over any path linking the points A and B. Any one of these three equations may be considered the defining equation for the potential difference; only the last one, however, may conveniently be used to calculate the potential difference in problems in which the electric field is known.

From the definition, it is clear that the unit of potential difference is the joule per coulomb, but since this concept is so frequently used, it is desirable to have a more concise name for the unit and the unit has been named the <u>volt</u>.

<u>Definition</u>. <u>Volt</u>: the potential difference across which one joule of work is required to move a charge of one coulomb.

Because the term potential difference itself is often clumsy to use, the concept is often called simply the voltage, although this term is not officially accepted as an SI name.

28.2 CONSERVATION OF ENERGY

As the definition of potential difference implies, when a charged object is forced to move against or allowed to move with an electric field, work is done on or by the object and thus it gains or loses electrical potential <u>energy</u>. While this electric energy has its origin in the interaction between the electric charge and the electric field, it has the same nature as any other kind of energy and thus may be transformed into other kinds of energy such as kinetic, thermal, etc. The work-energy principle and the law of conservation of energy apply to electrically charged systems in the same way they do to other systems. If a charged object is moved from one point to another between which there is a potential difference of ΔV, then its electric potential energy has been raised or lowered by an amount ΔVq; note that either ΔV or q (or both) may be negative yielding a reduction in energy if the <u>product</u> is negative.

As with other kinds of energy, only the change in electric potential energy is defined. If a specific value is to be assigned to the energy of an object, the point at which the energy is zero must be arbitrarily defined. For many problems in electrostatics, it is most convenient to define this zero level to be at infinity while for most problems involving electrical circuits it is more convenient to define it as the earth or "ground".

The fact that electric charge is quantized in units of the charge on an electron (see Chapter 26), allows the definition of a unit of energy which is particularly useful in problems dealing with electrostatic interactions, such as the interactions between atomic nucleii, which are composed of elementary charged particles. This unit is the electron-volt which is a small unit of energy and which may, of course, be used to describe other kinds of energy as well as electric potential energy.

<u>Definition</u>. <u>Electron-volt</u>: the electrostatic potential energy gained or lost by an electron as it is moved through a potential difference of one volt.

28.3 ELECTRIC POTENTIAL

All of the concepts introduced so far in this chapter are based on the potential difference between two points in an electric field. It is often convenient to describe the potential of a single point in "absolute" terms, i.e. without specific reference to a second point. We may do this by first defining a zero point for potential and then defining the potential of a given point as the potential difference between that point and the arbitrarily chosen zero point. The zero point for potential is defined in the same way that the zero point for potential energy was defined in the previous paragraph.

The definition given above shows that the potential difference between two points may be calculated by

$$\Delta V = -\int E \cdot ds \qquad (28.14)$$

If this dot product is expressed in vector notation in terms of cartesian coordinates,

$$E = iE_x + jE_y + kE_z$$
$$E\,ds = E_x dx + E_y dy + E_z dz$$

giving

$$V = -\int E_x dx - \int E_y dy - \int E_z dz$$

where the integration is carried out from the chosen zero point to the point at which the potential is to be calculated.

The equation above allows us, by the process of integration, to find the electric potential of a point in a known electric field. This equation may be inverted to find the electric field from a known potential. Thus, if the potential is known as a function of the position coordinates, x, y, and z, the electric field may be found by

$$E = -i\frac{\partial V}{\partial x} - j\frac{\partial V}{\partial y} - k\frac{\partial V}{\partial z}$$

28.4 POINT CHARGE POTENTIAL

We may use the electric field produced by a point charge both as an example of the sort of calculation described in the paragraph above and as a starting point for the calculation of the potential at a point in the field produced by a given distribution of charge. In Chapter 27, the expression for the electric field in the vicinity of a point charge was derived and this expression may be used in conjunction with the equations above to find the potential difference between two points in the field of such a charge; the result is Eq. (28.21) in the textbook. If this equation is modified by setting the initial point at infinity, thus establishing the zero point of potential, Eq. 28.22 results,

$$V(r) = Q/4\pi\epsilon_0 r \qquad (28.22)$$

The simplicity of this equation belies its importance. While it appears only to yield the potential of a point in the vicinity of the simplest of all charge distributions, the single point charge, any charge distribution, no matter how complicated, may be considered to be made up of an appropriate arrangement of point charges and the principle of superposition may be applied using the above equation for each of the charges.

Furthermore, hidden within the derivation of the equation is the fact that the path over which we choose to integrate between the two points is irrelevant; the same result is obtained for the potential for any possible path of integration. Were this not true, the concept of potential difference as we have defined it would be totally useless. A field for which this is true is said to be a conservative field and may always be used to define a potential function. Mathematically, the criterion for a field, F, to be conservative is

$$\oint F \cdot ds = 0$$

where the integral sign with the circle superimposed on it means that the integration must be carried out over a __closed__ path.

28.5 MULTIPLE CHARGE POTENTIALS

Equation 28.22 in the textbook was developed to yield the potential at a point in the vicinity of a single point charge. However, it may be extended by use of the principle of superposition to __any__ distribution of charges, no matter how complicated, and offers an alternative to using the defining equation (28.12) in the solution of problems; more often than not, this alternative is easier to use because it involves the integration of the scalar quantity, potential, rather than the vector quantity, electric field.

If the charge distribution is composed of a set of recognizably discrete charges, the potential at a point near such a set may be calculated simply by adding (scalar addition) the contributions of each member of the set.

$$V = 1/4\pi\epsilon_0 \Sigma q_i/r_i \qquad (28.27)$$

where the q_i is the magnitude of the ith charge (and may be either positive or negative) and the r_i is the distance from the ith charge to the point at which the potential is to be evaluated.

Actually, since as we saw in Chapter 26, charge is quantized, any charge distribution is made up of discrete charges. However, if we are dealing with charges distributed over a macroscopic region of space, the scale of the problem is so large that we may, for all practical purposes, consider the charge to be continuously distributed. In such a case, we must integrate over the continuous distribution rather than sum over the discrete distribution as above.

$$V = 1/4\pi\epsilon_0 \int dq/r$$

where dq is an infinitesimal element of the charge distribution and r is the distance from this element to the point at which the potential is to be evaluated. The limits on the integral must be chosen so that the entire distribution is covered by the integration.

28.6 EQUIPOTENTIAL SURFACES

In Chapter 27 we were introduced to the concept of the electric field line to provide a pictorial description of the electric field. A complementary and equally useful picture may be developed using the concept of equipotential surfaces.

> <u>Definition.</u> <u>An equipotential surface</u>: a mathematical surface which is the locus of all points which have the same value of electric potential. An equipotential surface may or may not coincide with the surface of a material object.

From the definition of potential difference, it is clear that in any region of space, the electric field lines intersect the equipotential surfaces at right angles. There is one important consequence of this fact that deserves mention. According to our model of a good conductor such as a metal, there is an almost unlimited amount of free charge (electrons) distributed throughout the volume of a conductor including the region immediately beneath the surface. Thus, if such a material is subjected to an electric field which has a <u>tangential</u> component at the surface, this free charge will move under the influence of the field until it is distributed in such a way that there is no longer a tangential component to the electric field at the surface and indeed, no field at all inside the bulk of the material. As a consequence, in an electro<u>static</u> situation the electric field must be perpendicular to the surface of a conductor and the surface of the conductor must be an equipotential surface.

28.7 CORONA DISCHARGE

The corona discharge observed at sharp points on a highly charged metallic object is an interesting, and in many cases important, phenomenon; however, the main thrust of this section is to show that when a charge is placed on a conducting object, the charge will distribute itself on the surface of the object in such a way that the portions of the surface which have the smallest radius of curvature, such as sharp points, will bear higher charge densities than portions which have more gentle curvature. This fact will often be helpful in forming a qualitative picture of the field in the immediate vicinity of a charged conductor as a starting point in the solution of problems involving such a conductor.

EXAMPLE PROBLEMS

Example 1

Two large parallel metal plates having dimensions much larger than the distance between them are filled with a dielectric material bearing a uniformly distributed positive electric charge of density

$$\rho = 2 \times 10^{-5} \text{ coulombs/m}^3.$$

Such a distribution produces an electric field which is zero midway between the two plates and is everywhere else perpendicular to the plates, increasing uniformly with distance away from the midplane (see Example Problem 4, Chapter 27). If the two plates are 2 cm apart, what is the potential difference between one of the plates and a) a point midway between the two plates? b) the other plate?

Solution

From the answer to Problem 4, Chapter 27, we find that the electric field between the two plates is perpendicular to the midplane and is equal in magnitude to

$$E = \rho z/\varepsilon_0 .$$

To find the potential difference between any two points, we may use the definition of potential difference,

$$V = \int E \cdot d\ell$$

which becomes for this configuration

$$V = \rho/\varepsilon_0 \int z \, dz$$

where $z = 0$ at the midplane.

Evaluation of this integral gives

$$V = \rho/2\varepsilon_0 (z^2 - z_0^2) = 113 \text{ volts}$$

for the answer to (a) and $V = 0$ for the answer to (b).

Example 2

The "electron gun" in a television tube consists of two plane parallel plates with a source of electrons at one of the plates and a small hole drilled through the other to allow passage of the electrons into the main part of the tube. A potential difference of a few tens of thousands volts is maintained between the two plates to accelerate the electrons to a high speed. If the potential difference between the plates is 20 kilovolts, what is the speed of the electrons in the beam as they reach and pass through the hole?

Solution

Since in this device the potential energy of the electric field is transformed into kinetic energy of the electrons, the law of conservation of energy provides a means of solution. In

passing through the field, the electrons lose an amount of potential energy equal to Ve, and thus
$$Ve = 1/2\, mv^2$$
This equation may be solved for v and the appropriate numbers substituted to yield
$$v = 8.4 \times 10^7 \text{ m/s}$$
Are the electrons in this beam relativistic?

Example 3

A long since discarded model of the electron is a small hard sphere of radius
$$R = 1.4 \times 10^{-15} \text{ m}$$
bearing a charge e distributed uniformly over its surface. Using this outmoded model, determine the energy of the electron by calculating the work required to "build" it by bringing infinitesimal units of charge, dq, from infinity to the surface of the sphere until the total charge is built up.

Solution

The electric potential of the sphere when it has a charge q on it is
$$V = \frac{1}{4\pi\epsilon_0} \frac{q}{R}$$
and the work required to bring an additional charge dq to the surface is
$$dW = V\, dq$$
Thus, the total amount of work required is found by integrating this last expression from q = 0 to q = e.
$$W = \frac{1}{4\pi\epsilon_0 R} \int q\, dq$$

$$W = \frac{e^2}{8\pi\epsilon_0 R} = 8.24 \times 10^{-14} \text{ joules}$$

How does this compare with the rest energy of the electron, .511 Mev?

Example 4

Two concentric metal spheres bear equal and opposite charges of .01 μC and have radii of 9.5 cm and 10.5 cm. What is the potential difference between the spheres?

Solution

Because of the spherical symmetry, the electric field between the spheres is identical to that of a point charge located at their common center. Thus, the potential at the surface of each sphere is given by
$$V = 1/4\pi\epsilon_0 R$$
and the potential difference between the two spheres is
$$V = V_1 + V_2 = 90.1 \text{ volts}.$$

Example 5

Two identical point charges are located a distance d from each other. Sketch the equipotential surfaces in the neighborhood of these charges.

Solution

A healthy portion of intuition is required for this problem. If we first consider points quite near one (either one) of the charges, the equipotential surfaces will be almost concentric spheres centered on the individual charge because the other charge is so far distant that its effects cannot "be felt". On the other hand, for points quite far from either charge, the two charges will seem to merge into a single charge and the equipotential surfaces will again be almost concentric spheres, now centered on the midpoint of the line joining the two charges. If we use y as the coordinate along the perpendicular bisector of the line joining the two charges, the transition between these two regions occurs at the point for which $(\partial V/\partial y) = 0$ which is $y = .707a$. It is left to your imagination to actually sketch these surfaces.

PROBLEMS

1. Potassium bromide is an ionic molecule in which the positive potassium ion and the negative bromine ion each bear a net charge equal to that of the electron. The ions are bound together to form the molecule by the coulomb attraction and thus form an electric dipole having a dipole moment which has been measured as
$$\mu = 3.035 \times 10^{29} \text{ CM}.$$
Calculate the distance of separation of the two ions and, from this, the amount of work required to break the ionic bond. Consider the ions to be point charges.

2. In a certain region of space, the electrical potential is described by the equation,
$$V = \frac{Q}{2\pi\varepsilon_0 a} (1 + (x/a)^2).$$
Using the unit vector notation, write the equation describing the electric field in this region.

3. A small rifle bullet has a mass of 200 gm and a speed of 1000 ft/s. What is the kinetic energy of this bullet in electron-volts?

4. Two positive point charges each of magnitude Q are separated by a distance 2a. Taking the x axis to lie along the line joining the charges and the origin to be at the point midway between them, derive the expression for the electric potential at a point on the x axis such that $x > a$. Show that for $x \gg a$, this expression is consistent with the potential for a single charge of magnitude 2Q.

5. Two straight parallel wires with radii small compared to their distance of separation bear equal and opposite charges (see problem 41 in textbook). Show that the equipotential surfaces in the vicinity of these wires are cylinders such that the ratio of the distances from the two wires to any point on the equipotential surface is a constant.

Chapter 29
Capacitance and Capacitors

PREVIEW

The capacitor is a simple and extremely useful device not only as a circuit element in electronic circuits but in other applications which call for the storage of electric charge. All electrical circuits contain capacitors whether intentionally included or not and, particularly at high frequencies, unwanted capacitors may be exceedingly troublesome. For both of these reasons, it is essential that those who use electronic circuits for measurement understand the characteristics of these devices.

SUMMARY

29.1 CAPACITANCE AND CAPACITORS

Qualitatively, a capacitor might be defined as a device upon which electric charge can be stored and from which the stored charge can be retrieved. Any conductor can satisfy this definition and single large conductors are occasionally used for this purpose. However, the more generally accepted definition involves two conductors:

Definition. Capacitor: two conductors located sufficiently close to but insulated from one another that, when they bear equal and opposite charges, an electric field exists between them, originating on the positively charged one of the pair and terminating on the other.

Definition. Capacitance(of a capacitor): the ratio of the magnitude of the charge placed on one of the conductors (note that the two conductors bear equal and opposite charges) to the potential difference between them.

The letter, C, is usually used to denote the capacitance of a capacitor and thus, from the above definition,
$$C = Q/V , \tag{29.1}$$
we see that the units of capacitance are Coulombs/Volt which has been given a name of its own, the farad. A one farad capacitor, however would be so large that the unit, microfarad (µf) is more frequently used. It should not be inferred from this that one farad capacitors are not available; they are, however, not very common and are used primarily to provide standby power for low voltage appliances such as computers in the event of power outages.

By far, the most common configuration for a capacitor is two parallel metal plates separated by a very thin insulating material, in some cases, air or a vacuum. In cases in which a large capacitance is needed and space is at a premium, these plates with the intervening insulator may be rolled tightly into a small cylinder. The capacitance of such a capacitor is given by $C = \varepsilon(A/d)$ where A is the area by which the two plates overlap one another, d is the distance of separation between them, and ε is the permittivity (see section 29.4) of the insulating material. While other configurations yield more complicated equations for the capacitance, all of them depend on these factors in essentially the same way.

<u>MAXIMUM VOLTAGE</u>: Since the electric field between the conductors is given by E = V/d, for very thin insulators, even moderate voltages may produce sufficiently large electric fields to break down the insulation and arc between conductors. Thus, commercial capacitors are always marked with two numbers, one specifying the capacitance and the other specifying the maximum voltage at which it may be safely used. In addition, electrolytic capacitors are marked with an indication of which terminal must be connected to the higher potential; failure to observe this polarity may cause the capacitor to overheat and fail, sometimes explosively.

29.2 CAPACITORS IN SERIES AND PARALLEL

Two connection configurations are considered here, the series and parallel connections. An important application of these connections is that capacitors may be combined to yield values of capacitance and/or voltage not available in a single unit.

When two capacitors, or other circuit elements, are connected together end to end so that any charge flowing out of one of them must flow into the other, they are said to be connected in series. If the two ends of one of the capacitors are connected to the two ends of the other so that the same potential difference appears across each of them, they are said to be connected in parallel. Using these descriptions and the definition of capacitance, it can be shown that for any number, N, of capacitors connected in series, the capacitance of the combination is given by

$$\frac{1}{C} = \sum_N \frac{1}{C_n} , \qquad (29.11)$$

while for any number, N, of capacitors connected in parallel, the capacitance of the combination is given by
$$C = \sum_N C_n . \qquad (29.12)$$

Thus, very large capacitors may be constructed by connecting several smaller capacitors in parallel; the voltage rating of the combination is then the same as the smallest voltage rating of the individual capacitors. On the other hand, a higher voltage rating may be achieved by connecting several capacitors in series, the voltage rating of the combination being equal to the sum of the voltage ratings of the individual capacitors; this series combination, however, yields a capacitance smaller than the smallest of the individual capacitors.

29.3 ELECTROSTATIC ENERGY OF A CHARGED CAPACITOR

When a capacitor is charged, the net result is the same as if this amount of charge were removed from one conductor and moved to the other and since a potential difference exists between the conductors, work must be done to effect the move. If the conductors already bear charges of magnitude q so that there is a potential difference V = q/C between them, then the work required to move an additional charge dq from one to the other is dW = Vdq and the total amount of work required to charge the device to a charge Q is found by integration to be

$$W = \int \frac{q\,dq}{C} = 1/2 \, \frac{Q^2}{C} . \qquad (29.13)$$

Since this work has not been dissipated (e.g. by conversion into heat), it must be stored as energy in the capacitor; it is useful to consider the electric field between the conductors as the residence of the energy stored in the capacitor. Thus, using the definition of capacitance, we have three different but completely equivalent expressions for the energy stored in the electric field between the plates of a charged capacitor:

$$U = \tfrac{1}{2}\frac{Q^2}{C} = \tfrac{1}{2}CV^2 = \tfrac{1}{2}QV \,. \qquad (29.15, 16, 17)$$

We interpret this energy as "belonging" to the electric field and indeed, experimental evidence indicates that any electric field, whether associated with a capacitor or not, has associated with it an energy density (energy per unit volume) given by

$$u = U/V = \tfrac{1}{2}\,\varepsilon_0 E^2 \,. \qquad (29.19)$$

29.4 EFFECT OF AN INSULATOR ON CAPACITANCE

If two otherwise identical capacitors have different insulating materials separating the conductors, it is found that they have different capacitances. The smallest capacitance is obtained by using a vacuum as the insulator, all other ordinary insulating materials yielding larger values. Quantitatively, the effect of the insulator on the capacitance may be described by assigning a dielectric constant, denoted by κ, to each material so that the capacitance of a capacitor containing such an insulating material is κ times the capacitance of a capacitor with the same dimensions but with a vacuum as an insulator.

Thus, from the definition of capacitance, it is clear that if a dielectric material is inserted between the plates of a capacitor which is isolated so that the charge cannot change, the voltage (and therefore the electric field) between the plates will be reduced. On the other hand, if the dielectric is inserted between the plates of a capacitor across which the voltage is held fixed, the charge on the plates must increase in order to keep the electric field between the plates at its original strength.

> **Definition.** **Dielectric constant:** a characteristic of an insulating material which describes its effect on an electric field in the region in which the material is situated.

Since the effects of the dielectric properties of a material are not confined to the use of that material in a capacitor, it is often useful to combine the dielectric constant with the permittivity constant. Thus,

$$\varepsilon = \kappa \varepsilon_0$$

where ε_0 is the permittivity of free space and ε is the permittivity of the dielectric material.

If a dielectric material only partially fills the space between the plates of a capacitor, the device may be treated as a combination of two or more capacitors in series or parallel. The operation of a number of useful transducers for position measurement is based upon this treatment.

29.5 ATOMIC VIEWPOINT OF THE EFFECT OF AN INSULATOR ON CAPACITANCE

The molecules which make up a dielectric or insulating material are electrically neutral in that they possess equal amounts of positive and negative charges however, the arrangement of these charges within the molecule may or may not be such that the center of positive charge coincides with the center of negative charge. Thus, while a molecule in a dielectric has no net charge, it may have a dipole moment.

Definition. <u>Polar molecule</u>: a molecule which has a permanent dipole moment because the center of negative charge does not coincide with the center of positive charge.

Definition. <u>Non-polar molecule</u>: a molecule which has no permanent dipole moment because the center of negative charge coincides with the center of positive charge.

The macroscopic effect of an electric field on materials constituted of these two types of molecule is the same even though there is a fundamental difference in the effects at the molecular level. The overall effect in both cases is that the electric field inside the material is smaller than it would be in the absence of the material. It is this reduction in the strength of the field which leads to the use of the prefix, "di", in the word, dielectric. The dielectric constant is a measure of the "ease" with which a material may be polarized.

In the context of its use as the insulating material between the plates of a capacitor, the most significant difference between the polar and non-polar materials is that in the polar materials, the dielectric constant is temperature dependent while in the non-polar materials, it is nearly independent of temperature.

EXAMPLE PROBLEMS

Example 1

The coaxial cable used for the antenna lead for a C.B. radio transmitter has an inner wire .028 inches in diameter and a concentric stranded sheath .125 inches in diameter, the two being separated by a polyethylene insulator, and is 3 m long. What is the capacitance of this cable?

Solution

The solution to this problem is a straightforward application of Eq. (29.8) in the textbook modified for the non-vacuum insulation between the two conductors.

$$C = 2\varepsilon L / \ln(r_2/r_1)$$

Given: $r_1 = .028$ in
          $r_2 = .125$ in
          $\kappa = 2.26$ (from Table 29.1)

Required: C

Since the radii of the conductors enter the equation only as a ratio, we may equally well use their diameters and furthermore, we need not convert to metric units. The length of the cable, however, must be specified in meters. From Table 29.1, we find that the dielectric constant for polyethylene is = 2.26 from which the appropriate value of may be calculated from

$$\varepsilon = \kappa \varepsilon_0$$

Substituting the appropriate numbers into these equations yields C = <u>252 picofarads</u>.

Example 2

What is the maximum voltage which the cable in Example Problem 1 can sustain between the two conductors without suffering a breakdown of the insulation?

Solution

From Table 29.2, we find that the dielectric strength of polyethylene is 18 kV/mm which is the maximum electric field strength which the insulation can sustain. The electric field in the vicinity of a long straight wire is given by

$$E = \frac{Q}{2\pi \varepsilon L r}$$

From this equation, it is clear that the electric field is strongest at the surface of the inner conductor and therefore the radius of this wire (in meters) must be used in the calculation. The charge on the wire may be found from $Q = CV$ and combining this with the equation on the previous page yields

$$V = 2\pi\varepsilon E L r / C = \underline{19152 \text{ volts}}$$

a voltage not likely to be encountered in this application.

Example 3

A parallel plane plate capacitor is charged to a voltage, V, and then isolated. A metallic sheet having a thickness one third the distance of separation between the plates is inserted between them. Describe the effect of inserting this sheet on the charge, the voltage, and the capacitance of the capacitor.

Solution

Since the capacitor has been isolated, the charge on the plates of the original capacitor cannot change; the metal sheet is polarized with equal and opposite charges appearing on opposite sides (but no net charge). The effect on the voltage and capacitance may be calculated in two different ways. Since the charge on the plates has not changed, the electric field in the remaining empty space must also not have changed; however, the electric field inside the metal sheet must be zero and the voltage between the plates, calculated by integrating the field from one plate to the other, must have been reduced to two thirds of its original value. Consequently, the capacitance is increased to three halves of its original value.

Alternatively, we may consider the capacitor with the metal sheet inserted to constitute two capacitors in series, each having one third the separation of the original capacitor, and thus each having three times the capacitance of the original. The series combination of two equal capacitors has a capacitance of one half the value of either of them, or three halves the original capacitance. Thus, the voltage has been reduced to two thirds of its original value.

Example 4

A 10,000 μf capacitor is charged to a potential difference of 25 volts. If all of the energy stored in this capacitor were recovered in the form of heat when the capacitor is discharged, how many calories would be available?

Solution

Given: $C = 10,000 \text{ μf} = .01 \text{ f}$
$V = 25 \text{ volts}$

Required: U (in calories)

The energy stored in the capacitor is given by

$$U = 1/2 \, CV^2 = 3.125 \text{ joules} = \underline{.747 \text{ calories}}$$

Example 5

A 220 pf capacitor and a 1000 pf capacitor are connected in series and the combination is connected to a 12 volt source. What is the potential difference across the 220 pf capacitor?

Solution

Given: $C_1 = 220 \text{ pf}$
$C_2 = 1000 \text{ pf}$
$V = 12 \text{ volts}$

Required: V_1

The equivalent capacitance of the series combination of the two capacitors is given by

$$C = \frac{C_1 C_2}{C_1 + C_2}$$

The two capacitors, being connected in series, bear the same charge, $Q = CV$, and the voltage across the 220 pf capacitor is

$$V_1 = (C/C_1)V = \frac{C_1 C_2}{C_1 + C_2} V = \underline{9.8 \text{ volts}}$$

PROBLEMS

1. Ordinary lamp cord consists of two (stranded) copper conductors each .8 mm in diameter and separated by 2.5 mm. They are insulated and held parallel to each other by a plastic coating which has a dielectric constant of $\varepsilon = 2.5$. The capacitance per unit length of such a pair of wires is given by

    $$C/\ell = \pi \kappa \varepsilon_0 [\cosh^{-1}(d/2R)]^{-1}$$

    where R is the radii of the two wires and d is their distance of separation. What is the capacitance of a two meter length of this lamp cord? Note: Do not be intimidated by this imposing looking equation; it is not at all difficult to work out on a hand calculator which includes hyperbolic functions.

2. A capacitor is constructed of two parallel plane plates of length L and width A, separated by a vacuum (or air). A slab of dielectric material of the same width as the plates and of thickness the same as the distance of separation of the plates is inserted to a distance, y, between the plates. By what fraction does the insertion of this dielectric cause the capacitance to change?

3. Using the same capacitor as in problem 2, a thin (very much thinner than the distance of separation between the plates) metallic sheet is inserted between the plates. What effect does the insertion of this sheet have on the capacitance of the capacitor?

4. A 7500 μf capacitor has a mass of 500 gm and is charged to a potential difference of 40 volts. If all of the energy stored in this capacitor could be recovered in the form of mechanical energy when the capacitor is discharged, to what height could the capacitor be lifted assuming an appropriate "lifting machine" were available?

5. Three capacitors of .47 μf, .01 μf, and 1 μf are connected in series and the combination is connected to a 24 volt source. What is the potential difference across the .47μf capacitor?

Chapter 30
Electric Current

PREVIEW

The last four chapters have been devoted to the study of electrostatics, electric charges at rest; here we pass to the study of charges in motion which constitute the electric current. The concepts of electric current, electrical conductivity, and electrical resistance are defined and applications of these concepts are described.

SUMMARY

30.1 ELECTRIC CURRENT

An electric charge in motion, often described as a "flow" of charge, constitutes an electric current.

> Definition. Electric current: the (time) rate at which electric charge passes by a given observation point. The unit of current is the ampere, one ampere being equivalent to one coulomb per second. Symbolically,
> $$I = \frac{dq}{dt}, \text{ or if } I \text{ is constant, } I = \frac{Q}{t}.$$ (30.1,2)

In this equation, dq (or Q) is the amount of charge which passes the observation point in the time interval dt (or t). This moving charge may be in the form of a stream of charged particles such as the electrons in the beam of a television picture tube or the flow of charge confined in a metal wire. While the moving charge may be of either sign, the definition given above for electric current is understood to be phrased in terms of positive charge; if the current is actually constituted of negative charge, the direction of the current is defined to be opposite to the direction of the motion of the charge.

When this definition is applied to the conduction electrons in a solid conductor, we obtain for the current, I,
$$I = enAv_D \qquad (30.3)$$
where:
- e is the electronic charge,
- n is the number of electrons per unit volume,
- A is the cross sectional area of the conductor, and
- v_D is the drift velocity of the electrons.

The drift velocity of the electrons is very slow, being, for typical values of the parameters above, of the order of a few hundredths of a centimeter per second.

For some applications, it is more convenient to describe the flow of charge in terms of the current density.

Definition. <u>Current density</u>: the current per unit area normal to the direction of current flow. Current density is a vector quantity with its direction parallel to the direction of the flow of charge.

$$J = I/A$$

In situations in which the flow of charge is not uniform over the cross section of the conductor, the current may be found by integrating the current density.

$$I = \int \mathbf{J} \cdot d\mathbf{A} \tag{30.6}$$

30.2 ELECTRICAL RESISTANCE AND OHM'S LAW

In most solid materials, the drift velocity of the conduction electrons is directly proportional to the applied electric field and, thus, the current is directly proportional to the potential difference applied between the ends of the conductor. This is known as <u>Ohm's law</u> and is written

$$I = V/R \quad \text{or} \quad I = GV$$

where R is the resistance in ohms (Ω) or G is the conductance in ohms. Clearly, $G = 1/R$.

Even in applications in which Ohm's law does not apply, this equation serves as the definition of resistance.

Definition. <u>Resistance</u>: the ratio of the potential difference across a device to the current through it.

$$R = V/I \tag{30.7}$$

30.3 ELECTRICAL CONDUCTIVITY AND ELECTRICAL RESISTIVITY

From Eq. 30.3, we see that the resistance of a conductor depends not only upon the characteristics of the material of which the conductor is made, but also upon the dimensions of the conductor. It is frequently necessary to describe the electrical characteristics of a material without reference to its dimensions, and to that end, resistivity and conductivity are defined.

Definition. <u>Resistivity</u>: the resistance per unit cross sectional area of a one unit length of a material.

$$\mathbf{R} = \rho\,(\ell/A) \text{ in } \Omega \text{ meters} \tag{30.8}$$

Definition. <u>Conductivity</u>: the reciprocal of the resistivity of a material.

<u>TEMPERATURE DEPENDENCE</u>: The resistivity of solid materials depends upon the temperature; for metals, the resistivity is nearly directly proportional to the absolute temperature, while for insulators and semiconductors, it decreases in a rather complicated manner as the temperature increases. This temperature dependence allows the use of both metals and semiconductors as temperature transducers. For metals, the effect may be described by

$$\rho = \rho_0[1 + \alpha(t - t_0)] \ . \tag{30.13}$$

Definition. The <u>temperature coefficient of resistivity</u>: the fractional change in resistivity per unit change in temperature. Its units are the reciprocal of the units in which the temperature is measured and it is symbolized by the greek letter α.

Definition. <u>Superconductor</u>: a member of a moderately large family of metals in which, for all temperatures below a certain critical temperature (different for each metal), the resistivity is identically zero.

A rather remarkable fact may be discovered from the last column in Table 30.1. All of the elementary metals, with the exception of iron, listed in that table, have about the same value for the temperature coefficient of resistance and this value is quite close to 1/273. This implies that extrapolation would yield zero resistance at absolute zero for these metals.

30.4 DYNAMIC RESISTANCE

Definition. Non-ohmic material: a material for which the relationahip between the voltage and current is non-linear and for which Ohm's law does not apply.

Definition. Dynamic resistance: the slope (dV/dI) of the curve representing the voltage as a function of the current. For an ohmic material, the dynamic resistance is equal to the static resistance: dV/dI = R.

Many non-ohmic devices have important applications in electronic technology.

30.5 ENERGY CONSERVATION AND ELECTRIC POWER

The flow of electric current through a device is usually accompanied by a conversion of electrical energy into some other form of energy. As with any other form of energy, the rate of conversion to or from electrical energy is termed power and may be calculated by

$$P = VI \tag{30.18}$$

This equation derives directly from the definitions of current and voltage and is thus valid in any application. For resistive materials in which the electrical energy is transformed specifically into heat, Ohm's law may be substituted into this equation to give two alternate forms,

$$P = I^2 R \quad \text{and} \quad P + \frac{V^2}{R} \tag{30.19, 30.20}$$

EXAMPLE PROBLEMS
Example 1

A constant current is used to charge a 10,000 µf capacitor from the uncharged state to a voltage of 12 volts. What current is required to reach this final voltage in a time of 5 s ?

Solution

Given: $C = 10{,}000$ µf $= .01$ farads
$V = 12$ volts
$t = 5$ s

Required: I

In order to find the current, we must calculate the amount of charge which has flowed onto the capacitor during the specified time.

$$Q = CV$$

We may then use the definition of current to solve the problem.

$$I = Q/t = C(V/t) = .024 \text{ amp} = \underline{24 \text{ ma}}$$

It should be noted that special provisions are required to assure a constant current into the capacitor during charging.

Example 2

A length of aluminum wire 1.5 mm in diameter carries an electric current of 10 amperes. What is the drift velocity of the electrons which constitute this current?

Solution

Given: $d = 1.5$ mm → $r = .75$ mm
$I = 10$ amperes
$e = 1.602 \times 10^{-19}$ coulombs

Required: v_D

In order to use Eq. 30.3 to solve this problem, we must know the density of free electrons within the aluminum. This value may be obtained by looking up the value for the molecular weight, M, the density, , and Avogadro's number in the available tables; then

$$n = N_A \frac{\rho}{M}$$

We find from the tables,

$N_A = 6.02 \times 10^{23}$ particles per mole
$M = 26.98$ gm per mole
$\rho = 2.7$ gm per cubic cm

Using this expression and $A = \pi r^2$, in Eq. 30.3, we have

$$v_D = \frac{IM}{eN_A \rho \pi r^2}$$

Converting all of these values to mks units as we must in problems involving electrical units, and substituting them into the final equation, we obtain

$$v_D = \underline{5.86 \times 10^{-4} \text{ m/s}}$$

Example 3

A carbon resistor of the type used in electronic circuits has a resistance of 2200Ω, a diameter of 3 mm, and a length of 6 mm. If a voltage of 15 volts is impressed across this resistor, what is the current through it? What is the resistivity of the carbon used in the manufacture of this resistor?

Solution

Given: $R = 2200$ Ω
$D = 3$ mm $= .003$ M
$\ell = 6$ mm $= .006$ M
$V = 15$ volts

Required: I, ρ

The first part of the problem is a straightforward application of Ohm's law.

$I = V/R = 15$ volts$/2200$ Ω $= .0068$ amp
$I = \underline{6.8 \text{ ma}}$

The second part of the problem is solved by use of Eq. (30.8).

$\rho = R(A/\ell) = .00259$ Ω mm
$\rho = \underline{2.59 \text{ ΩM}}$

Note that this is a much higher value than that given for carbon in Table 30.1. This results from the use of finely granulated carbon mixed with a filler in the manufacture of the resistor.

Example 4.

A 1 m length of copper wire having a diameter of .1 mm is to be used as a resistance thermometer. If the thermometer is to be capable of detecting temperature differences of one degree, what is the smallest change in resistance which the resistance measuring device must be able to detect?

Solution

Given: $\ell = 1$ M
$d = .1$ mm $= .0001$ M
$\Delta t = 1°C$

From Table 30.1
$\rho = 1.724 \times 10^{-8}$ m
$\alpha = .00393 \, /°C$

Required: R

The change in resistivity of copper for a one degree change in temperature is, from Eq. (30.13),

$\Delta \rho = \rho_0 \alpha$

This equation may be used along with Eq. 30.8 to calculate the resulting change in resistance.

$\Delta R = \Delta \rho (\ell/A) = \rho_0 \alpha (\ell/A)$

Substituting numerical values into this equation gives

$R = \underline{.0086 \, \Omega}$

Example 5.

An electric motor is a device which transforms electrical energy into mechanical energy. A certain small motor contains a coil of wire having a resistance of .4 and is designed to operate on 12 volts. When operating, it draws a current of 6 amperes. How much electrical power is it using? How much of this power is being wasted (transformed into heat)?

Solution

Given: $R = 0.4 \, \Omega$
$V = 12$ volts
$I = 6$ amperes

Required: P
$P_{(heat)}$

The total power being used by the motor is given by the product of the voltage and current

$P = VI = 12 \text{ volts} \times 6 \text{ amperes} = \underline{72 \text{ watts}}$

The portion of this power being used to heat the coil (resistance) is given by

$P_{(heat)} = RI^2 = \underline{14.4 \text{ watts}}$

PROBLEMS

1. An iron water pipe 2 m long has an outside diameter of 2 cm and a wall thickness of 3 mm. If a current of 100 amperes passes through this pipe, what is the potential difference between the two ends?

2. A thin rectangular sheet of brass 0.1 mm thick, 6 cm wide, and 20 cm long has a wire welded at the midpoint of each end and a potential difference of 6 volts applied to the two wires. Because of the small size of the contacts, the flow of current through the strip is non-uniform

and the current density through a cross section of the strip midway between the two ends is given approximately by

$$J = \frac{\sigma V}{\ell}\left[1 - 4\left(\frac{y}{\ell}\right)^2\right],$$

where y is measured from the midpoint of the line joining the two contacts and perpendicular to this line. What is the current through the strip? <u>Note</u>: The conductivity, σ, of brass may be found in Table 30.1 in the textbook.

3. A 10 m length of silver wire .02 cm in diameter is to be used as the basis for a resistance thermometer. If the resistance measuring instrument to be incorporated into the thermometer is capable of measuring resistance changes of $0.1\,\Omega$, what is the smallest temperature change which can be measured?

4. The current-voltage characteristics of a junction diode are shown in the graph of FIG. 1. It is clearly a non-ohmic device. Using data from the graph, calculate both the static and dynamic resistance of this device when the current is 0.4 amperes.

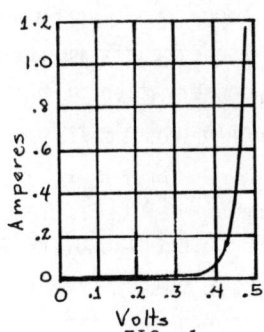

FIG. 1

5. A small electric heater designed to heat a single cup of water has a power rating of 300 watts when connected to a 120 volt source. What is the resistance of the heater? What current will the heater draw when connected to such a source?

Chapter 31
Direct-Current Circuits

PREVIEW

In this chapter, the sources of potential difference, the capacitors, and the resistors which have been encountered in the last three chapters are combined into electrical circuits. The rules by which these circuits may be analysed are derived from the fundamental laws of conservation of energy and conservation of charge and are applied to the analysis of instruments and circuits designed to measure electrical quantities.

SUMMARY

31.1 SOURCE OF ELECTROMOTIVE FORCE

The term, electromotive force is an obsolete, often misleading, term which has become so firmly imbedded in our scientific vocabulary that it is nevertheless used to refer to a device which maintains a potential difference between two points (terminals). Because the term may be misleading, we usually substitute the letters, emf, for the term.

> Definition. Source of emf: a device which separates negative from positive charge, thereby creating the electrical potential difference required to drive electric charge through a circuit. Emf is measured in volts and is symbolized by $\mathcal{E}$.

When a source of emf is connected to an external circuit, charge will flow (current) from the positive terminal of the source through the circuit to the negative terminal so that, if the potential difference between the terminals of the source is to be maintained, the process of charge separation must be continuous.

> Definition. Circuit: a conductive path containing one or more circuit elements such as resistors and capacitors through which an electrical current may flow from the positive terminal of a source of emf to the negative terminal.

A real voltage source may always be modelled as an ideal source of emf with an internal resistor connected in series with it. Thus, the terminal voltage of a source, the potential difference between its terminals, is

$$V = \mathcal{E} - Ir, \tag{31.4}$$

where $\mathcal{E}$ is the emf of the ideal source, r is the internal resistance, and I is the current which the source is supplying to the external circuit.

> Definition. Terminal potential difference: the actual potential difference between the terminals of a voltage source. It is always smaller than the emf by an amount proportional to the current being supplied to the external circuit.

Clearly, the emf of a source can be measured only when the source is delivering <u>no</u> current to an external circuit, including the measuring instrument.

In voltage sources designed to deliver large amounts of power to an external circuit, the internal resistance is small and may frequently be ignored.

31.2 KIRCHOFF'S LOOP RULE

Kirchoff's two rules, which are derived from fundamental conservation laws, provide powerful and convenient methods of analysing electrical circuit.

> <u>Definition</u>. <u>Circuit analysis</u>: the process of determining
> mathematically the currents through and the voltages across
> some or all of the elements which comprise an electrical circuit.

Kirchoff's loop rule, sometimes called Kirchoff's second rule, is derived from the law of conservation of energy and declares that, around any closed path in a circuit, the algebraic sum of the voltage changes is equal to zero;

$$\Sigma \Delta V = 0 \ .$$

In applying this rule to circuit analysis, one may assign positive values to either voltage increases or decreases; it is only necessary to be consistent.

For the analysis of a complex circuit, it is necessary to apply the rule to a sufficient number of different closed paths to cover all of the circuit elements and voltage sources included in the circuit.

31.3 APPLICATION OF KIRCHOFF'S LOOP RULE

> <u>Definition</u>. <u>Resistors in series</u>: two or more resistors connected together in series so
> that the same current must pass through each of them. This combination may be replaced,
> for purposes of analysis, by a single equivalent resistor whose resistance is equal to the
> sum of the resistances of the individual resistors.

$$R_{eq} = \Sigma R_n \qquad (31.9)$$

From this equation, it is easy to show that if N <u>equal</u> resistors are connected in series, the equivalent resistance is simply N times the resistance of the individual resistances. In any event, the equivalent resistance is always larger than the largest of the individual resistances.

31.4 KIRCHOFF'S JUNCTION RULE

Kirchoff's Junction Rule, sometimes called Kirchoff's first rule, is derived from the law of conservation of charge and declares that for any junction, or point at which two or more current paths are connected together, in a circuit, the algebraic sum of the currents entering (or leaving) the junction is zero.

$$\Sigma I = 0$$

In applying this rule to circuit analysis, one may assign positive values either to currents entering the junction or to currents leaving the junction; it is only necessary to be consistent.

For a complete analysis of a circuit containing N junctions, it is necessary to write the Kirchoff junction equation for each of (N-1) of the junctions, normally omitting the junction at ground potential.

Definition. <u>Circuit ground</u>: the point in a circuit which is arbitrarily assigned a potential of zero volts and to which all other voltages in the circuit are referred.

31.5 APPLICATION OF KIRCHOFF'S JUNCTION RULE

<u>Resistors in parallel</u>: When two or more resistors are connected in parallel so that the voltages across each of them are the same, the combination may be replaced, for purposes of circuit analysis, by a single equivalent resistor whose reciprocal value is equal to the sum of the reciprocals of the values of the individual resistors.

$$1/R_{eq} = \Sigma(1/R_n) \tag{31.22}$$

From this equation, it is easy to show that, if the parallel combination consists of N <u>equal</u> resistors, the value of the equivalent resistor is 1/N times the value of the individual resistors. In any event, the value of the equivalent resistor must be smaller than the value of the smallest of the individual resistors.

Another special case worth remembering is that for which the parallel combination consists of just two resistors. In this case, Eq. 31.22 yields an equivalent resistor which is equal to the product of the two resistances divided by their sum.

$$R_{eq} = \frac{R_1 R_2}{R_1 + R_2}$$

Kirchoff's rules are quite general and can be applied to any circuit, no matter how complex. There are some circuits, however, in which all of the resistors may be recognized as being contained in either a series or parallel combination with other resistors; these circuits are susceptible to reduction, through a series of steps, to a single equivalent resistor, a technique usually simpler and more convenient than the direct application of Kirchoff's rules.

31.6 CHARGING A CAPACITOR: THE RC CIRCUIT

In applying Kirchoff's loop rule to circuits, the voltage drop across a resistor may be calculated by Ohm's Law,

$$V = R I .$$

However, Ohm's Law may not be applied to a capacitor since the voltage across the capacitor is not proportional to the current through it, but rather is proportional to the charge.

$$V = Q/C$$

If this equation is differentiated with respect to time,

$$\frac{dV}{dt} = \frac{1}{C} \frac{dQ}{dt} = (1/C) I$$

from which we can see that the <u>time derivative</u> of the voltage, rather than the voltage itself, is proportional to the current.

Definition. <u>Capacitor</u>: a device across which the voltage cannot change instantaneously. An instantaneous change in the voltage would require an infinite current, which is impossible.

If a series combination of a capacitor and a resistor is connected to a voltage source, the source will charge the capacitor at a rate determined by the size of the resistor (and, or course, the size of the capacitor). In such a circuit, the voltage across the capacitor as a function of time is given by

$$V_C = V(1 - e^{-t/RC})$$

where V is the source voltage. The current through (actually into one end and out of the other) it is given by

$$I_C = I_0 e^{-t/RC} \qquad (31.31)$$

where I_0 is the initial current; $I_0 = V/C$.

> Definition. <u>The time constant</u> of an RC circuit is the time required for the current to fall to $1/e$ (about 37%) of its initial value or for the voltage to rise to $(1-1/e)$ (about 63%) of its final value. From Eq. 31.31, the time constant may be calculated by $T = RC$.

31.7 CURRENT AND POTENTIAL DIFFERENCE MEASUREMENTS

Current and voltage measurements may be made by use of a variety of instruments, some of which are sensitive to current and some of which are sensitive to voltage. Since current and voltage are related to each other by Ohm's law, either may be measured by any of these instruments by properly combining it with a carefully selected resistor. An instrument designed to measure current is called an ammeter, while one designed to measure potential difference is called a voltmeter. The terminals of a voltmeter must be connected between the two points across which the voltage is to be measured. An ammeter must be connected in series with the circuit element through which the current is to be measured.

> Definition. <u>Moving coil</u> (or d'Arsonval meter): a current sensitive instrument in which a coil of wire is mounted so that it will rotate between the poles of a permanent magnet when a current passes through it, thus moving a pointer across a calibrated scale.

The d'Arsonval meter may be converted into a voltmeter by connecting a large resistor in series with it.

> Definition. <u>Cathode ray oscilloscope</u>: a voltage sensitive instrument in which a beam of high speed electrons is deflected from their straight path by the electric field produced when a voltage is applied to two parallel plane plates. The amount of deflection is observed as a spot of visible light produced when the electron strikes the fluorescent screen on the front of the instrument.

The oscilloscope may be used to measure current by connecting a low resistance in parallel with it.

> Definition. <u>Digital meters</u>: actually pulse counting instruments. By use of an analog to digital converter (ADC), the number of pulses counted may be made proportional to either the current or the voltage and thus be used as either a voltmeter or ammeter.

To minimize the error caused by connecting (the internal resistance of) a voltmeter across a circuit element, the internal resistance should be much higher than the resistance of the circuit element.

$$R_M \gg R_{(circuit\ element)}$$

To minimize the error caused by connecting (the internal resistance of) an ammeter in series with a circuit element, the internal resistance of the ammeter should be much lower than the resistance of the circuit element.

$$R_M \ll R_{(circuit\ element)}$$

EXAMPLE PROBLEMS

Example 1

A laboratory technician, upon being assigned the task of determining the internal resistance of a source of emf, devised the following experiment. She connected a 50 Ω resistor to the source and measured the current through it. She then added a second, variable, resistor in series with the first and found that when the second resistor was set at 54.6 Ω, the current was reduced to one half its original value. From these data, calculate the internal resistance of the source.

Solution

Given: $R_1 = 50\ \Omega$
$R_2 = 54.6\ \Omega$
$I_2 = 1/2\ I_1$

Required: r

With the single 50 Ω resistor in the circuit, Ohm's law gives

$$\mathcal{E} - I_1 r = R_1 I_1$$

while with two resistors in the circuit, the corresponding equation is

$$\mathcal{E} - I_2 r = (R_1 + R_2) I_2$$

Solving each of these equations for $\mathcal{E}$ and equating them gives

$$I_1(R_1 + r) = (R_1 + R_2 + r) I_2$$

and since $I_1 = 2I_2$, this yields

$$r = R_2 - R_1 = \underline{4.6\ \Omega}$$

Example 2

In the circuit of FIG. 2-1, use the equivalent resistance method to find the current delivered to the circuit by the source.

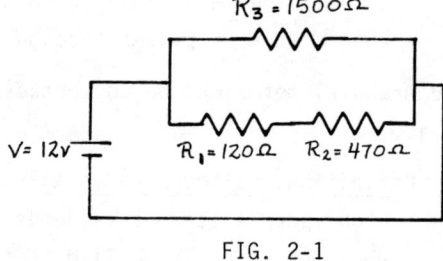

FIG. 2-1

Solution

Given: V = 12 volts
$R_1 = 120\ \Omega$; $R_2 = 470\ \Omega$; $R_3 = 1500\ \Omega$

Required: I

The series combination of R_1 and R_2 is in parallel with R_3. Thus

$$R_{eq} = \frac{(R_1 + R_2)R_3}{R_1 + R_2 + R_3} = 423.4\ \Omega$$

The current, I, may now be found by use of Ohm's law.

$$I = V/R_{eq} = .0283 \text{ amperes} = \underline{28.3\text{ ma}}$$

Example 3

Use Kirchoff's rules to find the current I_3 in FIG. 3-1.

Given: $R_1 = 330\ \Omega$; $R_2 = 4700\ \Omega$; $R_3 = 560\ \Omega$
V = 24 volts

Required: I_3

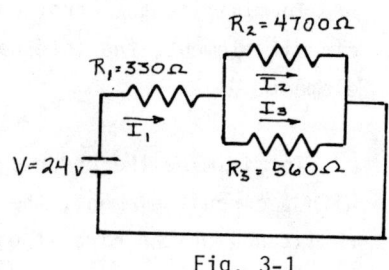

Fig. 3-1

Traversing the loop through the source, R_1 and R_2, we have

$$V - R_1 I_1 - R_2 I_2 = 0 .$$

Traversing the loop through the source, R_1 and R_3, we have

$$V - R_1 I_1 - R_3 I_3 = 0 \ .$$

From Kirchoff's junction rule, $I_1 - I_2 - I_3 = 0 \ .$

Eliminating I_1 and I_2 from these equations gives

$$I_3 = (R_1 + R_2 + R_3)/(R_1 R_2 + R_1 R_3 + R_2 R_3) = \underline{.031 \text{ amp}}.$$

Example 4

An initially uncharged 200 μf capacitor is connected in series with a resistor and the combination is connected to a voltage source which has a negligible internal resistance. From the charging curve shown in FIG. 4-1, calculate the value of the resistor and the voltage of the source.

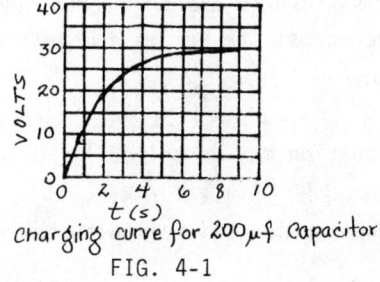

Charging curve for 200 μf Capacitor
FIG. 4-1

Solution

Given: C = 200 μf

 charging curve

Required: V

 R

The voltage across the initially uncharged capacitor will start at zero and gradually increase, approaching the source voltage after a "long time". From the graph, we can determine that this final voltage is 30 volts. Thus, the source voltage must have been V = 30 volts.

Also from the graph, we see that the capacitor voltage rose from zero to 63% of its final value, 18.9 volts, in a time of 2 s. Thus, the time constant, which is the product of the resistance and the capacitance, is 2 s.

$$T = RC = 2 \text{ s}$$

From this, the value of the resistance may be calculated.

$$R = T/C = 2 \text{ s}/.0002 \text{ f}$$

$$R = 10,000 \ \Omega = \underline{10 \text{ k}\Omega}$$

Example 5

The most direct way to measure the resistance of a resistor is to place it in a circuit so that simultaneous measurements of the voltage across it and the current through it may be measured. Ohm's law can then be used to calculate the resistance from the meter readings. FIG. 5-1 shows a circuit which might be used for this purpose. If the two meters were ideal, Ohm's law would give R = V/I where V and I are the voltmeter and ammeter readings. Derive the equation which must actually be used in view of the finite resistances of the meters.

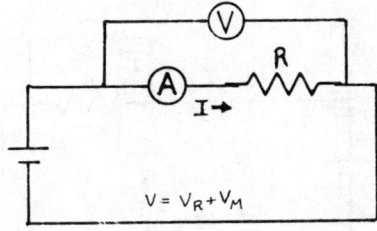

FIG. 5-1

230 Chapter 31

Solution

Given: V = Voltmeter reading
         I = Ammeter reading
         R_V = Voltmeter resistance
         R_A = Ammeter resistance

Required: R

The ammeter correctly reads the current through the resistor; however, the voltmeter reads the voltage across the series combination of the resistor and the ammeter. Thus, the voltmeter reading will be
$$V = (R + R_A)I \ .$$

This equation may be solved for R to yield
$$R = V/I - R_A \ .$$
Note that, using this circuit, the resistance of the voltmeter has no effect on the final result.

PROBLEMS

1. Suppose that in Example Problem 1 the current measured with the single 50Ω resistor in the circuit is 150 milliamperes. Find, without prior knowledge of the internal resistance of the source, the emf of the source.

2. Use the equivalent resistance technique to find the current in the 560Ω resistor in FIG. 3-1.

3. Use Kirchoff's laws to find the current delivered to the circuit by the source in FIG. 2-1.

4. An initially uncharged capacitor is connected in series with a 10 kΩ resistor and the combination is connected through a switch to a voltage source which has a negligible internal resistance. The graph of FIG. 1 shows the current through the resistor as a function of time after the switch is closed. From data taken from the graph, calculate the voltage of the source and the capacitance of the capacitor.

5. An alternative (to the connection in FIG. 5-1) connection for the measurement of resistance is shown in FIG. 2. Derive the equation by which the resistance may be calculated from the voltmeter and ammeter readings from this circuit.

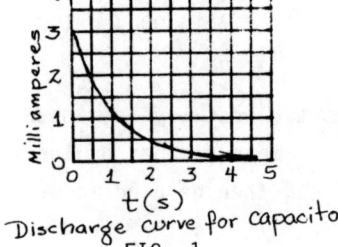
Discharge curve for capacitor
FIG. 1

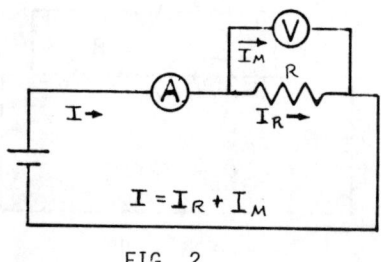
FIG. 2

Chapter 32
Magnetism and the Magnetic Field

PREVIEW

In this chapter, the phenomenon of magnetism is described and related to the phenomenon of electricity. The magnetic field is defined in terms of its effects on electric charges in motion, both linear and circulatory. A method for measuring the magnetic field and other applications of these effects is described in detail.

SUMMARY

32.1 MAGNETISM

Some magnetic effects can be described in terms of magnetic poles and a quantitative relationship almost identical to Coulomb's law for electric charges. However, magnetic poles cannot be localized as can electric charges, and indeed, there is no evidence that an isolated magnetic pole can even exist. A more fundamental explanation is that magnetism has its origin in moving electric charges, either in the form of electric currents in wires or in the circulatory motion of the charged particles which make up atoms.

> Definition. Magnetism: a natural phenomenon which results in a force which acts upon electric charges in motion but not upon charges at rest.

32.2 MAGNETIC FIELD

Experiment shows that when a magnetic field is impressed upon a moving electrically charged particle, the particle experiences a force which would be absent if the particle were at rest. These experimental facts are used to frame an operational definition of the magnetic field.

> Definition. Magnetic field: (officially termed the magnetic induction) a vector concept symbolized by B and defined in terms of the force it exerts on a moving charged particle. The direction of B is defined as the unique direction in which the particle can move without experiencing any magnetic force. The magnitude of B is defined as
>
> $$B = \frac{F}{qv \sin\theta}$$

where θ is the angle between the direction of the field and the direction of the velocity. The sense of the vector representing B is defined so as to satisfy the vector equation

$$F = q(v \times B) . \qquad (32.1)$$

This vector equation shows that the direction of the magnetic force is perpendicular to both the direction of the velocity and the direction of the magnetic field. The same right hand rule which is used to describe the directions in the vector cross product is used to describe the directions of the magnetic field, magnetic force, and charge velocity.

The unit of B is called the tesla and it is this unit which must be used in calculations involving magnetic fields. However, one tesla is a quite strong magnetic field and in conversation, a more conveniently sized unit, the gauss, is frequently used; one gauss is equal to one ten thousandth of a tesla.

> Definition. Magnetic field lines: imaginary lines drawn in space in such a way that they are everywhere tangent to the direction of the magnetic field and that their density is directly proportional to the strength of the field.

As in the case of electric field lines, the magnetic field lines can be very useful in visualizing the effects of magnetic fields, but it must be remembered that they are intellectual constructs which must not be taken too literally.

> An important difference between electric and magnetic field lines is that whereas electric field lines originate and terminate on electric charges, isolated magnetic charges (poles) do not exist and thus, magnetic field lines are continuous; they neither originate nor terminate.

The absence in nature of isolated magnetic poles is taken as one of the fundamental laws of electromagnetism and is stated quantitatively in the magnetic analog of Gauss' law,

$$\int \vec{B} \cdot d\vec{A} = 0 \tag{32.6}$$

where the integration is performed over a closed surface.

> Definition. Magnetic flux: the normal component of the magnetic field vector, $\vec{B}$, passing through a surface area integrated over that area. It may be visualized as the number of magnetic field lines passing through the surface. The unit of magnetic flux is called the weber and the concept is symbolized by
>
> $$\Phi_B = \int \vec{B} \cdot d\vec{A} \tag{32.3}$$

In terms of the concept of magnetic flux, the important Eq. (32.6) may be stated - the magnetic flux passing through any closed surface is zero. In terms of magnetic field lines, this means that any line which enters a closed surface must eventually leave the surface.

32.3 APPLICATIONS OF MOVING CHARGES IN A MAGNETIC FIELD

For an understanding of the effects of a magnetic field, it is best to develop two separate "mental pictures", one involving individual atomic sized charged particles and the other involving the moving charges which constitute the electric currents carried by conductors. In this latter case, the charged particles (electrons) are confined to the conductor and the magnetic force experienced by the individual electrons is effectively transferred to the conductor.

> Trajectories of moving charged particles in a magnetic field. Since the magnetic force is always perpendicular to the velocity of the particle, the force can cause only a change in the direction of the velocity, never in its magnitude. Further, if the velocity is not perpendicular to the field, only that component of the velocity which is perpendicular to the field will be affected; the component which is parallel to the field will remain unchanged.

The magnetic force will thus be a centripetal force acting on the component of velocity which is perpendicular to the magnetic field and the particle will travel a circular path or, if the velocity also has a component parallel to the field, a helical path. The radius of curvature of the circle or helix can be calculated by setting the centripetal force equal to the magnetic force.

$$\frac{mv^2}{r} = qvB \tag{32.7}$$

If the path of the particle can by some means be made visible, Eq. (32.7) can be used, by measuring the radius of curvature, to determine the dynamical characteristics of the particle such as momentum, kinetic energy, or mass.

In the case in which the moving charges constitute the current in a segment of wire, the magnetic forces on the individual charges is transferred to the wire and so the wire experiences a force perpendicular both to itself and to the direction of the magnetic field. The magnitude of the force is given by

$$dF = Bi\, dl\, \sin\theta, \tag{32.19}$$

where θ is the angle between B and the current in the wire segment. It must be remembered that an isolated segment of wire cannot support an electric current; it must be a part of a complete circuit and so Eq. 32.19 must be taken to represent the <u>contribution</u> of this small segment to the force on the total circuit. The total force may be found by integrating Eq. 32.19 over the complete circuit, or at least, that part of the circuit which is subjected to the magnetic field. This force can be quite strong and, indeed, is the basis of operation of many devices, such as electric motors, which are used to transform electrical energy into mechanical energy.

If the conductor carrying a current is a relatively wide but very thin flat sheet, then we observe a combination of the two effects described above. Not only is the force on the individual charge carrier transferred to the conductor, but the carriers are deflected along a curved path within the conductor so that a transverse electric field is established perpendicular to the direction of current flow. This electric field produces a potential difference across the width of the conductor, the polarity of which depends upon the sign of the charge constituting the current.

$$V_H = \frac{B\,I}{tnq} \tag{32.28}$$

where t is the thickness of the strip, n is the number per unit volume of charge carriers, and q is the magnitude (including sign) of the charge on each carrier. This phenomenon, called the Hall effect, thus can distinguish between currents constituted of negative and positive charges. Note that in many semiconductors, while the charge carriers are electrons, they are not free electrons, and may behave as if they were positively charged particles.

32.4 MAGNETIC DIPOLE IN A MAGNETIC FIELD

If a loop of wire of any arbitrary shape is placed in a <u>uniform</u> magnetic field and an electric current is passed through the wire, careful analysis of the forces on the various segments of the loop shows that the net magnetic force on the loop is zero. Further analysis shows, however, that there is a torque on the loop given by

$$\tau = IB \cdot A$$

where the sense of the vector A representing the area enclosed by the loop is determined by applying the right hand rule to the direction of the current. This behavior is identical to the behavior of an electric dipole in an electric field if we define the magnetic dipole moment to be

$$\mu = iA \tag{32.32}$$

$$\tau = \mu \times B \tag{32.29}$$

Definition. The <u>magnetic dipole moment of a loop current</u>: a vector having a magnitude equal to the product of the current and the area enclosed by the loop and having a direction perpendicular to the plane of the loop. The sense of the vector is determined by applying the right hand rule to the direction of the current.

This definition applies not only to currents in wire loops but also to free charged particles moving around closed paths, or simply spinning on an axis.

As in the case of an electric dipole, the magnetic dipole in a non-uniform magnetic field experiences not only a torque tending to align it with the direction of the field, but also a net force in the direction of increasing field strength.

$$F = \mu \frac{dB}{dx} \qquad (32.35)$$

where the x axis is oriented in the direction of most rapidly increasing **B**.

Because the magnetic field exerts a torque on a dipole, work is required to rotate the dipole in a magnetic field and there will be a potential energy associated with the orientation of the dipole. If we take the zero of this potential energy to be when the dipole moment is perpendicular to the field, then this energy is given by

$$U = -\mu \cdot B \qquad (32.33)$$

The magnetic moments of the dipoles arising from the circulating charges in atoms will be invoked in a later chapter to explain the magnetic properties of matter.

EXAMPLE PROBLEMS

Example 1

A small aluminum sphere bearing a positive charge of .001 coulomb and having a mass of 50 gm is moving horizontally near the surface of the earth with a speed of 300 m/s and is, of course, subject to the gravitational force (its weight). What is the strength of the horizontal magnetic field directed perpendicular to the direction of the velocity required to just balance the gravitational force and keep the sphere from falling? If the sphere is moving from south to north, what must be the direction of the field?

Solution

Given: m = .05 Kgm
Q = .001 coulomb
v = 300 m/s from south to north

Required: B (magnitude and direction)

The gravitational force on the sphere is its weight, W = mg, acting downward. The magnetic force, F = Q(v x B), must therefore act upward. Applying the right hand rule to this vector cross product, we find that if v is from south to north, and F is to be upward, then B must be directed from east to west. Also, the forces must be equal in magnitude. Thus

$$mg = QvB$$

from

$$B = mg/Qv = \underline{1.63 \text{ tesla}}$$

This is a quite strong magnetic field which would require a large iron core electromagnet or a small superconducting magnet.

Example 2

At a certain place in the northern hemisphere, the earth's magnetic field has a magnitude of .35 gauss and is directed toward the north and downward at an angle of 71° below the horizontal. A power transmission line at that location is carrying an electric current of 1000 amperes horizontally from east to west. What magnetic force is exerted on each meter of the power line by the earth's field?

Solution

Given: $B = .35$ gauss $= 3.5 \times 10^{-5}$ tesla
         $\theta = 71°$ below the horizontal
         $I = 1000$ amperes

Required: F/ℓ

THe force on the current carrying conductor is given by

$$F = I(\vec{\ell} \times \vec{B}).$$

In this case, the magnetic field is perpendicular to the conductor and thus, the magnitude of the force is given by

$$F = BI\ell$$

or

$$F/\ell = BI = \underline{.035 \text{ newtons}}$$

The direction of the force is the direction of the cross product, $\vec{\ell} \times \vec{B}$, which is toward the south and 20° below the horizontal.

Example 3

An electron traveling at a speed of 2×10^7 m/s enters a region in which there is a magnetic field of 3 gauss directed perpendicularly to the path of the electron. If the region containing the field is 10 cm wide, by what angle has the electron been deflected from its original path upon leaving the field region?

Solution

Given: $v = 2 \times 10^7$ m/s
         $B = 3$ gauss $= .0003$ tesla
         $d = 10$ cm $= .1$ m

Required: θ

After entering the magnetic field, the electron travels along a circular arc with a radius of curvature given by

$$r = mv/Be$$

The angle of deflection will be the same as the angle subtended by this arc which can be calculated by

$$\sin\theta = d/r$$

Substituting the expression for r above into this equation yields

$$\theta = \arcsin(dBe/mv) = \underline{15.3°}$$

Example 4

A rotary solenoid consists of a rectangular coil of n turns of wire mounted on a shaft between the poles of a permanent magnet so that the magnetic field is parallel to the plane of the coil. When a current is passed through the coil, a torque is exerted on the shaft, causing it to execute some desired mechanical function. If the magnet produces a field of 1000 gauss and the coil has 250 turns, is 1.5 cm wide and 6 cm long and if the particular application requires a torque of .3 cm newtons, what current must be used to energize the solenoid?

Solution

Given: n = 250 turns
B = 1000 gauss = .1 tesla
L = 6 cm = .06 m
w = 1.5 cm = .015 m
τ = .003 MxN

Required: I

The torque on the coil is given by

$$\tau = nBIA = nBILw$$

from which

$$I = \tau/nBLw = \underline{.133 \text{ amperes}}$$

Example 5

A thin, flat, circular disc 10 cm in radius bears a total charge of +.001 C which is uniformly distributed throughout its volume. The disc is spinning at a rate of 3600 rpm about an axis through its center and perpendicular to its plane. What is the magnitude and direction of the magnetic moment created by the rotating charge?

Solution

Given: R = 10 cm = .1 m
Q = .001 C
f = 3600 rpm = 60 rps

Required: μ (magnitude and direction)

Since magnetic moment is defined in terms of ring currents, we may divide the disc into elements of area, dA, each of which constitutes a ring of rotating charge as shown in FIG. 5-1. The amount of charge, dq, contained on each ring is found by multiplying the charge density, ρ, by dA.

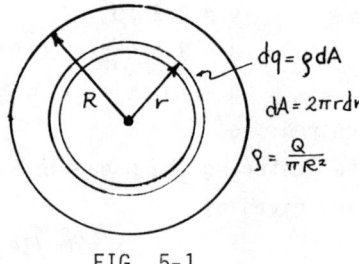

FIG. 5-1

$$dq = \rho dA = (Q/\pi r^2)\, 2\pi r\, dr$$

Each of these rotating rings is equivalent to a current of

$$dI = dq/t = f\, dq$$

and thus produces a magnetic moment

$$d\mu = dI\, A = f\, dq\, \pi r^2$$

The total magnetic moment of the disc may be found by integrating the contributions of these ring currents over the whole disc. Thus

$$\mu = (2\pi fQ/R^2) \int_0^R r^3 dr .$$

$$\mu = (1/2)\pi fQR^2 = \underline{.00094 \text{ A m}^2}$$

The vector representing the magnetic moment is directed along the axis of rotation; its sense is determined by applying the right hand rule to the direction of rotation.

PROBLEMS

1. An alpha particle is moving with a speed of 100 km/s in the positive x direction through a magnetic field of 1000 gauss in the positive y direction. What electric field (magnitude and direction) is required to balance the magnetic force on the alpha particle so that it will continue to travel in a straight line?

2. If the power transmission line in Example Problem 2 is reoriented so that it is carrying current from south to north, what is the magnetic force on the line?

3. An electron traveling with a speed of 2×10^7 m/s enters a region in which there is a magnetic field of 30 gauss directed 10° away from the electron's path. Since the component of the electron's velocity parallel to the field will not be affected, the electron will travel a helical path through the field. If the region containing the field is 25 cm long, how many turns of the helix will the electron execute before leaving the field?

4. A length of very flexible wire is formed into a flat but irregularly shaped loop and its ends are fixed to a pair of twisted lead wires so that a current may be passed through it. The loop is then placed in a uniform magnetic field. By careful analysis of the torque on the loop and of the forces on each segment of the wire, deduce the behavior of the loop when a current is passed through it.

5. The disc in Example Problem 5 also has a certain amount of mass and therefore possesses an angular momentum. If the mass of the disc is 100 gm, find the ratio of its magnetic moment to its angular momentum.

Chapter 33
Magnetic Field of Electric Current

PREVIEW
   The last chapter alluded to the fact that the origin of the magnetic field is a moving charge or an electric current. The present chapter develops that idea quantitatively with two fundamental laws which allows the calculation of the magnetic field for any given distribution of electric currents. These laws are used to describe the magnetic field produced by a particularly useful device, the solenoid, and finally to describe the modern method of defining the unit of and making precise measurements of electric current.

SUMMARY
33.1 BIOT AND SAVART'S LAW
   The Biot and Savart law serves the same function in magnetostatics (steady currents) as does Coulomb's law in electrostatics. It is more difficult to use because an isolated current carrying segment of wire cannot exist; if it is to carry a current, it must be a part of a complete circuit. Therefore, the Biot and Savart law can yield only the <u>contribution</u> to the magnetic field of a small segment of a complete circuit. The total field at a given point may only be found by integrating over all nearby currents.

   <u>Biot and Savart's Law</u>: the magnetic field at a point a (vector) distance r from a small segment of wire of length ds which carries an electric current I is given by

$$dB = \frac{\mu_0}{4} \frac{I \, ds \times \hat{r}}{r^2}. \tag{33.1}$$

Analysis of the vector cross product reveals that this magnetic field is perpendicular to the plane which contains the vector, ds and the vector, r.

   In Eq. (33.1), the proportionality constant is defined and so determines the size of the magnetic field unit.

$$\mu_0 = 4\pi \times 10^{-7} \text{Tm/A}$$

33.2 MAGNETIC FIELD OF ELECTRIC CURRENT
   Two examples serve to illustrate the use of the law of Biot and Savart. In the first case, the magnetic field near a long straight current carrying wire is considered. Although the wire must be part of a complete circuit, if the application at hand requires that we consider only points quite close to the wire, then the other portions of the circuit will be so far away from the point that, because of the inverse square nature of the field, we may safely ignore the contributions from these more distant portions. The magnitude of the magnetic field a (perpendicular) distance r from an infinitely long straight wire which carries a current I is given by

$$B = \frac{\mu_0 I}{2\pi r} \, . \tag{33.11}$$

The magnetic field lines are circles coaxial with the wire and the sense of B is given by the right hand rule:

> Right hand rule: if the wire is grasped with the right hand, the thumb pointing in the direction of the current, the curled fingers will show the direction of the magnetic field.

Note that to find this result, we were able to integrate the Biot and Savart equation only because we could safely ignore the "troublesome" parts of the circuit.

The second example is the problem of finding the magnetic field at points on the axis of a circular loop of wire in which there is a current. In this case, we are able to integrate the Biot and Savart equation because of the symmetry of the problem; there are no "troublesome" parts of the circuit. The result is

$$B = \frac{\mu_0 I}{2} \frac{a^2}{(a^2+x^2)^{3/2}} \tag{33.14}$$

where x is the distance from the field point to the center of the loop.

The direction of the field is along the axis of the loop and its sense may be determined by applying the right hand rule to each segment of the wire. Note that this equation is valid only for points on the axis of the coil; for points off the axis, the Biot and Savart equation becomes extremely difficult to integrate.

> Magnetic dipole field: the current carrying loop to which Eq. 33.14 applies is just the configuration described in Chapter 32 as a magnetic dipole and becomes particularly important in the region for which $x \gg a$. In this region, the equation becomes
>
> $$B = \frac{\mu_0}{2\pi} \frac{\mu}{x^3} \tag{33.16}$$

where $\mu = \pi a^2 I$, and must not be confused with μ_0, the permeability of free space.

33.3 AMPERE'S LAW

Ampere's law is stated in terms of the magnetic field at points along any arbitrary <u>closed</u> path in space and the electric currents encircled by that path.

$$\oint \mathbf{B} \cdot d\mathbf{s} = \mu_0 I \tag{33.19}$$

where the integral is evaluated over the closed path and I is the total electric current piercing the surface bounded by the path. The positive direction for I is determined by applying the right hand rule to the path of integration. It should be noted that it is not necessary for the surface bounded by the path of integration to be a plane surface; any one of the infinite number of surfaces bounded by a given closed path will suffice. In applying Ampere's law to the solution of problems, one must select the path of integration so that it conforms with the geometry of the magnetic field and the integration may be easily performed.

<u>Criteria for selecting the path of integration</u>

If B is zero, the integral is zero.

If B is perpendicular to the path, the integral is zero.

If B is constant along the path, the integral is easily evaluated.

33.4 MAGNETIC FIELD OF A SOLENOID

The solenoid is an electromagnetic device which has many important applications both in the scientific laboratory and in industry.

> **Definition.** <u>Solenoid</u>: a coil of wire in the form of a cylindrical helix and arranged so that an electric current may be passed through it. It is usually wound so that the turns of wire are quite close together and there may be several layers of turns. The interior may be empty or may be filled with a magnetic material.

If the length of the solenoid is considerably larger than its diameter, it may be approximated by an infinitely long solenoid, in which case the magnetic field inside is strictly axial and is uniform throughout the interior.

$$B_{axial} = \mu_0 n I \tag{33.27}$$

where n is the number of turns per unit length. This equation, while strictly valid only for a closely wound, infinitely long solenoid, is a very good approximation for a solenoid of finite length at points not too close to either end where a radial component of the field develops as the field lines curve outward around the solenoid. In applications in which these "end effects" must be avoided altogether, the solenoid may be bent into a toroid (doughnut shape) so that there are no ends. In this configuration, the internal field is no longer uniform but is easy to calculate.

There is also a small azimuthal field immediately outside the solenoid, similar in form to that for a long straight wire, whose strength decreases as the turns of wire are wound more closely together.

$$B_{azimuthal} = \mu_0 I / 2\pi r \tag{33.28}$$

Because the solenoid consists of a number of closely spaced loop currents, there is a magnetic moment associated with solenoid of finite length which is just the sum of the magnetic moments of the individual loops.

$$\mu_{solenoid} = n \ell (I \pi a^2) \tag{33.29}$$

where a is the radius of the solenoid.

33.5 DEFINITION OF THE AMPERE

Since a current carrying conductor both produce a magnetic field and is affected by a magnetic field, there will be a magnetic interaction between two neighboring conductors; they will exert forces on one another. In the case of two long straight parallel wires separated by a distance, d, this force per unit length is given by

$$F/\ell = \mu_0 I_1 I_2 / 2 \pi d \tag{33.31}$$

If the two currents are parallel, these forces will be attractive while if they are antiparallel, the forces will be repulsive.

This phenomenon forms the basis for the operational definition of the unit of electric current, the ampere, and for the current balance, the instrument used for the precise measurement of current.

EXAMPLE PROBLEMS

Example 1

Two very long straight wires are connected to the midpoints of two opposite sides of a wire square as shown in FIG. 1-1 and a current I is passed through the wires. What is the strength of the magnetic field at the center of the square?

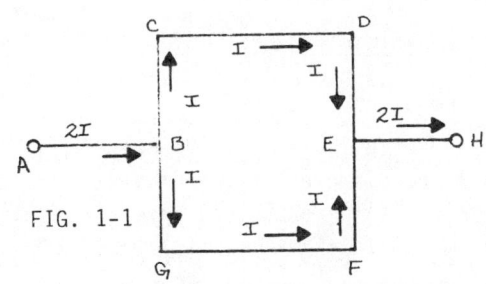

FIG. 1-1

Solution

When the integral in the Biot and Savart law is evaluated, the contributions of the segments A-B and E-H are zero because the cross product vanishes. The contributions of the segments B-C and B-G cancel each other as do the contributions of the segments C-D and G-F. So too do the contributions of the segments D-E and F-E. The magnetic field at the center of the square is zero. All that work for nothing!

Example 2

Two infinitely long straight parallel wires are situated a distance 2a apart and carry the same current in the same direction. Derive the expression for the magnetic field at points between the two wires.

Solution

If we place the origin of our coordinate system at a point midway between the two wires and orient the Y axis perpendicular to the wires, the magnitude of the portion of the field due to the first wire is, by Eq. 33.11

$$B_1 = \frac{\mu_0 I}{2\pi(a+y)}$$

while the portion of the field due to the second wire is, by the same equation

$$B_2 = -\frac{\mu_0 I}{2\pi(a-y)}$$

Using the principle of superposition, the total field is the sum of these two.

$$B = B_1 + B_2 = \frac{\mu_0 I y}{\pi(a^2 - y^2)}$$

This result may be checked intuitively; the field should be zero midway between the two wires and infinite at the position of either wire. Is it? A graph of the function is shown in FIG. 1-2.

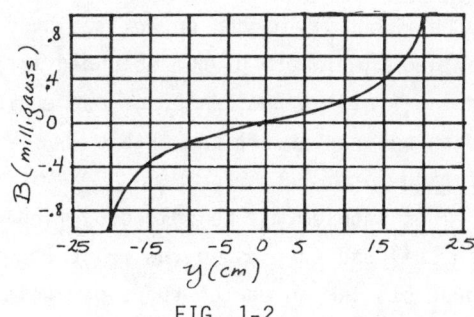

FIG. 1-2

Example 3

A closely spaced, single layer coil is to be wound on a toroidal form which has a rectangular cross section. The outside diameter (OD) of the toroid is 30 cm, the inside diameter (ID) is 28 cm, and the thickness is 2 cm. The diameter of the wire is 0.5 mm. Calculate the total number of turns on the coil and the length of wire required.

Solution
Given: OD = 30 cm
         ID = 28 cm
         t = 2 cm
         d = 0.5 mm
Required: N, L

The winding pitch, the distance between the centers of consecutive turns, can be no smaller than the diameter of the wire.

$$P = \ell/N = \pi D/N = d$$

Since the ID yields the smaller of the two circumferences, it must be used for D in this equation. Thus

$$N = \pi(ID)/d = \underline{1759 \text{ turns}}$$

Note that N should be an integer and because of mechanical restrictions, may not be "rounded up".

The length of wire required for each turn is the perimeter of the cross section with the length of each side increased by the diameter of the wire. Thus the total length of wire needed is

$$L = 2((1/2)(OD-ID)+t)N = \underline{10,554 \text{ cm}}$$

To this 105.54 m of wire must be added whatever length of wire is necessary to connect the coil to the external circuit.

Example 4

Use Ampere's law to find the expression for the magnetic field at points inside the coil of Example Problem 3 as a function of r, the distance from the center of the toroid to the field point.

Solution

For the path of integration, we choose a circle with radius r concentric with the toroid center. Because of the symmetry of the toroid and the absence of end effects, the magnetic field will be parallel to the path of integration at all points and thus, Ampere's law gives

$$\oint \mathbf{B} \cdot d\mathbf{s} = \mu_0 NI = 2\pi r B$$

from which

$$B = \mu_0 NI/2\pi r$$

Example 5

Eq. 33.27 is a good approximation for the field at the center of a solenoid of finite length provided the length of the solenoid is considerably greater than its radius. Find the expression for the field at the center of one end of a solenoid of length L and radius a for which the length and radius are comparable.

Solution

This problem may be solved by finding contribution to the field at the desired point due to each coil and then using the principle of superposition to add these contributions together. If n represents the number of turns per unit length, n=N/L, then an element of length, ds, contains nds turns and, according to Eq. (33.14), produces a field

$$dB = \frac{\mu_0 n I dx}{2} \frac{a^2}{[a^2+x^2]^{3/2}}$$

at a point on the axis a distance x from the coil. To find the field at one end of the solenoid, we must integrate this expression from x=0 to x=L. The time honored method of evaluating such an integral is to "look" it up in the tables"; this procedure yields

$$B = \frac{\mu_0 n I L}{2[L^2+a^2]^{1/2}}$$

PROBLEMS

1. A section at the center of a very long straight wire is formed into a semicircle as shown in FIG. 1 and a current I is passed through the wire. Derive the expression for the magnetic field at the center of the semicircle.

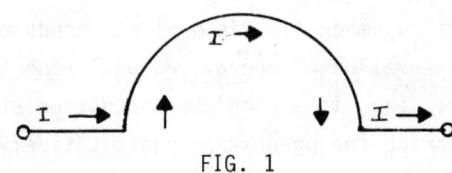

FIG. 1

2. Two infinitely long parallel straight wires, each carrying the same current in the same direction, are situated a distance 2a from each other. Orienting the X axis parallel to the wires with the origin midway between them and the Y axis perpendicular to the plane in which the wires lie, derive the expression for the magnetic field, B, at points along the Y axis.
Note: The perceptive student will note that this expression vanishes both for y=0 and for y=infinity thus implying some value of y for which B is a maximum. He will be unable to eat or sleep until he finds the position of this maximum.

3. From the results of Example Problem 4, find the expression for the magnetic flux passing through any cross section of the core of the toroid. Compare this expression with that derived by assuming that the magnetic field within the core is uniform and has the value calculated for the center line of the core.

4. Eq. 33.28 describes the tangential (azimuthal) magnetic field just outside a solenoid due to the helical nature of the windings and shows that this field is independent of the spacing of the wires or the total number of turns; it does, however, depend upon the number of layers of wire on the solenoid, a factor which has not been considered in Eq. 33.28. The normal method of winding a solenoid is to wind the first layer from right to left, the second from left to right and so on until the winding is finished. Describe, using Ampere's law, how the tangential field depends upon the number of layers.

5. A straight 50 cm length of wire is placed parallel to and 2 cm directly above an identical length of wire. The two wires are connected in series so that the same current of 5 amperes flows through both of them but in opposite directions. A third identical wire is placed midway between these two to form a current balance and is connected in series with the other two so that the current through it is in the same direction as the current through the lower wire. What is the magnitude and direction of the force on the central wire?

Chapter 34
Electromagnetic Induction

PREVIEW

The phenomenon described in this chapter is essentially the inverse of that studied in the previous chapter where we found that a steady electric current produces a magnetic field. Although a steady magnetic field cannot produce an electric current without violating the law of conservation of energy, we will find that a <u>changing</u> magnetic field does produce an electromotive force and, if a complete circuit exists, an electric current. Faraday's law is introduced to describe the phenomenon quantitatively. Lenz's law describes the polarity of this emf and is an application of the law of conservation of energy. Among the applications of Faraday's law are the electric generator, the transformer, and eddy current devices.

SUMMARY

34.1 MOTIONAL ELECTROMOTIVE FORCE

A conductor has been described as a material which contains a large number of free charge carriers, usually negatively charged electrons. If a conductor moves through a magnetic field, these free charges will experience a force

$$F = q(v \times B) \qquad (34.1)$$

and, being free, will move in response to this force, redistributing the charge and creating an electric field within the conductor. If the conductor is in the form of a straight wire of length ℓ, as a result of this electric field, there will be an emf between the two ends of the wire given by

$$\mathscr{E} = B\ell v \qquad (34.3)$$

if the wire is perpendicular to both the direction of the velocity, v, and the magnetic field, B. If this is not the case, then the right side of Eq. 34.3 must be multiplied by the sine of the angle between the vectors v and B and by the cosine of the angle between the wire and the normal to the plane defined by v and B. Thus, in general,

$$\mathscr{E} = (v \times B) \cdot \ell$$

> Definition. <u>Motional emf</u>: the potential difference induced across a conductor which is moving through a magnetic field.

If the wire forms part of a complete circuit, an electric current will flow in response to this induced emf. The polarity of the emf or the direction of the current is determined by Lenz's law which is given in section 34.2.

34.2 FARADAY'S LAW OF INDUCTION

The emf induced in a moving wire is a special case of a much more general phenomenon described by Faraday's law which applies to any region of space whether there is a conductor present or not.

Faraday's Law: In any region of space in which there is a changing
magnetic field, an electric field is induced such that

$$\oint \vec{E} \cdot d\vec{\ell} = -\frac{d\Phi}{dt}, \tag{34.11}$$

where the integral is to be evaluated around any closed path and Φ
is the magnetic flux passing through the surface bounded by that path.

Note that the path of integration may, but need not, conform to a conducting path.

In the form given in Eq. 34.11, Faraday's law is accepted as one of the fundamental laws of electromagnetism. The integral on the left side of Eq. 34.11 will be recognized as the potential drop around the closed path and the magnetic flux is defined as

$$\Phi = \int \vec{B} \cdot d\vec{A} \tag{34.9}$$
$$\Phi = \int B \cos\theta \, dA.$$

Thus, for many circuit applications, it is convenient to write Faraday's law as

$$\mathcal{E} = -\frac{d\Phi}{dt} = -\frac{d}{dt}(BA\cos\theta).$$

Here, B represents either the strength of a uniform field in the region of space considered, or the average value of a non-uniform field. Thus, we see from this equation that if a magnetic field passes through a conducting circuit, there are three ways (or any combination of these three) to induce an emf in that circuit:

An emf may be induced in a circuit by:
   a) changing the magnitude, B, of the field ,
   b) changing the area, A, of the circuit affected by the field ,
   c) changing the orientation, θ, of the circuit relative to the field.

There are numerous technological applications of each of these methods.

34.3 LENZ'S LAW

Lenz's law is an application of the law of conservation of energy to Faraday's law and accounts for the negative sign in Eq. 34.11.

Lenz's Law: the polarity of the induced emf or the direction
   of the induced current if there is a complete circuit, is such
   as to oppose whatever effect is causing it to be induced.

Since it is generally easier to apply Lenz's law to the induced current in a circuit than to the emf, it is convenient in those cases in which there is not a complete circuit, to establish an imaginary circuit for purposes of analysis.

Thus, if a current is induced as a result of an increasing magnetic field strength, the direction of the current will be such as to produce an opposing magnetic field. If the current is induced as a result of rotating a coil in a magnetic field, the current will be in such a direction as to produce a torque to oppose the rotation of the coil.

34.4 APPLICATIONS OF FARADAY'S LAW

Applications of Faraday's law may be analysed in terms of one, or a combination, of the three effects listed in section 34.2; they include:

The transformer in which a continuously changing magnetic field
produced by an alternating current in one coil (the primary)
threads through and induces an emf in a second nearby coil
(the secondary). There being no moving parts involved, the
transformer can be made to have an extremely high efficiency.

The rotating coil generator in which a multiturn coil of wire
rotates in a constant magnetic field, thereby generating an
alternating emf. By use of a commutator, a switch synchronized
with the rotation of the coil, the emf may be made unidirectional.

The flip-coil magnetometer in which the current induced in a small
coil of wire as it is rotated through 180° is used as a measure of
the magnetic field. This same measurement may be made by moving
the coil into or out of the field rather than rotating it. The
instruments used in conjunction with the flip-coil are usually
integrating instruments so that the analysis is done in terms of
charge rather than current; this obviates the necessity of
controlling the speed with which the coil is moved and yields an
instrument indication proportional to the field itself rather
than the time derivative of the field.

Electromagnetic braking, or eddy current braking, in which motion
opposing forces are produced (Lenz's law) by the very large
circulating (eddy) currents induced as large area conductors
are moved through a magnetic field.

EXAMPLE PROBLEMS
Example 1

A flat circular coil containing 200 turns of fine wire has a diameter of 15 cm and is in a magnetic field which is inclined at an angle of 35° to the axis of the coil. The resistance of the coil is 35 Ω. If the strength of the field is changing at the rate of 100 gauss per second, what is the magnitude of the current induced in the coil?

Solution

Given: $N = 200$ turns
$D = 15$ cm $= .15$ m
$\theta = 35°$
$dB/dt = 100$ gauss/s $= .01$ T/s
$R = 35\Omega$

Required: I

Since the strength of the field is the only component of Φ which is time dependent, Faraday's law for this case becomes

$$V = NA (dB/dt) \cos \theta,$$

and by Ohm's law, the current through the loop is

$$I = V/R = .00083 \text{ amperes} = \underline{.083 \text{ ma}}$$

Example 2

FIG. 2-1 shows two parallel conducting rails, separated by a distance of 20 cm and connected together at one end by a 2Ω resistor. There is a magnetic field of 1000 gauss perpendicular to the plane containing the rails and a straight length of wire, marked A-B in the diagram, lies across and is perpendicular to the rails. If the wire is pushed to the right along the rails at a speed of 30 m/s, what force is needed to keep it moving at this constant speed? What is the polarity of the emf induced in the wire?

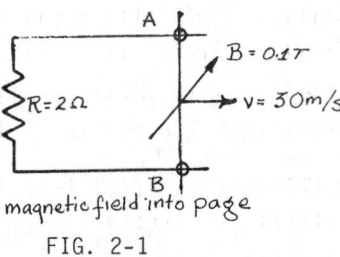
FIG. 2-1

Solution

Given: B = 1000 gauss = 0.1 tesla
R = 2 Ω
v = 30 m/s
ℓ = 20 cm = 0.2 m

Required: F, polarity

Since the wire, the field, and the velocity are mutually perpendicular to each other, the emf induced in the moving wire is

$$\mathcal{E} = B\ell v$$

As a result of this emf, a current flows in the circuit, including the moving wire,

$$I = \mathcal{E}/R = B\ell v/R$$

From Chapter 32, the wire carrying this current in a magnetic field must experience a force given by

$$F = BI\ell = \frac{B^2 \ell^2 v}{R}$$

which must, according to Newton's first law, be balanced if a constant speed is to be maintained. Substituting the appropriate numbers into this equation yields a force of F = $\underline{.006 \text{ newtons}}$ in the direction of motion. The magnetic force must, according to Lenz's law, be in the direction to oppose this driving force, which means that the current must be from B to A in the diagram, which in turn means that the point A must be positive relative to the point B. Note that the moving wire is acting as a <u>source</u> of emf for the rest of the circuit.

Example 3

A long solenoid wound with 20 turns/cm carries a 1000 hz alternating current described by

$$I = I_0 \sin (2\pi f t)$$

with a maximum value of 2 amperes. If the diameter of the coil is 2 cm, what is the rate of change of magnetic flux passing through the plane at the center of the solenoid and oriented perpendicular to the solenoid axis?

248 Chapter 34

Solution

Given: $I_0 = 2$ A
$D = 2$ cm $= .02$ m $\rightarrow R = .01$ m
$n = 20$ turns/cm $= 2000$ turns/m
$f = 1000$ hz

Required: $d\Phi/dt$

The magnetic field at the center of a long solenoid is
$$B = \mu_0 nI = \mu_0 I_0 \sin(2\pi ft)$$
and the magnetic flux is
$$\Phi = BA = \pi R^2 B$$

Thus, the rate of change of flux is found by substituting for B in this equation and differentiating.
$$d\Phi/dt = 2\pi^2 R^2 \mu_0 nI f \cos(2\pi ft)$$
$$d\Phi/dt = \underline{9.9 \times 10^{-3} \cos(2\pi ft) \text{ webers/s}}$$

Example 4

A small superconducting magnet consists of a solenoid 10 cm long and 3 cm in diameter and wound with 8,000 turns of very fine superconducting wire. Because of the small wire and close packing, the self induced voltage across the leads of the magnet must not exceed one volt. How many seconds will be required to energize the magnet from 0 to 5 tesla without exceeding this maximum voltage?

Solution

Given: $B = 0$ to $B = 5$ tesla
$L = 10$ cm $= .1$ m
$d = 3$ cm $\rightarrow r = 1.5$ cm $= .015$ m $N = 8,000$ turns
$V = 1$ volt

Required: Δt

As the magnetic field in the solenoid is increased, the induced voltage is given by Faraday's law. Assuming that the field is increased at a uniform rate, the technique leading to the lowest induced voltage,
$$V = N \frac{d\Phi}{dt} = NA \frac{\Delta B}{\Delta t}$$

Solving this equation for Δt gives
$$\Delta t = 2\pi N r^2 \frac{\Delta B}{V} = \underline{56.5 \text{ s}}$$

Note that since the resistance of the coil is zero, there is no voltage other than the self induced voltage across it.

Example 5

A 50 cm length of cylindrical copper pipe has an outside diameter of 25 mm and an inside diameter of 20 mm. A length of copper wire is threaded through the inside of the pipe, bent around the wall and brought back along the outside and the two ends connected together forming a rectangle with length equal to that of the pipe and width equal to the wall thickness of the pipe. A current of 100 amperes is then caused to flow through the pipe from end to end. If a time of

.5 millisecond was required for the current to rise from 0 to 100 amps, what was the average voltage induced in the wire loop during this time?

Solution

Given: ℓ = 50 cm = .5 m
OD = 25 mm
ID = 20 mm
I = 100 amp
Δt - .0005 s

Required: V

The magnetic field in the conductor as a result of the current flowing through it is azimuthal and, in magnitude, is given by Eq. (33.11).

$$B = \mu_0 I/2\pi r$$

Since this field is perpendicular to the plane of the coil, the flux through the coil may be calculated by dividing the coil into elements of area of length ℓ and width dr; thus,

$$\Phi = \frac{\mu_0 I \ell}{2\pi} \int_{r_1}^{r_2} \frac{dr}{r}$$

$$\Phi = \frac{\mu_0 I \ell}{2\pi} \ln \frac{r_2}{r_1}$$

The only component of Φ which changes with time is the current, I, and thus, the voltage induced in the loop is

$$V = \frac{\mu_0 \ell}{2\pi} \ln \frac{r_2}{r_1} \frac{dI}{dt} = \underline{.0045 \text{ volts}}$$

PROBLEMS

1. FIG. 1 shows two parallel straight wires separated by a distance of 50 cm and connected together at one end so that they carry the same current but in opposite directions. Midway between these two wires and coplanar with them, lies a rectangular coil of width 40 cm and length 60 cm. If the current, I, in the wires is increasing at a rate of 2 amperes/s, what is the emf induced in the rectangular coil?

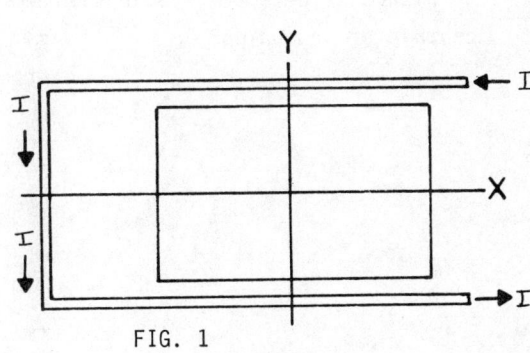

FIG. 1

2. If in Problem 1, the current instead of being a slowly varying direct current is an alternating current having the form,

$$I = I_0 \cos(2\pi f t),$$

where I_0 = 2 amps and f = 100 khz, what is the maximum value of the alternating voltage induced in the rectangular coil?

250   Chapter 34

3. An electric generator consists of a large electromagnet with pole faces machined into concentric cylinders so that the magnetic field in the gap between them is radial, i.e. directed outward from the common axis of the cylinders. Rotating within the gap is a rotor constructed of a number of conducting rods lying parallel to the cylindrical axis with their ends welded together in a parallel connection. The two common ends of the rods are connected through sliding contacts to an external circuit. The design specifications of the generator are:
    length of each rod - 130 cm
    radius of circle of rotation - 50 cm
    speed of rotation - 3600 rpm
    strength of magnetic field at rod position - 8000 gauss
    What emf is generated by this machine?

4. An earth inductor is a simple flip-coil designed to measure the earth's magnetic field. The device is a flat circular coil, 30 cm in diameter, wound with 300 turns of fine wire having a total resistance of 200 Ω. It is connected to a ballistic (charge measuring) galvanometer which has an internal resistance of 800 Ω. In use, it is placed so that the plane of the coil is horizontal and then rapidly rotated through 180° so that, after rotation, the plane of the coil is again horizontal. As a result of this operation, the galvanometer indicates that a charge of 1.5 µC flowed through the circuit. The operation is repeated beginning with the plane of the coil vertical and oriented east-west, giving a galvanometer indication of .52 µC flowing through the circuit. From these data, calculate the magnitude and direction of the earth's field.

5. A geophysicist wishes to measure the time variation of the earth's magnetic field. He constructs a square coil of 100 turns of wire mounted on fence posts around a horizontal field 50 m on each side. The coil is connected to an oscilloscope with which he can monitor the voltage induced in the coil. If the vertical component of the earth's field is .35 gauss and he wishes to detect variations as small as .1%/s, what minimum voltage must his instrument be capable of measuring?

Chapter 35
Inductance and Inductors

PREVIEW

The inductor is introduced as a two-terminal circuit element and its electrical characteristics are compared with those of the other two-terminal elements, the resistor and capacitor, encountered in previous chapters. The concept of inductance is developed from Faraday's law to describe the relationship between the voltage across and the current through the inductor. The energy of a magnetic field is derived from a consideration of the work required to establish the field in a solenoid. Finally, the oscillatory nature of the current and voltage in circuits containing capacitors and inductors is described.

SUMMARY

35.1 SELF-INDUCTANCE AND INDUCTORS

When an electric current flows in a circuit, a magnetic field will be generated in accord with Amperes's law. According to Faraday's law, if the current, and therefore the magnetic field, is changing with time, there will be an emf induced in the circuit. This induced emf will have the same effect on the circuit as an emf introduced by a battery or other source. While these induced emfs will be present to at least a small extent in any circuit, a circuit element may be devised to enhance and take advantage of the effect; such a device is called an inductor.

> Definition. An inductor: a two-terminal circuit element in which an emf appears across the terminals as a result of a changing (with time) current passing through it. According to Lenz's law, the polarity of the emf is such as to oppose the change in the current from which it results.

The law of Biot and Savart assures us that, whatever the geometry, the magnetic flux passing through the device or the circuit will be proportional to the current. According to Faraday's law, the magnitude of the induced emf is proportional to the rate of change of this flux and, therefore, to the rate of change of the current. The proportionality factor describing these relationships is called the self-inductance of the inductor and is symbolized by the letter, L.

$$\Phi = LI \tag{35.4}$$
$$\mathcal{E} = -L(dI/dt) \tag{35.6}$$

> Definition. The self-inductance of an inductor: the ratio of the magnitude of the emf induced across the terminals of the inductor to the rate of change of the current through it.

From this definition, we see that the unit of self-inductance is the volt/ampere/second, a rather clumsy combination of units which is given the name, henry, and abbreviated, H.

Qualitatively, an inductor may be described as a device through which the current cannot change instantaneously.

> **Definition.** **A one henry inductor**: an inductor for which a current changing at the rate of one ampere per second will induce an emf of one volt.

In making experimental measurements of the inductance of an inductor, Eq. 35.6 is used. In calculating the inductance of an inductor from its geometry, Eq. 35.4 is used.

By applying Eq. 35.4 to the solenoid described in Chapter 34, we find that the self-inductance of the solenoid is given by

$$L = \mu_0 N^2 A/\ell, \tag{35.8}$$

while application of this same equation to the coaxial cable yields a self-inductance per unit length of

$$L/\ell = (\mu_0/2\pi) \ln(r_1/r_2) \tag{35.11}$$

35.2 CIRCUIT ASPECTS OF INDUCTORS

When an inductor is included as one of several elements in an electrical circuit, it is, like each of the other elements, subject to Kirchoff's laws. The emf described by Eq. 35.6 is one of the voltage drops to be summed in applying Kirchoff's loop rule. Since an inductor is always some kind of a coil wound of wire, there is some resistance associated with it, a resistance which often is comparable with other resistances in the circuit and therefore cannot be ignored. When analysing a circuit containing an inductor,

> **a real inductor** may always be characterized as an ideal (resistanceless) inductor in series with the resistance of the wire of which it is wound.

When an inductor is connected in series with a resistor, either its own intrinsic resistance or an external resistor, and the circuit is energized with a constant emf such as from a battery, since the current through the inductor cannot change instantaneously, the current will increase gradually from zero to a final constant value determined by the magnitude of the emf and the resistance. The equations governing the current and voltage in such a circuit are

$$I = \frac{\mathcal{E}}{R}[1 - e^{-Rt/L}] \tag{35.13}$$

and

$$V = \mathcal{E}e^{-Rt/L} \tag{35.14}$$

In these equations, the time, t, is measured from the instant at which the circuit is first energized, for example by closing a switch. Eq. 35.14 indicates that the voltage across the (ideal) inductor starts out at a maximum value of $\mathcal{E}$ and gradually decreases to zero as the current through the inductor changes less and less rapidly and finally becomes constant. As in the case of an RC circuit, a time constant may be defined for this LR circuit.

> **Definition.** **The time constant**: time required for the current through the inductor to increase from zero to 63% of its final value. Alternatively, the time constant may be defined as the time required for the voltage across the inductor to decrease from its initial value to 37% of that initial value. Either of these definitions indicate that the time constant may be calculated

by T=L/R where R is the resistance in series with the inductor and must include the intrinsic resistance of the inductor itself.

<u>WARNING</u>: The description of an inductor as a device through which the current cannot change instantaneously reveals a potential hazard in the operation of a device such as a large electromagnet which has a large inductance associated with it. If the current through such a device is suddenly interrupted, as by opening a switch, very large voltages will be induced in the circuit. At the very least, these voltages will cause arcing across the switch and may very well be hazardous to either the equipment or to nearby personnel.

35.3 ENERGY STORED IN A MAGNETIC FIELD

When an inductor is first energized, the current and magnetic field build up gradually. Since the current must flow against the emf induced by the changing magnetic field, energy is required to establish the field. For a given (ideal) inductor, the amount of energy required to establish the final constant field is

$$U_M = 1/2 \, LI^2 \tag{35.20}$$

This energy is not transformed into thermal energy (heat) as in the case of a resistor, but rather is stored in the magnetic field surrounding the inductor and is retrieved by the circuit when the current and the magnetic field are again reduced to zero. It is instructive to think of this energy as residing in the magnetic field independently of the inductor and, indeed, experiments show that any magnetic field has associated with it an energy density (energy per unit volume) given by

$$u_M = \frac{1}{2\mu_0} B^2$$

35.4 OSCILLATIONS IN A CIRCUIT CONTAINING A CAPACITOR AND AN INDUCTOR

If an initially charged ideal capacitor and an ideal inductor are connected together to form a circuit, neither the voltage nor the current in the circuit can change instantaneously. Analysis of the circuit using Kirchoff's laws shows that the charge initially stored on the plates of the capacitor will flow through the inductor, inducing a sufficient voltage to recharge the capacitor with the opposite polarity. The circuit will oscillate with a frequency given by

$$\omega = 2\pi\nu = 1/(LC)^{1/2}$$

In such a circuit, the energy is transferred alternately from the electric field in the capacitor to the magnetic field in the inductor and back to the electric field. Any real circuit will, of course, contain some resistance - at least the resistance of the connecting wires and the intrinsic resistance of the inductor. The resistance in the circuit will cause the frequency of oscillation to be reduced and also cause the oscillations to gradually decrease in amplitude and eventually die out. This effect may be conveniently described by introducing a "damping factor", d=R/2L, such that

$$I = I_0 \, e^{-dt}$$

The frequency of oscillation is then given by

$$\omega = [(1/LC) - d^2]^{1/2}$$

While ideal capacitors and inductors cannot be made, it is possible to electronically simulate a negative resistance which may be connected in series with the real inductor to cancel the unavoidable circuit resistance; in this way, it is possible to achieve the continuing oscillations described above in terms of the ideal components.

EXAMPLE PROBLEMS

Example 1

A flat circular coil is wound of N turns of wire and has radius a. Making the (incorrect) assumption that the magnetic field in the plane of the coil is uniform, derive the expression for the inductance of the coil.

Solution

According to Eq. (33.15), the magnetic field at the center of the loop is given by
$$B = \mu_0 NI/2a$$
Since we are assuming that the field over the plane of the coil is uniform, the magnetic flux passing through the plane of the coil is given by
$$\Phi = (\mu_0 IN^2/2a) \int 2\pi r \, dr$$
where the integral is evaluated from r = 0 to r = a.

The integration yields
$$\Phi = 1/2 \, \mu_0 \, \pi N^2 Ia$$
From Eq. (35.4)
$$L = \Phi/I = 1/2 \, \mu_0 \, \pi N^2 a$$

Example 2

In the circuit of FIG. 2-1, the inductor has a negligible intrinsic resistance. Suppose that the switch has been closed for a sufficiently long time for the current to stabilize in the circuit and then is suddenly opened. What is the voltmeter reading five milliseconds after the switch is opened? Assume an ideal voltmeter.

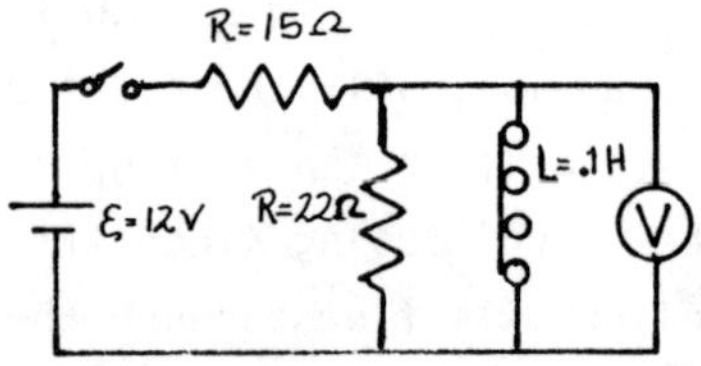

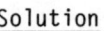

FIG. 2-1

Solution

Given: $\mathcal{E}$ = 12 volts
  r = 15 Ω
  R = 22 Ω
  L = 0.1 henry
  t = .005 s

Required: V

For a steady current, the resistance of the parallel combination of the resistor, R, and the inductor is zero and therefore, the current in the circuit just before the switch is opened is
$$I_0 = \mathcal{E}/r = \underline{0.8 \text{ amps}}$$
all of it passing through the inductor. The voltmeter reading at this time is zero. Since the current in the inductor cannot change instantaneously, immediately after the switch is opened, this same current must pass through the resistor, R, and the voltmeter reading at this time is
$$V_0 = I_0 R = \underline{17.6 \text{ volts}}$$

As the magnetic field in the inductor collapses, the voltage across the inductor decays according to
$$v_L = v_0 e^{-Rt/L}$$
and when t = .005 s, the voltmeter will read
$$V_L = \underline{5.9 \text{ volts}}$$

Example 3

A toroidal coil has a rectangular cross section with an ID of 20 cm, an OD of 30 cm and a width of 5 cm. If the coil contains 120 closely spaced turns of wire and carries a current of 2 amperes, what is the magnetic energy within the coil?

Solution

Given: N = 120 turns
h = 5 cm = .05 M
ID = 20 cm -> r_i = .1 M
OD = 30 cm -> r_2 = .15 M
I = 2 A

Required: U

Using Ampere's law, we find that the magnetic field in the toroid is
$$B = \frac{\mu_0 NI}{2\pi r}$$

and, thus, the energy per unit volume (energy density) is
$$u = \frac{\mu_0}{8} \left(\frac{NI}{\pi r}\right)^2$$

The total energy may be found by integrating the energy density over the full volume of the toroid.
$$U = \frac{\mu_0}{8} \left(\frac{NI}{\pi}\right)^2 \int dr/r$$
$$U = \frac{\mu_0}{8} \left(\frac{NI}{\pi}\right)^2 \ln(r_2 r_1)$$
$$U = \underline{3.72 \times 10^{-4} \text{ Joules}}$$

Example 4

An inductor having an inductance of 100 mh and a resistance of 200 Ω is connected to a 0.1 μf capacitor. What is the natural frequency of oscillation of this combination? What would be the natural frequency if the resistance of the inductor were zero? What is the time constant of the circuit?

Solution

Given: L = 100 mh = 0.1 henry
R = 200 Ω
C = 0.1 μf = 1 × 10^{-7} farads

Required: ν ; ν_0 ; T

The damping factor for this circuit is d = R/2L = 1000/s .

Thus, the natural frequency of oscillation is
$$\nu = \frac{1}{2\pi} [(1/LC) - d^2]^{1/2} = \underline{1584 \text{ hz}}$$

If the inductor were ideal and had no resistance associated with it, the natural frequency of the circuit would be

$$\nu_0 = (1/LC)^{1/2} = \underline{1584 \text{ hz}}$$

The time constant of the circuit is

$$T = L/R = \underline{.0005 \text{ s}}$$

Note that there is only a small difference in the frequencies even with a time constant as small as one half millisecond.

Example 5

An inductor which has an inductance of 1 millihenry and a resistance of $15\,\Omega$ is connected to an initially charged .47 µf capacitor so that the circuit oscillates with an exponentially decreasing amplitude. How many cycles of oscillation must be completed before the amplitude has decayed to one tenth of its initial value?

Solution

Given: $L = 1 \text{ mH} = .001 \text{ H}$
         $R = 15\,\Omega$
         $C = .47\;\mu f = 4.7 \times 10^{-7} \text{F}$
         $I = .1\, I_0$

Required: t/T = number of cycles

The amplitude of the current oscillations is given by

$$I = I_0\, e^{-Rt/2L}$$

and the frequency of the oscillations is given by

$$\nu = (1/2\pi)\left\{\frac{1}{LC} - \left(\frac{R}{2L}\right)^2\right\}^{1/2}$$

while the period of the oscillations is

$$T = 1/\nu$$

Solving the first of these equations for t yields

$$t = -(2L/R)\,\ln(I/I_0)$$

Thus,

$$t/T = -\nu(2L/R)\ln(I/I_0) = 2.2 \text{ cycles}$$

Three full cycles must be completed.

PROBLEMS

1. What is the self-inductance of the superconducting magnet described in Example Problem 4 of Chapter 34?

2. In Example Problem 1 above, it was assumed that the magnetic field is uniform over the plane of the coil. A slightly better approximation is

$$B = \frac{\mu_0 IN}{2a}\left[1 + \frac{3}{4}\left(\frac{r}{a}\right)^2\right]$$

Using this expression for the magnetic field in the plane of the coil, derive the expression for the self-inductance of the coil.

3. For the circuit shown in FIG. 1, use Kirchoff's two laws to show that the equation governing the current through the inductor is

$$\mathcal{E} = rI_L + (1 + \frac{r}{R})L \frac{dI_L}{dt}$$

and from this equation, deduce the time constant for the circuit.

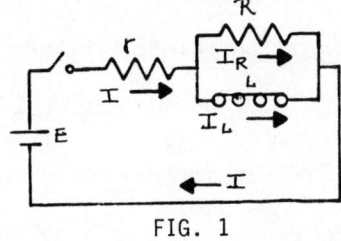

FIG. 1

4. The pole faces of a large iron core electromagnet are parallel to each other. They are each 10 cm in diameter and are separated by a distance of 2 cm. If the magnetic field between the pole faces is 1.5 tesla, how much magnetic energy is contained in the space between the pole faces?

5. Eq. 35.11 gives the inductance per unit length and Eq. 29.8 the capacitance per unit length of a coaxial cable. If one end of a ten meter length of such a cable is shorted (the inner and outer conductors connected together), and a potential difference is suddenly applied to the other end, electrical oscillations will occur in the cable. Assuming negligible resistance in the cable, what is the frequency of the oscillations?

Chapter 36
Magnetic Properties of Matter

PREVIEW

In the previous four chapters, we dealt with the magnetic fields in the empty space near moving electric charges and current carrying conductors. In this chapter, we study the way in which the magnetic field interacts with the moving charges which constitute the atomic structure of bulk materials and thereby affects the characteristics of that material. The related concepts of permeability and susceptibility are defined to describe these characteristics and we find that materials may be divided into three distinct classes in terms of their magnetic properties. The chapter ends with a discussion of the magnetic field of the earth.

SUMMARY

36.1 BEHAVIOR OF MATERIALS IN A NONUNIFORM MAGNETIC FIELD

Careful magnetic field measurements in the vicinity of a piece of material will show, in most cases, that the field is zero if the material has been properly shielded from the influence of external fields such as that of the earth. A few materials, most notably iron, may be pretreated in such a way that these measurements will show a quite strong magnetic field associated with them.

All materials, when placed in a non-uniform magnetic field will experience a force due to that field. This force will, for some materials, be directed toward the region of stronger field and for other materials, be directed toward the region of weaker field. In most cases, the force is so weak that it can be detected only by very sensitive instrument systems; for a few materials, however, the force can be very strong. These same forces are not experienced in a uniform field, although, in the case of the pretreated materials described at the end of the previous paragraph, the uniform field will produce a torque on the material.

36.2 CLASSIFICATION OF MAGNETIC MATERIALS

The measurements described qualitatively in Section 36.1 suggest that bulk matter normally has no magnetic field associated with it, but that exposure to an external magnetic field will <u>induce</u> a magnetic dipole moment in the material. In some few cases, the induced magnetic moment will remain after the external field has been removed. These effects may be described quantitatively in terms of the induced dipole moment per unit volume. Thus, the magnetic field inside a material sample may be described as

$$B = B_E + B_I$$

where B_E is the field which would be present at that point in space in the absence of the sample, and B_I is the field due to the induced dipoles within the sample. Thus, if the net magnetic moment induced in the sample by the external field is represented by μ_I, then

$$B_I = (\mu_0/V)\, \mu_I$$

and
$$B = B_E + (\mu_0/V)\, \mu_i \qquad (36.12)$$

WARNING: Note that the symbol, μ, is used here in one instance for the permeability of free space and in another instance for the magnetic moment. Be careful to not confuse them.

Since for not too large external fields, the induced dipole moment is directly proportional to the external field, it is often convenient to write

$$\mu_I = \chi_M (B_E V/\mu_0) \qquad (36.13)$$

where χ_M is called the magnetic susceptibility.

The field inside the sample may then be written as

$$B = B_E (1 + \chi_M) \qquad (36.14)$$

In cases in which the susceptibility is not constant (independent of the external field), it is often more convenient to define a different coefficient, the relative permeability, as

$$\kappa_M = 1 + \chi_M \qquad (36.16)$$

so that

$$B = \kappa_M B_E$$

Three classes of magnetic materials may be defined as follows:

36.3 DIAMAGNETISM

Definition. <u>Diamagnetic material</u>: a material which experiences a force toward the weaker region of a non-uniform magnetic field. This force is very weak and its magnitude is independent of the temperature of the material.

Careful measurements show that <u>all</u> materials exhibit diamagnetism but that in many materials, the diamagnetism is almost completely hidden by one of the other types of magnetism which are, in general, stronger. For these materials, χ is negative so that the field in the region occupied by the sample is <u>smaller</u> than it would be in the absence of the sample and, since $\chi \ll 1$, κ is a number between 0 and 1.

The superconductor is a special case of diamagnetism. When a metal is in the superconducting state, there can be no magnetic field inside it. If a magnetic field is impressed upon a sample while it is in the superconducting state, the field cannot penetrate the sample; if a magnetic field exists inside a normal sample at a higher temperature, then when the temperature is lowered and the sample becomes superconducting, the field will be expelled from the sample. Thus, a superconductor must have a relative permeability of zero and a susceptibility of minus one.

36.4 PARAMAGNETISM

Definition. <u>Paramagnetic materials</u>: a material which experiences a force toward the stronger region of a non-uniform magnetic field. This force is very weak and its magnitude depends in a predictable way on the temperature of the material; in general, as the temperature decreases, the force becomes stronger.

For these materials, χ is positive so that the magnetic field in the region occupied by the sample is <u>larger</u> than it would be in the absence of the sample and, since $\chi \ll 1$, κ is a number just slightly larger than 1. Many paramagnetic materials become ferromagnetic at sufficiently low temperatures.

Paramagnetism in non-metals results from the tendency of the magnetic moments of the individual atoms or molecules of the sample to align with the external magnetic field. This tendency toward alignment is opposed by thermal agitation which always produces a tendency toward randomness. Clearly, as the strength of the field increases, the fraction of the atoms aligned with the external field also increases but just as clearly, this fraction cannot exceed 100%. Thus, as the externally applied field increases in strength until nearly all of the atomic dipoles are aligned with it, further increases become less and less effective.

> Definition. Magnetic saturation: a characteristic of paramagnetic and ferromagnetic materials resulting from the nearly complete alignment of the atomic dipoles of the material. In the magnetically saturated state, the magnetic susceptibility of the material is zero.

> Curie's Law: For paramagnetic materials, if the external field is not too strong nor the temperature too low, the susceptibility is inversely proportional to the absolute temperature.

$$X_M = (\mu_0 C/V)/T \tag{36.17}$$

where the quantity enclosed in the parentheses is called the Curie constant. Note that, while at normal temperatures, the paramagnetic susceptibility is quite small, it can become significant at low temperatures because of Curie's law. This fact makes certain paramagnetic materials useful for low temperature thermometry. Since no material can support an infinite susceptibility, it is clear from Eq. 36.17 that no material can follow Curie's law to the ultimate low temperature of absolute zero. For each paramagnetic material, there is a certain cricital temperature below which the material becomes either ferromagnetic or antiferromagnetic; this temperature is called the Curie temperature for ferromagnetic materials or the Neel temperature for the antiferromagnetic materials.

In metals, except for some of the transition group metals, the paramagnetism is a result of the alignment of the conduction electrons rather than the orbital electrons and the susceptibility is temperature independent; metals generally do not follow Curie's law.

36.5 FERROMAGNETISM

> Definition. Ferromagnetic material: a material which experiences a force toward the stronger region of a non-uniform magnetic field. This force is quite strong and its magnitude does not depend sensitively upon the temperature except in that if the temperature is raised above a certain critical temperature, different for each material and called the Curie temperature, the ferromagnetism will revert to parmagnetism.

For these materials, the susceptibility is positive and quite large, having values of several thousand and is strongly dependent upon the strength of the applied external field. As do the paramagnetic materials, these materials exhibit the phenomenon of magnetic saturation when subjected to sufficiently strong fields. Unlike the paramagnetic materials, these materials also exhibit the phenomenon of hysteresis.

> Definition. Magnetic hysteresis: a characteristic of ferromagnetic materials in which the response to a decreasing applied field is different in magnitude from the response to an increasing field. As a result of hysteresis, if the external field is first increased

to near saturation and subsequently decreased to zero, a significant fraction of the induced dipole moment will remain in the sample. This remanent field can be eliminated only by the application of a sufficiently strong external field in the opposite direction.

Definition. Remanent field: the value of the magnetic field, B, remaining in a ferromagnetic sample after the magnetizing field has been reduced to zero.

Definition. Coercive force: the strength of the reverse magnetizing field, B_E/μ_o, required to reduce the magnetic field in a ferromagnetic sample to zero after initial magnetization.

If a ferromagnetic material is subjected to a cyclical magnetic field, as might be produced by an alternating current in a coil, a plot of B versus the applied field yields a curve such as that shown in Fig. 36.13e in the textbook. The importance of this hysteresis curve lies in the fact that during each cycle traversed, magnetic energy is transformed into thermal energy in an amount proportional to the area enclosed by the curve.

$$U = \frac{V}{\mu_o} \oint B_E dB$$

where V is the volume of the sample and the integration is carried out around the closed path represented by the hysteresis loop.

This dissipation of energy is responsible for a significant portion of the heat generated in alternating current machinery such as transformers and electric motors.

36.6 THE EARTH'S MAGNETIC FIELD

There exists in the vicinity of the earth and other members of the solar system a magnetic field which closely resembles the field of a magnetic dipole. In the case of the earth, measurements suggest a dipole centered near the center of the earth and having its axis tilted about 11° from the axis of rotation and directed more or less toward the south geographic pole. Thus, at the surface of the earth, there is a horizontal component of the field directed generally toward the north and a vertical component directed downward.

Definition. Angle of declination: the angle between the horizontal component of the earth's field and geographic north. This angle varies from about 20° on the west coast of the United States through zero on a line through western Ohio and Kentucky to about -20° on the east coast.

Definition. Angle of inclination (often called the dip angle): the angle measured downward from the horizontal to the direction of the magnetic field lines at the surface of the earth. This angle varies from about 55° to 75° in the United States.

Both the direction and strength of the earth's field change with time and indeed, recent evidence suggests that the direction of the field has actually reversed on several occasions in the geologic past.

Celestial objects outside the solar system are known to have magnetic fields associated with them also.

EXAMPLE PROBLEMS

Example 1

Using a method developed by Michael Faraday to measure the magnetic susceptibility of a metal, a cylindrical rod of the metal 20 cm long and 2 cm in diameter is suspended by one end from an equal arm balance so that the lower end is in the uniform magnetic field of 0.5 T produced by a permanent magnet and the upper end is well outside of the field. With the magnet removed, the balance is balanced; then the magnet is replaced and it is found that an additional 38 milligrams is required to rebalance the balance. From these data, find the susceptibility of the metal.

Solution

Given:
$L = 20$ cm $= .2$ M
$D = 2$ cm $= .02$ M
$B = .5$ T
$\Delta m = .038$ gm $= .000038$ Kgm

Required: χ

Over the length of the rod, the magnetic field varies from B at the lower end to zero at the upper end. A magnetic force is experienced by that part of the rod which is in the non-uniform field near the edges of the magnet pole faces. The rod may be divided into disc shaped elements of volume,

$$dV = A\, dy = \pi D^2/4\, dy$$

each of which will have an induced dipole moment of

$$d\mu_I = \frac{\chi B A}{\mu_o} dy$$

and experience a force either toward or away from the region of stronger field given, according to Eq. 32.35, by

$$dF = d\mu\, (dB/dt) = \frac{\chi B A}{\mu_o} dB$$

This equation may be integrated from one end of the rod to the other to find the total force. Thus,

$$F = mg = \chi \frac{A}{\mu_o} \int_0^B B\, dB = \frac{\chi A}{2\mu_o} B^2$$

This equation may be solved for χ to yield

$$\chi = \underline{11.8 \times 10^{-6}}$$

Since weights had to be added rather than subtracted to rebalance the balance, the metal must have been paramagnetic.

Example 2

Oxygen is paramagnetic and, in the liquid form, has a magnetic susceptibility of

$$\chi = 4.67 \times 10^{-6}$$

If a test tube containing 25 milliliters of liquid oxygen is placed in a uniform magnetic field of 1000 gauss, what is the magnitude of the dipole moment induced in the liquid?

Given:
$V = 25$ ml $= 2.5 \times 10^{-5}$ M^3
$\chi = 4.67 \times 10^{-6}$
$B_E = 1000$ gauss $= 0.1$ T

Required: μ_I

The solution to this problem is a straightforward application of Eqs. 36.11 and 36.13. Combining these equations gives

$$\mu_I = B_E V/\mu_0$$

Substitution of the numerical values from the given data yields

$$\mu_I = \underline{1.99 \text{ J/T}}$$

Example 3

Hydrated copper sulfate is paramagnetic and has a Curie constant of .051 K. What is the magnetic susceptibility of this compound at 300 K? What is its susceptibility at 10 K?

Solution

Given: Curie Constant = .051

$T_1 = 300$ K

$T_2 = 10$ K

Required: X_1, X_2

The Curie constant is defined in Eq. 36.17 so that

$$X = \frac{\text{Curie constant}}{T}$$

Substituting the given data into this equation yields

$$X_1 = 1.7 \times 10^{-4}$$

and

$$X_2 = 5.1 \times 10^{-3}$$

Example 4

The graph in FIG. 4-1 shows the hysteresis loop for a sample of iron. From this graph, determine the amount of energy per unit volume transformed into heat each time the sample traverses the loop.

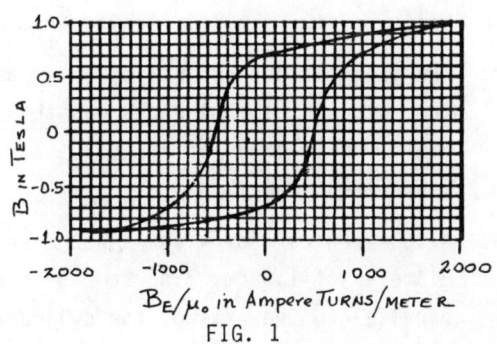
FIG. 1

Solution

The energy loss per unit volume per cycle is equal to the area enclosed by the hysteresis loop. Thus, the problem is one of measuring the area. A number of methods are available for such a measurement, but the best method in the present case is to count the squares on the graph. There are approximately 186.5 squares enclosed by the loop. The length of the vertical side of each square is 0.1 tesla and the length of the horizontal side of each square is 100 amperes per meter; thus, the area of each square is (0.1 x 100) tesla-amperes/m or 10 Joules/m^3. Multiplication of this scale factor by the number of squares yields an area of 1865 Joules/m^3/cycle.

Example 5

In Puerto Rico, the angle of inclination (dip) is 50°. The strength of the horizontal component of the earth's field at this location is 0.3 gauss. What is the full strength of the field?

Solution
    Given: θ = 50°
                B_H = 0.3 gauss
    Required: B
    The angle of inclination is the angle between the field lines and the horizontal plane; thus
    $$B = \frac{B_H}{\cos\theta} = \underline{.47 \text{ gauss}}$$

PROBLEMS
1. The magnetic susceptibility of a sample of copper sulfate is to be measured by the Gouy method in which a small spherical sample is suspended from one arm of an equal arm balance in a non-uniform field so that the magnetic force on the sample can be measured. The field is provided by a permanent magnet with pole faces specially shaped so that B varies uniformly from 5000 gauss at the center to zero at a distance of 10 cm above the center line. The sample is 1 cm in diameter and is placed precisely in the center of the region of non-uniform field. The balance is balanced with the magnet removed and it is found that an additional 9.04 milligrams is required to rebalance the balance when the magnet is replaced. What is the susceptibility of the sample? Is it paramagnetic or diamagnetic?

2. The magnetic susceptibility of a paramagnetic sample is measured at 1.4 K. The temperature of the sample is then lowered and the susceptibility is again measured and found to have a value 3.5 times as large as its value at the higher temperature. At what temperature was the second measurement made?

3. A 300 turn coil of wire is wound snugly around a cylindrical sample of indium metal having a diameter of 1.5 cm. The sample is placed in a uniform magnetic field of 200 gauss which is parallel to the axis of the cylinder. While the magnetic field remains constant, the temperature of the sample is reduced while the voltage across the coil is monitored. At a temperature of 1.7 K, the indium becomes superconducting and the magnetic flux is expelled from the sample producing a momentary voltage across the coil. If the average value of this voltage was 0.6 volts, what time interval was required for the sample to become perfectly diamagnetic?

4. From the hysteresis loop shown in FIG. 4-1, determine the values of a) the remanent field, b) the coercive force and c) the relative permeability, κ, for small values of B.

5. In Cimmarron, NM, the angle of declination is +12.5°, which is to say, the compass needle points 12.5° east of geographic north. If the pilot of a small plane wishes to fly due west to the village of Taos on a day on which there is no wind, what must be his compass heading?

Chapter 37
Alternating Currents

PREVIEW

Up to this point, we have dealt almost exclusively with direct current (dc) in which the polarity of the source of emf and the direction of the current in a circuit does not change. In an alternating current circuit, the polarity of the source and, therefore, the direction of the current does change periodically. It is this changing voltage and current which give capacitors and inductors a special importance in ac circuits and which makes possible a very simple and relatively inexpensive device, the transformer, by use of which ac voltages can readily be raised or lowered. The transformer, in turn, makes practical the transmission of electrical power over long distances without which our modern technological society could not easily function.

SUMMARY

37.1 ALTERNATING CURRENTS

While the term, ac, is an acronym for alternating current, it has by convention come to be applied to a current or a voltage which varies sinusoidally with time. Although there are other periodically changing forms of current and voltage which are of considerable importance, it is more important by far to study and understand the sinusoidal form because:

1. The sinusoidal form is easily produced by both mechanical generators and electronic circuits, is easy and inexpensive to transmit over long distances, and is easy to use in many appliances. For these reasons, ac is used almost exclusively for commercial and industrial power the world over.

2. Periodically varying currents and voltages which are not sinusoidal can, according to a theorem developed by Fourier, always be represented by a superposition of a series of sinusoidal functions, and so, if a circuit can be analysed for sinusoidal currents and voltages, it can be analysed for <u>any</u> periodic function.

For an ac circuit, the instantaneous values of the current and voltage may be represented by the equations,

$$V = V_m \cos(\omega t + \Phi) \quad \text{and} \quad I = I_m \cos(\omega t + \Phi)$$

where the subscript m denotes maximum, or peak, values, and $\omega = 2\pi\nu$ is the frequency measured in radians per second. The phase angle is called Φ and is determined by the (arbitrary) selection of the zero point for time measurements. Clearly, the sine function could be as easily used in the above equations. Obviously, it is convenient, whenever possible, to set $\Phi = 0$.

Since the instantaneous value of the current or voltge varies from positive maximum to negative maximum over each cycle, it is necessary to have some agreed upon parameter to characterize the magnitude of these functions. For some purposes, the peak values are satisfactory; however, two other values are often used.

Definition. <u>The average value</u> of a periodic function is found by integrating the function over one cycle and dividing by the period. This is the value to which an ordinary dc meter of the d'Arsonval type responds. For an ac (sinusoidal) current or voltage, the average value is zero and therefore, this value is useless for specifying the magnitude of an ac function.

$$I_{ave}(\text{or } V_{ave}) = (1/T) \int_0^T I_m(\text{or } V_m)\, dt$$

Definition. <u>Root mean square (rms) value</u>: the square root of the average value of the square of the periodic function integrated over one cycle. For most purposes, the rms value is the most convenient and meaningful value with which to specify the magnitude of an ac voltage or current and most ac meters are calibrated in terms of the rms value.

$$I_{rms}(\text{or } V_{rms}) = \sqrt{(1/T) \int_0^T I^2(\text{or } V^2)\, dt}$$

For ac currents and voltages, the rms values are

$$I_{rms} = I_m/\sqrt{2}, \text{ and } V_{rms} = V_m/\sqrt{2}$$

The instantaneous power delivered to an ac circuit is, of course, given by the product of the instantaneous voltage and the instantaneous current and is of little importance. Of more importance is the average power per cycle which is given by

$$P_{ave} = V_{rms} I_{rms} \cos \phi$$

where ϕ is the angle by which the current is out of phase with the voltage. The factor, $\cos \phi$, is called the power factor.

37.2 BEHAVIOR OF A RESISTOR CONNECTED TO AN AC GENERATOR

When an electric current flows through a resistor, Ohm's law may be used to calculate the magnitude of the voltage across the resistor. Ohm's law is valid for instantaneous, peak, or rms values of the current and voltage.

Magnitude: $V = RI$

Phase: The voltage is in phase with the current

Power: $P_{ave} = RI_{rms}^2$ (37.6)

For ac circuits, resistors in series or parallel are treated just as they are for dc circuits.

37.3 BEHAVIOR OF A CAPACITOR CONNECTED TO AN AC GENERATOR

If a source of ac voltage, $V = \mathcal{E}\cos \omega t$, is connected across a capacitor, the current through (actually, a flow of charge into one plate and out of the other) the capacitor is governed by an Ohm's law like equation. If we define $X_C = 1/\omega C$, we obtain

Magnitude: $V = X_C I = I/\omega C$

Phase: The voltage lags the current by 90°

Power: $P_{ave} = 0$

where V and I may be either maximum or rms values. X_C is called the capacitive reactance.

Definition. <u>Reactance</u> (either capacitive or inductive): the ratio of the magnitude of the voltage across the capacitor or inductor to the magnitude of the current through it.

The values of the current or voltage used in this definition may be either peak or rms values and, from this definition, it is clear that reactance is the analog of resistance for these two circuit elements.

Two or more capacitors connected in series or parallel may be combined into an equivalent capacitor in the same way as in Chapter 29.

37.4 BEHAVIOR OF AN INDUCTOR CONNECTED TO AN AC GENERATOR

If a source of ac voltge, $V = \mathcal{E} \cos \omega t$, is connected across an (ideal) inductor, and if we define the inductive reactance as $X_L = \omega L$, the relationship between the magnitude of the voltage across the inductor to the current through it is given by an Ohm's law like equation.

Magnitude: $V = X_L I = I \omega L$
Phase: The voltage leads the current by 90°
Power: $P_{ave} = 0$

<u>Warning</u>: Both real capacitors and real inductors have a certain amount of resistance associated with them and, indeed, a real resistor may have a capacitance and/or an inductance associated with it. However, except for very precise work or high frequencies, only in the inductor are these "stray" resistances and capacitances large enough that they may not be safely ignored. The intrinsic resistance of a real inductor must usually be included as a series resistor.

Two or more inductors connected in series or parallel may be combined into an equivalent inductor in the same way that resistors are combined.

37.5 THE RC CIRCUIT

Kirchoff's laws, so useful in circuit analysis, strictly apply only to the instantaneous values of voltage and current in ac circuits. However, by using vector-like methods, we may apply these laws to the rms or peak values as well. Although voltage and current are both scalars and have no direction associated with them, they do have an angle, the phase angle, associated with them and we may use the methods of vector addition to apply Kirchoff's laws. The angle between two "vectors" (called phasors by some authors) used to represent voltages or currents is a measure of the difference in phase between them. Using this method, a "useful" triangle may be used for the addition of currents or voltages.

A circuit containing a series combination of a resistor and a capacitor connected to a source of ac voltage can be analysed by Kirchoff's loop rule to give the equation,

$$\mathcal{E}/I = \sqrt{R^2 + X_C^2} = Z$$

<u>Definition</u>. <u>Impedance</u> (of a circuit to an ac current): the ratio of the magnitude of
the ac voltage of the source to the magnitude of the ac current delivered to the circuit by
the source. Impedance plays the same role for the circuit as does reactance for individual
circuit elements. Impedance is represented by the symbol Z and has units of ohms.

Using this definition, if the source voltage is

$$V = \mathcal{E} \cos \omega t,$$

the current in the circuit is

$$I = (\mathcal{E}/Z) \cos (\omega t + \Phi)$$

where

$$\omega = 2 \pi \nu \; ; \; \mathcal{E}/Z = I_m \; ; \; \Phi = \arctan(X_C/R)$$

is the angle by which the current leads the source voltage.

37.7 THE RLC CIRCUIT

If a series combination of a resistor, a capacitor, and an inductor is connected to a source of ac voltage, application of Kirchoff's loop rule gives for the impedance of the circuit,

$$Z = \sqrt{R^2 + (X_L - X_C)^2} = \sqrt{R^2 + (\omega L - 1/\omega C)^2}$$

In this circuit,

$$\Phi = \arctan[(X_L - X_C)/R] = \arctan[(\omega L - 1/\omega C)/R]$$

and may be either positive, negative, or zero depending on the values of the two reactances. The current may either lead, lag, or be in phase with the source voltage.

37.8 RESONANCE IN A SERIES RLC CIRCUIT

In the series RLC circuit, the condition in which the capacitive reactance and the inductive reactance are equal in magnitude is of particular importance; in this case, the effects of the capacitor are just cancelled by the effects of the inductor and the circuit behaves as if it contained only a resistor. Since the values of the reactances depend upon the frequency, this condition can be met at only one frequency, called the resonant frequency, for given values of the components.

> Definition. Resonance: the condition in a circuit containing reactive elements in which the circuit impedance is purely resistive and the current delivered to the circuit is in phase with the source voltage.

In the series RLC circuit, resonance occurs at a frequency given by

$$X_L = X_C \rightarrow \omega = \sqrt{\frac{1}{LC}}$$

At this frequency, $Z = R$ and $\Phi = 0$. Since R is the minimum value the impedance may have in a series RLC circuit, the current in the circuit is a maximum (for a fixed source voltage) at this frequency.

37.9 TRANSFORMERS

> Definition. A transformer, often called a mutual inductance, consists of two electrical circuits, usually coils of some sort, positioned so that some or all of the magnetic flux produced by a changing current in one of them will thread through the other, producing therein an emf.

Whichever coil is connected to the ac voltage source is called the primary coil while the one in which the voltage is induced is called the secondary coil. In general, the voltage induced in the secondary coil is

$$V_s = \pm M \frac{dI_p}{dt},$$

where the + sign is used if the coils are wound in the same sense and the - sign if they are wound in the opposite sense. Note that either coil could be used as the primary and that the value of the mutual inductance, M, would be the same in either case. If both coils are wound on the same iron core, the ferromagnetism of the core will both increase the magnitude of the flux and act as a magnetic circuit to direct <u>all</u> of the flux produced by the primary through the secondary. This construction is used for transformers in which a significant amount of power must be transferred efficiently from the primary circuit to the secondary circuit. Iron core transformers are routinely made with efficiencies well above 90%. In such a transformer where there is no flux "leakage", the relationship between the primary and secondary coils is particularly simple:

$$\frac{V_S}{N_S} = \frac{V_P}{N_P} \; ; \; \frac{I_S}{N_P} = \frac{I_P}{N_S} \; ; \; P_S = P_P$$

A transformer may also be used to improve the efficiency with which power is delivered from a source to a load. Maximum efficiency requires that the intrinsic resistance of the source be equal to the resistance of the load. A mismatch may be corrected by inserting a transformer between the source and load so that

$$R_{source} \, N_S^2 = R_{load} \, N_P^2$$

EXAMPLE PROBLEMS
Example 1

At some places in the USA early in this century, electric power was generated at a frequency of 25 Hz. If the voltage specified for the generator output was 2300 volts, write the equation which describes the instantaneous voltage.

Solution

Given: ν = 25 Hz
V_{rms} = 2300 V

Required: V_m, ω, $v(t)$

Since the voltage specified for the generator output is, by convention, the rms value, we need to calculate the peak value.

$$V_m = \sqrt{2} \, V_{rms} = \underline{3252.7 \text{ V}}$$

The frequency in Hz must be converted to radian frequency.

$$\omega = 2\pi\nu = \underline{157.1 \text{ rad/s}}$$

Using these values, the instantaneous voltage is given by

$$V = \underline{3252.7 \cos 157.1 \, t}$$

Note that a sine function or a combination of sine and cosine functions could equally well have been used.

Example 2

An electronic signal generator produces a "sawtooth" voltage such as that shown in FIG. 2-1. What are the average and rms values of this voltage?

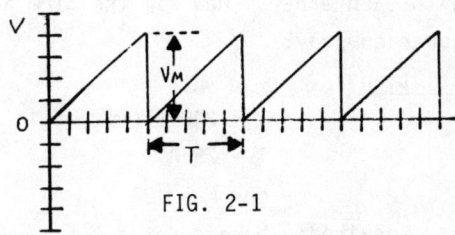

FIG. 2-1

Solution

The equation describing the sawtooth voltage function is

$$V = \frac{V_m t}{T}$$

The average value is found by integrating over one cycle and dividing by period.
$$V_{ave} = \frac{V_m}{T^2} \int_0^T t\, dt = 1/2\, V_m$$
The rms value is found by integrating the square of the function over one cycle, dividing by the period, and taking the square root.
$$V_{rms} = \sqrt{\frac{V^2}{T^3} \int_0^T t^2\, dt} = \frac{V_m}{\sqrt{3}}$$

Example 3

A series combination of a resistor and a capacitor are connected to a source of 60 Hz ac voltage. A voltmeter, an ammeter, and a wattmeter are also connected to the source so as to measure its voltage as well as the current and power which it delivers to the circuit. The meter readings recorded are: V = 110 volts ; I = 1.2 amperes ; P = 120 watts. What are the values of the resistor and capacitor?

Solution

Given: V = 110 V
          I = 1.2 A
          P = 120 W
          ν = 60 Hz

Required: R , C

The meter readings will yield values of the impedance, Z = V/I, and the power factor, cos Φ = P/(VI). From the trigonometry of the "useful" triangle,
$$R = Z \cos \Phi = \frac{P}{I^2} = \underline{83.3\, \Omega}$$

Also, from the trigonometry of the "useful" triangle,
$$X_C = Z \sin \Phi = (V/I) \sin (\arccos P/VI) = 38.19\, \Omega$$
and
$$C = 1/(2\pi\, X_C) = \underline{69.5\, \mu f}$$

Example 4

A small industrial plant operates a number of ac machines which have large inductances associated with them. These machines operate on 240 volts and draw a total of 25 amperes with a power factor (cos Φ) of 80%. The power company offers the firm lower rates if they will install a device to compensate for this small power factor so that the voltage and current will be in phase with each other. How can the firm accomplish this, thus saving money both for the firm and the power company?

Solution

Given: V = 240 V (rms)
          I = 25 A
          cos Φ = .80

Required: Make cos Φ = 1

The fact that the machines are inductive implies that the current lags behind the voltage by an ammount

ϕ = arccos .80 = 36.9°

The firm can install a capacitor in parallel with the machines which will draw additional current (but not additional power), the additional current leading the voltage by 90°. The "useful" triangle representing the "vector" addition of this current to the current drawn by the machines is shown in FIG. 4-1 where we see that

$$I_C = V/X_C = 2\pi\nu CV = I/\sin\phi$$

From this equation, we may calculate the required capacitance,

$$C = I/(2\pi\nu V \sin\phi) = 461 \text{ }\mu f$$

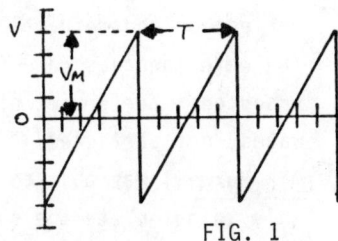

FIG. 4-1

Example 5

The transformer in an electronic power supply steps down the voltage from the 120 VAC provided by the power company to the 24 VAC required for the operation of the instrument of which it is a component. If the instrument requires a current of 3 amperes, what current will flow in the primary windings of the transformer?

Solution

Given: V_S = 24 V
V_P = 120 V
I_S = 3 A

Required: I_P

Using Eq. 37.84, which is actually a statement of the law of conservation of energy,

$$I_P = (V_S/V_P)I_S = .6 \text{ A}$$

PROBLEMS

1. The output of an ac voltge source is described by the equation,

   $$V = 36 \cos 1000t + 48 \sin 1000t \text{ volts}$$

   What are the peak and rms values of the voltage, frequency and the phase angle?

2. An electronic signal generator produces a "sawtooth" voltage such as that shown in FIG. 1. Calculate the average and rms values of this voltage function in terms of the peak (maximum) value.

3. A series combination of a 150 Ω resistor and a 50 μf capacitor are connected to a 120 V(rms), 60 Hz source of ac voltage. A wattmeter, a voltmeter, and an ammeter are also connected in such a way as to measure the electrical characteristics of the source as it delivers current to the circuit. Assume that the source has negligible internal resistance. What are the readings on the meters?

4. Eq. 37.75 in the textbook describes the "full width at half power" of the RLC resonance peak. Show that this peak is not symmetric by deriving the expression for the difference between

   $\omega_+ - \omega_r$ and $\omega_r - \omega_-$

5. The large transformers used in the transmission and distribution of electrical power are rated in kilovolt-amperes rather than in kilowatts. Recalling that the heat generated within a transformer is due to resistance heating, eddy currents and hysteresis, deduce a reason for the use of this seemingly strange and cumbersome unit.

Chapter 38
Electromagnetic Waves and Maxwell's Equations

PREVIEW

The existence of electromagnetic waves was predicted in 1864 by Maxwell as a "surprise" result of his attempt to bring some organization to the then chaotic and separate disciplines of electricity and magnetism. He found that four of the familiar laws could be used as a foundation for a unified theory of electromagnetism after modifying one of them in order to make the four mathematically consistent. Furthermore, he found that, in addition to explaining all of the then known electrical and magnetic phenomena, the equations describing these laws had a wave like solution, implying that electric and magnetic fields could be propagated as waves. The four equations are now called Maxwell's equations and form the basis for our knowledge not only of electricity and magnetism, but also of optics.

SUMMARY

38.1 INTRODUCTION

For many years, controversy surrounded theories concerning the nature of light. Many scientists, following Newton's lead and citing the "straight line propagation", held that light consisted of a stream of particles. Others, following Huygens and citing his explanation of reflection and refraction, held that light was a wave phenomenon. The most convincing evidence for the wave theories came from the experiments of Young and Fresnel who were able to show that light produced interference effects and could be polarized; these phenomena are strictly associated with waves, not particles.

Maxwell was able to show that electric and magnetic fields could be propagated as waves and was able to calculate the speed of these waves. His calculated value turned out to be the same as the measured speed of light. Unable to believe that this was a coincidence, he boldly pronounced that light is an electromagnetic wave; although Maxwell did not live to see his theory confirmed by experiment, it now forms the basis not only for our knowledge of optics, but also of fields as diverse as radio, microwaves, X-rays and many others.

38.2 ELECTROMAGNETIC WAVES

> Definition. Electromagnetic wave: a transverse wave consisting of oscillating electric and magnetic fields. The electric and magnetic fields are perpendicular to each other and both are perpendicular to the direction of propagation. The speed of the wave (in a vacuum) is independent of the frequency or wave length. The direction in which the wave travels is the direction of the vector cross product, (ExB). According to Maxwell's theory, E-M waves are produced by accelerating electric charges.

As with any wave, the frequency, period, wave length, and speed are related by
$$c = \lambda\nu \; ; \; T = 1/\nu \; ; \; c = \lambda/T. \tag{38.1,2,3}$$
The symbol, c, is used to denote the speed of an E-M wave because this value is an unchanging fundamental constant of nature.

The E and B fields of a plane wave which is propagating in the positive z direction may be represented by
$$E_x = E_0 \sin[(2\pi/\lambda)(z-ct)] \tag{38.4}$$
and
$$B_y = B_0 \sin[2\pi/\lambda)(z-ct)] \tag{38.5}$$

It is often convenient to express these equations in terms of the wave number, k, rather than the wave length; the wave number is defined as $k = 2\pi/\lambda$ so that the equations become
$$E_x = E_0 \sin(kz-\omega t) \tag{38.8}$$
and
$$B_y = B_0 \sin(kz-\omega t) \tag{38.9}$$
where, as usual, $\omega = 2\pi\nu$ \hfill (38.7)

> **Definition**. A <u>plane wave</u>: a wave for which the locus of all points having the same phase is a plane. Usually, E-M waves which have traveled a long distance from their source are very nearly plane waves.

38.3 MAXWELL'S EQUATIONS

The four equations which have come to be known as Maxwell's equations have all been encountered in earlier chapters, although one of them must, as Maxwell discovered, be modified. They are:

<u>Faraday's law</u>
$$\oint \mathbf{E} \cdot d\mathbf{s} = -(d/dt)\int \mathbf{B} \cdot d\mathbf{A} \tag{38.12}$$

<u>Gauss' law</u>
$$\int_{cs} \mathbf{E} \cdot d\mathbf{A} = q/\varepsilon_0 \tag{38.13}$$

the non-existence of magnetic monopoles
$$\int_{cs} \mathbf{B} \cdot d\mathbf{A} = 0 \tag{38.11}$$

and <u>Ampere's law</u>
$$\oint \mathbf{B} \cdot d\mathbf{s} = \mu_0 I + \mu_0\varepsilon_0 (d/dt)\int \mathbf{E} \cdot d\mathbf{A} \tag{38.17}$$

The last term on the right hand side of this last equation was not part of Ampere's law as it was stated in Chapter 33; this is the term which Maxwell found that he must add to Ampere's law to achieve consistency among the four equations. This term (exclusive of the permeability) is called the displacement current and is, according to Ampere's modified law, effective in producing a magnetic field in the same way that a conduction current is.

This modification of Ampere's law introduces a philosophically pleasing symmetry to Maxwell's equations; according to Faraday's law, a changing magnetic field creates an electric field and according to Ampere's law, a changing electric field creates a magnetic field. It is this coupling of the changing electric and magnetic fields which creates electromagnetic waves.

38.4 THE SPEED OF ELECTROMAGNETIC WAVES

When Maxwell's equations are solved for E and B, the results describe a wave traveling with a speed given by

$$c = \frac{1}{\sqrt{\mu_0 \epsilon_0}} \tag{38.19}$$

Substituting the numerical values of the permeability and permittivity of free space into Eq. 38.19 yields a value of

$$c = 2.99793 \times 10^8 \text{ m/s} \tag{38.22}$$

for the speed of Maxwell's waves. It was the fact that this value, calculated from the basic physical constants, was so near to the measured speed of light that led Maxwell to conclude that light is an electromagnetic wave.

The amplitudes of the oscillating electric and magnetic fields which constitute an E-M wave are related by

$$E_0 = B_0 c \tag{38.27}$$

and these fields oscillate in phase with each other.

> NOTE that while Maxwell's theory predicts correctly that the speed of E-M waves in a vacuum is independent of their wave length, if these waves are traveling in a material medium, their speed depends upon the permeability and permittivity of that medium (the same equation without the subscripts) and may, therefore, depend upon the wave length.

Electromagnetic waves stimulate the nerve endings in the retina of the human eye only over a very small range of frequencies (less than one octave) or wave lengths. The spectrum of E-M waves extends very far on either side of this small range from the long wavelength broadcast radiowaves to the very short gamma rays emitted by radioactive nuclei.

38.5 ENERGY TRANSFER VIA ELECTROMAGNETIC WAVES

Electric and magnetic fields both having energy associated with them, it is clear that an electromagnetic wave must carry energy with it as it travels through space.

> Definition. Intensity of an E-M wave: the amount of energy per unit time passing through a unit area oriented perpendicular to the direction of travel. The SI units of intensity are watts/meter^2.

The intensity of an E-M wave is proportional to the square of the amplitude of the wave; the amplitude of the wave is, by convention, taken as the amplitude of the oscillating electric field, but since the electric and magnetic fields are related by Eq. (38.27), the amplitude of the magnetic field could equally well be used. The intensity of an E-M wave may be calculated by

$$I = c \frac{\epsilon_0 E_0^2}{2} = c \frac{B_0^2}{2\mu_0} \tag{38.44}$$

38.6 POLARIZATION

Since E-M waves are transverse waves, the electric field may oscillate in any direction which is perpendicular to the direction of propagation, the specific direction being determined by the direction of the acceleration of the electric charge which acts as a source of the wave. Since many waves, particularly in the case of light, arise from the superposition of a very large number of independently oscillating charges, the fields composing such waves are uniformly distributed over all of these allowed directions. Such a wave is said to be unpolarized.

Definition. Polarization: a wave phenomenon peculiar to transverse waves in which the oscillation is confined to a single direction. In the case of electromagnetic waves, the direction of polarization is, by convention, taken to be the direction in which the electric field oscillates. The magnetic field is, of course, oscillating in a direction perpendicular to the direction of polarization.

Radio waves which radiate from antennas in the form of straight wires or loops, are polarized in the direction of the antenna. Although most sources of light are unpolarized, we seldom look directly at a source of light and much of the light entering our eyes is at least partially polarized since light waves may be polarized by reflection, refraction, scattering or absorption. We are unaware of this polarization because the human eye is insensitive to the state of polarization.

When light is reflected from a non-conducting surface, the reflected beam is at least partially polarized with the electric field parallel to the reflecting surface. Perhaps the most familiar polarizer is the synthetic polarizing material, polaroid, which absorbs all of the components perpendicular to its polarizing axis. If unpolarized light is passed through a polarizing material, its amplitude will be reduced to one half (actually less than one half because of imperfect transmission) of its original value and its intensity, being proportional to the square of the amplitude, will be reduced to one fourth of its original value. Polarized light passed through a polarizing material with its polarizing axis inclined at an angle, θ, to the direction of polarization, will have its intensity reduced to

$$I = I_0 \cos^2\theta \qquad (38.45)$$

EXAMPLE PROBLEMS
Example 1

The X-band microwaves frequently used in magnetic resonance research instruments have a wavelength of about 3 cm. What is the frequency of these microwaves?

Solution

Given: $\lambda = 3$ cm $= .03$ M

Required: ν

The relationship between frequency and wave length is $c = \lambda\nu$. Thus, the frequency is

$$\nu = c/\lambda = 9.99 \times 10^9 \text{ Hz}$$

Example 2

For a frequency of 100 MHz, the dielectric constant for water is 80 and the relative permeability is very nearly equal to 1. What is the speed of an E-M wave of this frequency in water?

Solution

Given: $\kappa = 80$

$\kappa_m = 1$

Required: v

The speed of the wave in a material medium is given by

$$v = \frac{1}{\sqrt{\mu\epsilon}}$$

since $\mu = \kappa_m \mu_o$ and $\varepsilon = \kappa \varepsilon_o$, the speed of the wave in water is given by

$$v = (\kappa_m \kappa)^{-1/2} c = 3.35 \times 10^7 \text{ m/s}$$

Example 3

A series combination of an initially uncharged capacitor and a resistor is connected to a source of emf. As the charge builds up on the capacitor, the electric field between the plates increases so that there is a displacement current "flowing" between the plates. Suppose that the plates are circular, of radius a, and separated by a distance, d. Derive the expression for the magnetic field between the plates at a radial distance, r, from their center.

Solution

At a time, t, after the charging begins, the voltage between the plates is

$$V = \mathcal{E}(1-e^{-t/RC})$$

and the electric field between the plates is

$$E = V/d = \frac{\mathcal{E}}{d}(1-e^{-t/RC})$$

We can find the magnetic field between the plates by use of Ampere's law. Since symmetry requires that the field be azimuthal, we use as a path of integration, a circle of radius, r, concentric with the center of the plates. Then the integrals in Ampere's law become

$$\oint B \cdot ds = 2\pi r B$$

and

$$\int E \cdot dA = \frac{\mathcal{E}}{d}(1-e^{-t/RC})\pi r^2$$

Since no conduction current passes through the area bounded by the path of integration, Ampere's law becomes

$$2\pi r B = \mu_o \varepsilon_o \frac{d}{dt}[\frac{\mathcal{E}}{d}(1-e^{-t/RC})]$$

Performing the indicated differentiation, we find

$$B = \frac{\mu_o V r}{2\pi a^2 R} e^{-t/RC}$$

Example 4

A laboratory laser produces a well collimated beam of red light with a diameter of 0.5 mm. If the power output of the laser is 0.1 milliwatt, what is the strength of the B field in the light wave?

Solution

Given: $d = 0.5$ mm $= .005$ M

           $P = 0.1$ mW $= .0001$ watt

Required: B

The intensity of the beam is the power per u;nit cross sectional area of the beam,

$$I = P\frac{4P}{\pi d^2} = 509.3 \text{ watts/square meter}$$

and, from Eq. 38.4),

$$B = \sqrt{2\mu_o I/c} = \underline{4.27 \times 10^{-12} \text{ T}}$$

Example 5

Problem 10 in the textbook gives the differential forms of Faraday's law and Ampere's law:

$$\frac{dE_x}{dz} = -\frac{dB_y}{dt} \quad \text{Faraday's law}$$

$$\mu_0 \epsilon_0 \frac{dE_x}{dt} = -\frac{dB_y}{dz} \quad \text{Ampere's law}$$

Combine these equations to eliminate E and obtain a single differential equation for B.

Solution

If we differentiate both sides of Faraday's law with respect to t so that

$$\frac{d^2 E_x}{dz\,dt} = -\frac{d^2 B_y}{dt^2}$$

and differentiate both sides of Ampere's law with respect to z so that

$$\mu_0 \epsilon_0 \frac{d^2 E_x}{dt\,dz} = -\frac{d^2 B}{dz^2}$$

then, since the order of differentiation is immaterial, the two equations may be solved for the derivative of E and equated, thus eliminating E from the equations. The result is

$$\frac{d^2 B_y}{dz^2} = \mu_0 \epsilon_0 \frac{d^2 B_y}{dt^2}$$

PROBLEMS

1. A citizen's band (CB) two way radio transmits at a frequency of 27 MHz. For correct impedance matching, the transmitting antenna should be one quarter wave length long. How long should the antenna be for this radio?

2. The CB antenna described in Problem 1 is vertical so that the radio waves radiating from it are polarized in the vertical direction. At a point due north of the antenna when the electric field is directed vertically upward, in what direction is the magnetic field pointing?

3. At optical frequencies, the dielectric constant of most materials is significantly lower than at radio frequencies. The speed of light in water is about three fourths of its speed in a vacuum. Assuming the relative permeability to remain near one at optical frequencies, what is the dielectric constant of water?

4. The laser in Example Problem 1 produces light with a wave length of 632.8 nanometers. This light is reflected back and forth in the laser tube which is 30 cm long. How many wave lengths of the light are contained within the length of the tube?

5. Use the procedure outlined in Example Problem 5 above to eliminate B from the differential forms of Faraday's law and Ampere's law and so obtain a single differential equation for E. Does the plane wave form described by Eq. 38.8 satisfy this equation?

Chapter 39
Reflection, Refraction, and Geometric Optics

PREVIEW

Optics is the study of light, the name generally given to the portion of the electromagnetic spectrum perceived by human sight. This chapter introduces optics and treats the branch known as geometric optics. The chapters that follow cover the branch known as physical optics. Geometric optics represents light as light rays and assumes that light travels in straight lines. The chapter begins by introducing Huygens' principle and using it to derive the laws of reflection and refraction. These laws are then applied in the ray theory of geometric optics to analyze image formation by mirrors and lenses. The chapter also discusses a number of optical systems, including the human eye, magnifiers, microscopes, and telescopes.

SUMMARY

39.1 HUYGENS' PRINCIPLE, REFRACTION, AND DISPERSION

Wave fronts are defined in Chapter 18 as surfaces over which the phase of a wave is constant. Huygens' principle describes how wave fronts move.

<u>Huygens' Principle</u>: each point on a wave front acts as a source of secondary wavelets, and the wave front at any subsequent moment is the envelope of the secondary wavelets.

A Huygens' construction allows us to follow the progress of a light wave through a medium or from one medium into another, as determined by the speed of light in the media involved. The speed of light in a medium differs from that in a vacuum and is specified by a quantity known as the index of refraction.

<u>Definition</u>. <u>Index of refraction</u>: the ratio of the speed of light in a vacuum (c) to the speed of light in a material medium (v), given by $n = c/v$ (Eq. 39.1).

The index of refraction is always greater than unity, since the speed of light is larger in a vacuum than in any medium. The indices of refraction for several substances are given in Table 39.1. The index of refraction of air is usually assumed to be unity, as it is actually only very slightly above this value.

If the speed of light in a medium depends on the wavelength, the medium is said to exhibit dispersion. For a dispersive medium, then, the index of refraction varies with the wavelength, and the dispersion is determined by the difference in the indices of refraction for red and violet light.

<u>Definition</u>. <u>Dispersion</u>: the spreading of white light into a spectrum of colors in a medium whose index of refraction depends on the wavelength.

The graph of the index of refraction versus wavelength is called a dispersion curve. Dispersion in a lens produces a defect known as chromatic aberration. Perhaps the most obvious example of dispersion is the rainbow.

39.2 THE LAWS OF REFLECTION AND REFRACTION

In general, when a light wave is incident on the boundary or interface between two media, a portion of the wave is returned, or reflected, back into the first medium, and a portion is bent, or refracted, as it passes into the second medium. Huygens' principle can be used to derive the laws of reflection and refraction.

Law of Reflection: The angles of incidence and reflection are equal.

The angles of incidence and reflection are customarily measured between the normal to the boundary and the normals to the incident and reflected wave fronts. The latter two normals are in the direction of the incident and reflected rays, thus in a ray theory, the angles of incidence and reflection are measured between the normal to the boundary and the incident and reflected rays.

Snell's Law of Refraction: When light travels from one medium into another medium, the sines of the angles of incidence and refraction have a constant radio, given by the ratio of the indices of refraction according to $n_1 \sin \theta_1 = n_2 \sin \theta_2$ (Eq. 39.6).

The angles of incidence and refraction are measured between the normal to the boundary and the normals to the incident and refracted wave fronts, or equivalently, between the normal to the boundary and the incident and refracted rays.

39.3 LIGHT RAYS AND GEOMETRIC OPTICS

In geometric optics, light is represented by light rays and is assumed to travel in straight lines. Geometric optics can be used provided the wavelength of the light is much smaller than the characteristic dimensions of the optical system involved.

Definition. Total internal reflection: the complete reflection of light, originally in a more-dense medium, that is incident on the boundary with a less-dense medium at an angle that is equal to or greater than the critical angle, whose sine is given by the ratio of the indices of refraction of the two media.

The sine of the critical angle is given by Snell's law, with an angle of refraction of 90°:
$$\sin \theta_c = n_2/n_1 \tag{39.10}$$
where n_1 is the index of refraction of the more-dense medium and n_2 is that of the less-dense medium. The light must always be incident from the more-dense medium (n_1), since the sine of the critical angle must be less than unity.

The rainbow is an example of the dispersion of light and can be analyzed by applying the laws of reflection and refraction to a spherical raindrop. The light that forms the primary rainbow comes to us after undergoing two refractions and one internal reflection in the raindrop. The light of the primary rainbow is violet at the lower edge and red at the upper edge and appears between 40.6° and 42.2° above the direction of the incoming sunlight, called the antisolar direction. The secondary rainbow results from an additional internal reflection, has the colors reversed, and lies 50.7° - 53.6° above the antisolar point.

39.4 MIRRORS

The analysis of image formation by mirrors involves only the law of reflection. We use the ray theory of geometrical optics and consider only plane mirrors and spherical mirrors, which are formed from segments of spheres. The analysis is carried out in the paraxial-ray approximation, in which all rays are assumed to make a small angle with the symmetry axis of the mirror, resulting in negligible blurring of the image formed by spherical mirrors due to spherical abberation. The result is known as the spherical mirror equation:

$$\frac{1}{s} + \frac{1}{s'} = \frac{2}{r} \qquad (39.17)$$

where s, s', and r are measured along the symmetry axis and refer to the object distance, the image distance, and the radius of curvature, respectively.

For a plane mirror, $r = \infty$, and $s' = s$. For spherical mirrors, $r = 2f$ (Eq. 39.19), where f is the focal length, which is measured along the symmetry axis to the focal point, the point where incident rays parallel to the axis are converged. For spherical mirrors, the mirror equation becomes

$$\frac{1}{s} + \frac{1}{s'} = \frac{1}{f} \qquad (39.20)$$

The lengths s, s', and f carry no algebraic signs, and these are assigned according to the sign convention adopted for the mirror equation. The character of the image can be determined from the sign convention and is classified as upright or inverted, enlarged or reduced, and real or virtual.

> Definition. Real image: an optical image through which light rays actually pass, carrying light energy through the image position.

> Definition. Virtual image: an optical image from which light rays appear to diverge and through which no light energy passes.

It is assumed that the object is to the left of the mirror and that the object distance is positive for this situation. When the image distance is calculated, positive s' means the image is inverted and real and negative s' means the image is upright and virtual. When the image distance is assigned, s' is positive to the left of the mirror and negative to the right of (behind) the mirror. The focal length f is positive for concave mirrors and negative for convex mirrors. A summary of the character of the image for plane and spherical mirrors is given below:

Type of Mirror	Character of Image		
plane	upright	virtual	same size as object
convex spherical	upright	virtual	reduced
concave spherical	inverted	real	enlarged or reduced
	upright	virtual	enlarged

The image for a spherical mirror can also be located using a ray diagram. Out of the multitude of rays that leave every point on an object in all directions, we select a few useful rays, called "easy" rays, and use them to locate the images of a few key points on the object. The image is located by the intersection of any two rays emerging from the object, and the third can be used to check the result. The "easy" rays for spherical mirrors are given below:

Concave Mirror	Convex Mirror
(1) Ray parallel to axis reflected through focal point.	(1) Ray parallel to axis reflected along line whose backward extension passes through focal point.
(2) Ray through focal point reflected parallel to axis.	(2) Ray toward focal point reflected parallel to axis.
(3) Ray through center of curvature reflected back on itself.	(3) Ray toward center of curvature reflected back on itself.

For a convex mirror, it is the backward extensions of these reflected rays that appear to come from the virtual image. For a concave mirror the reflected rays either pass through a real image or appear to come from a virtual image, depending on the object distance. For a plane mirror, the "easy" rays are (1) the straight-in ray that is reflected back on itself and (2) the ray reflected from the mirror into the observer's eye. The backward extensions of these rays form a virtual image.

<u>Principle of Optical Reversibility</u>: if any ray is reversed, it will retrace its path through an optical system.

39.5 LENSES

By analyzing the rays entering and emerging from spherical surfaces using Snell's law of refraction, we obtain a quantitative description of image formation by spherical lenses. In the thin-lens approximation, in which the thickness of the lens is negligible compared to the other lengths involved, this equation is known as the thin-lens formula:

$$\frac{1}{s} + \frac{1}{s'} = \frac{1}{f} \tag{39.35}$$

where the expression for f is known as the lensmaker's formula:

$$\frac{1}{f} = \pm \left[\frac{n_2}{n_1} - 1\right]\left[\frac{1}{r_A} + \frac{1}{r_B}\right] \tag{39.34},(39.36)$$

The + sign is for double convex lenses, the − sign is for double concave lenses, and r_A and r_B denote the radii of the refracting surfaces of the lens. If one lens surface is convex and the other concave, the + sign is used, and r_A, the radius of the first surface, and r_B, the radius of the second surface, are given signs according to the sign convention that convex is positive and concave is negative when viewed from outside the lens.

Convex lenses are thicker in the center than at the edges, are sometimes called converging lenses, and produce images whose character is like those of concave mirrors. Concave lenses are thicker at the edges than at the center, are called diverging lenses, and have images whose character is like those of convex mirrors.

The thin-lens formula is mathematically identical to the mirror equation, and the sign convention is similar. For an object to the left of the lens, the object distance s is positive, and positive s means a real object. The image distance s' is positive to the right of the lens and negative to the left, and positive s' means a real image and negative s' a virtual image. The focal length f is positive for convex lenses and negative for concave lenses, as indicated by the lensmaker's formula. A summary of the image character for spherical lenses is given below:

Type of Lens	Character of Image		
concave spherical	upright	virtual	reduced
convex spherical	{ inverted	real	enlarged or reduced
	upright	virtual	enlarged

Images for thin lenses can be located by using a ray diagram and two of the three "easy" rays emerging from the object:

Convex Lens	Concave Lens
(1) Ray parallel to axis refracted through focal point.	(1) Ray parallel to axis refracted along line whose backward extension passes through focal point.
(2) Ray through focal point refracted parallel to axis.	(2) Ray toward focal point refracted parallel to axis.
(3) Ray through center of lens continues in straight line with no net refraction.	(3) Ray through center of lens continues in straight line with no net refraction.

39.6 OPTICAL SYSTEMS

In a normal human eye, light from a distant object is focused on the retina with the lens muscles relaxed. Accommodation is the ability of the eye to focus on closer objects by making the lens more convex and shortening its focal length. The closest point on which the eye can focus is called the near point, typically about 25 cm in young adults. The decrease in accommodation and the retreat of the near point with age is known as presbyopia.

Other common vision defects involve the inability of the eye to focus incoming rays on the retina. In myopia (nearsightedness), the rays are focused in front of the retina, the eye has a far point, or farthest point on which it can focus, that is less than normal, and the correction is a diverging or negative lens. In hyperopia (farsightedness), the focal point is behind the retina, the eye has a near point greater than normal, and the correction is a converging or positive lens.

The focal length of a corrective lens can be calculated using the expression for the effective focal length of two thin lenses separated by a distance that is small compared to their focal lengths:

$$\frac{1}{f} = \frac{1}{f_1} + \frac{1}{f_2} \tag{39.39}$$

where f_1 and f_2 refer to the focal lengths of the eye and the corrective lens and f is the focal length of the combination, designed to focus rays from distant objects on the retina.

The simple magnifier consists of a single converging lens that can produce an enlarged virtual image at the near point, allowing an object to be brought closer than the near point and still be in focus. Since the enlarged image subtends a larger angle than the object, the size of the retinal image is increased.

The angular magnification, or magnifying power (X), is the ratio of the angles subtended by the magnified image and the unmagnified image when the object is at the near point:

$$M = S_N/s = 1 + (S_N/f) \tag{39.43}$$

where S_N is the distance to the near point and f is the focal length of the lens. The magnification of a simple magnifier can be increased by reducing the focal length but is limited by aberrations to approximately M = 4X.

The compound microscope is used to obtain magnifications greater than those possible with a simple magnifier and consists of two lenses or lens systems. The lens closer to the object is called the objective, and that closer to the eye is called the eyepiece or ocular. The objective produces an intermediate, real, magnified image that serves as the object for the eyepiece. The eyepiece produces a final image that is often at a relatively large distance to provide for viewing by a relaxed or unaccommodated eye.

The magnification of the compound microscope is given by

$$M_{mic} = LS_N/f_o f_e \qquad (39.50)$$

where L is the distance between the focal points of the lenses, S_N is the near point, and f_o and f_e are the focal lengths of the objective and eyepiece. The distances L and S_N are usually fixed, so that the magnification is inversely proportional to the product of the focal lengths. A large magnification thus requires short focal lengths for both lenses.

The simple refracting telescope has two lenses that perform the same functions as the lenses in a microscope. The eyepiece views an intermediate image formed by the objective and produces a final image that is viewed by the eye. The magnification of the telescope is given by

$$M_{tel} = f_o/f_e \qquad (39.51)$$

The microscope and telescope differ in the dependence of the magnification on the focal length of the objective, reflecting the difference in the function of this lens in the two instruments. In a microscope the objective has a short focal length and provides magnification, while the telescope objective is designed to carry out the instrument's light-gathering function and has a large diameter and large focal length.

EXAMPLE PROBLEMS
Example 1

A student standing 80 cm from a plane dresser mirror judges the image of a brush on the dresser to be 1 m from her position. How far is the brush itself from the student?

Solution

Given: $s_{student}$ = 80 cm
distance from student to brush image = 1 m

Required: $s = s_{student} - s_{brush}$

The student sees the brush in the mirror at 1 m and is standing 80 cm from the mirror, therefore, the image of the brush is located 20 cm behind the mirror:

$$s'_{brush} = -20 \text{ cm}$$

where the minus sign indicates that the image of the brush is virtual.

The image of a plane mirror is located as far behind the mirror as the object is in front of it, therefore the image distance of the brush is given by

$$s_{brush} = -s'_{brush} = 80 \text{ cm} - 20 \text{ cm} = \underline{60 \text{ cm}}$$

In plane mirror problems, you should remember that the object and image distances are equal in magnitude and that the image and object are of the same size (magnification = 1).

Example 2

A beam of light enters a swimming pool of depth 3 m at its edge and illuminates a spot on the bottom of the pool 2 m from the edge. What angle does the beam make with the vertical a) in the water and b) in the air before entering the water? Assume that $n_{air} = 1$ and $n_{water} = 1.33$.

Solution

Given: $d = 3$ m
$x = 2$ m
$n_a = 1$
$n_w = 1.33$ (Table 39.1)

Required: a) θ_w, b) θ_a

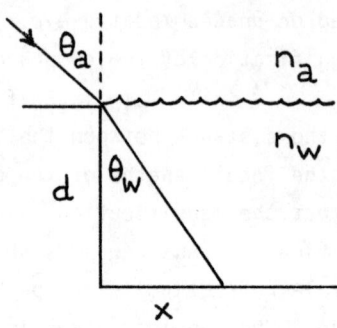

FIG. 2-1

a) The direction of the beam in the water is given to be from the edge to a spot on the bottom 2 m from the edge, as shown in FIG. 2-1. The angle θ_w the beam makes with the vertical in the water is given by

$$\tan \theta_w = 2 \text{ m}/3 \text{ m}$$
$$\theta_w = 33.7°$$

b) The relation between the angles the beam makes in air and water is given by Snell's law of refraction (Eq. 39.6) written for air and water:

$$n_a \sin \theta_a = n_w \sin \theta_w$$

where the angles must be measured with respect to the vertical.

Solving this equation for the angle θ_a the beam makes with the vertical in air and substituting for the angle in water θ_w from a) and the indices of refraction n_a and n_w:

$$\sin \theta_a = (n_w/n_a) \sin \theta_w = (1.33/1) \sin 33.7° = 0.738$$

$$\theta_a = \underline{47.6°}$$

The light beam is refracted toward the normal as it passes from the less dense air into the more dense water, with a bending of nearly 14°.

Example 3

Find a) the image distance and b) the character of the image for an object located 4 cm in front of a convex spherical mirror having a radius of curvature of -4 cm.

Solution

Given: $s = 4$ cm
$r = -4$ cm

Required: a) s', b) image character

a) The image distance can be found using the spherical mirror equation (Eq. 39.17):

$$\frac{1}{s} + \frac{1}{s'} = \frac{2}{r}$$

Although this form of the equation does not require the focal length f, it is given by $f = r/2$ for a spherical mirror. The sign convention for s and r is that s is positive when the object is in front of the mirror and r (also f) is negative for a convex mirror. Solving this equation for the image distance s' and substituting for the object distance s and the radius of curvature r:

$$\frac{1}{s'} = \frac{2}{r} - \frac{1}{s} = -\frac{2}{4 \text{ cm}} - \frac{1}{4 \text{ cm}} = -\frac{3}{4 \text{ cm}}$$

$$s' = \underline{-1.33 \text{ cm}}$$

According to the sign convention, the negative value of s' indicates that the image is virtual and located behind the mirror. In the spherical mirror equation, the object and image distances and the focal length are all distances, therefore any length unit can be used for these quantities provided that all are converted to the same unit prior to substituting in the equation.

b) The character of the image is the same for all convex mirrors. It is <u>upright</u>, <u>virtual</u> (behind the mirror), and <u>reduced</u>.

This can be verified by constructing a ray diagram using the "easy" rays for a convex mirror, as shown in FIG. 3-1. (1) Ray parallel to axis reflected along line whose backward extension passes through focal point. (2) Ray toward focal point reflected parallel to axis. (3) Ray toward center of curvature reflected back on itself.

The diverging rays indicate that the image is virtual, and the intersection of their backward extensions locates the image and verifies its character as virtual, upright, and reduced.

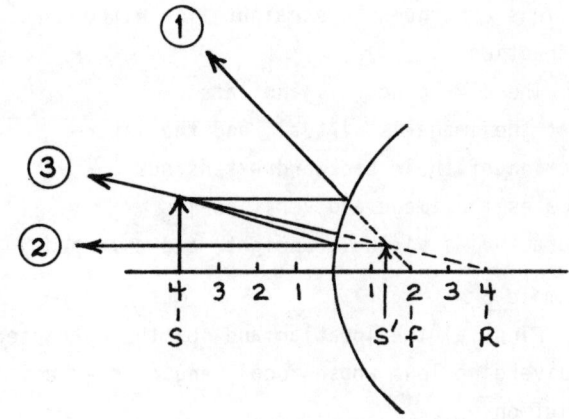

FIG. 3-1

Example 4

For an object located 1 cm from a converging thin lens of focal length 3 cm, find a) the location and b) the character of the image.

<u>Solution</u>

Given: $f = 3$ cm
         $s = 1$ cm

Required: a) s', b) image character

a) The image distance can be obtained from the thin-lens equation (Eq. 39.35):

$$\frac{1}{s} + \frac{1}{s'} = \frac{1}{f}$$

The sign convention is that f is positive for a convex (converging) lens and the side of the lens where the object is located is positive for the object distance and negative for the image distance. Solving this equation for s' and substituting for the focal length f and the object distance s:

$$\frac{1}{s'} = \frac{1}{f} - \frac{1}{s} = \frac{1}{3 \text{ cm}} - \frac{1}{1 \text{ cm}} = -\frac{2}{3 \text{ cm}}$$

$$s' = \underline{-1.5 \text{ cm}}$$

According to the sign convention, the negative value of s' indicates a virtual image located on the same side of the lens as the object. As noted in connection with the mirror equation, any consistent length unit is acceptable for the object and image distances and the focal length.

b) For a convex lens, the character of the image is not uniquely determined by the type of lens, but if the image is <u>virtual</u> as here, it is <u>upright</u> and <u>enlarged</u>.

The character of the image can be verified by constructing a ray diagram using the "easy" rays for a convex lens, as shown in FIG.4-1. (1) Ray parallel to axis refracted through focal point. (2) Ray through focal point refracted parallel to axis. (3) Ray through center of lens continues in straight line without refraction.

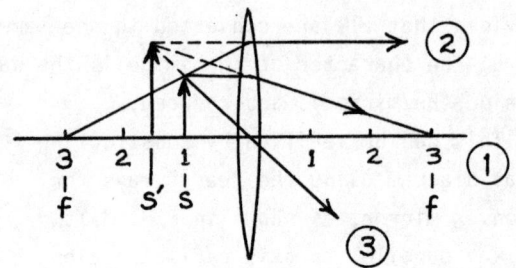

FIG. 4-1

The diverging rays indicate that the image is virtual, and the intersection of their backward extensions locates the image and verifies its character as virtual, upright, and enlarged.

Example 5

Find a) the location and b) the character of the image formed for an object located 3 cm from a diverging lens whose focal length is -2 cm.

Solution

Given: $f = -2$ cm
           $s = 3$ cm

Required: a) s' , b) image character

The image can be located using the thin-lens equation (Eq. 39.35):

$$\frac{1}{s} + \frac{1}{s'} = \frac{1}{f}$$

The sign convention is that f is negative for a concave (diverging) lens and the side of the lens where the object is located is positive for the object distance and negative for the image distance. Solving this equation for s' and substituting for the focal length f and the object distance s:

$$\frac{1}{s'} = \frac{1}{f} - \frac{1}{s} = -\frac{1}{2 \text{ cm}} - \frac{1}{3 \text{ cm}} = -\frac{5}{6 \text{ cm}}$$

$$s' = \underline{-1.2 \text{ cm}}$$

According to the sign convention, the negative value of s' indicates a virtual image located on the same side of the lens as the object. In using the mirror and lens equations and the equation for the focal length of a lens combination, the terms of the equation are the reciprocals of the object and image distances and the focal length and must be inverted to obtain the result.

b) The character of the image is the same for all diverging lenses. It is upright, virtual (same side as object), and reduced.

The character of the image can be verified by constructing a ray diagram using the "easy" rays for a concave lens, as shown in FIG. 5-1. (1) Ray parallel to axis refracted along line whose backward extension passes through focal point. (2) Ray toward focal point refracted parallel to axis. 3) Ray through center of lens continues in straight line without refraction.

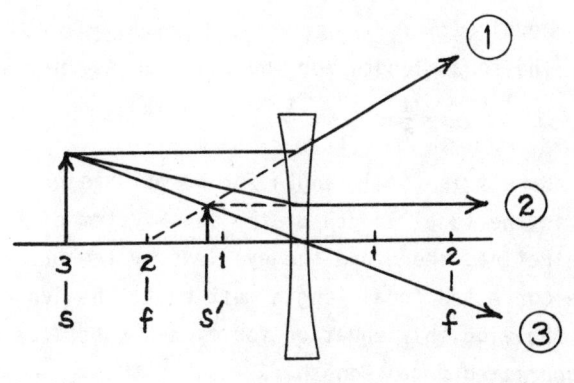

FIG. 5-1

The diverging rays indicate that the image is virtual, and the intersection of their backward extensions locates the image and verifies its character as virtual, upright, and reduced.

Example 6

Find the focal length in air of a diverging thin lens made of glass (n = 1.5) with radii of curvature 10 cm and 4 cm, as shown.

Solution

Given: r_A = 10 cm
         r_B = 4 cm
         n_1 = 1
         n_2 = 1.5

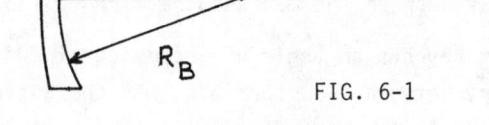

FIG. 6-1

Required: f

The focal length is given by the lensmaker's formula (Eq. 39.34):

$$1/f = [(n_2/n_1) - 1][(1/r_A) + (1/r_B)]$$

As shown in FIG. 6-1, the first surface is convex and the second surface concave when viewed from outside the lens. The signs are given according to the sign convention that r_A is the radius of the first surface and r_B the radius of the second surface and a convex surface is positive and a concave surface is negative. Therefore, substituting in this equation for n_2 and n_1 and r_A = +10 cm and r_B = -4 cm:

$$1/f = (1.5 - 1)[1/10 \text{ cm}) - (1/4 \text{ cm})] = -0.075/\text{cm}$$
$$f = \underline{-13.3 \text{ cm}}$$

Example 7

A farsighted person has a focal point that is 2 mm behind the retina. Find the focal length of the contact lens required to focus parallel rays on the retina, assuming an effective lens-retina distance of 2.1 cm.

Solution

Given: $f = 2.1$ cm
         $f_2 = (2.1 + 0.2)$ cm $= 2.3$ cm

Required: f_1

The focal length for two thin lenses in contact is given by Eq. 39.39:

$$\frac{1}{f} = \frac{1}{f_1} + \frac{1}{f_2}$$

Here f is focal length for normal vision, f_1 is the focal length of the contact lens, and f_2 is the focal length of the lens system of the eye. The uncorrected focal point is 2 mm behind the retina, therefore the eye lens system has a focal length of $f_2 = (2.1 + 0.2)$ cm $= 2.3$ cm. The corrected focal length must equal the lens-retina distance, so that $f = 2.1$ cm.

Solving this equation for f_1 and substituting for the corrected focal length f and the uncorrected focal length f_2:

$$\frac{1}{f_1} = \frac{1}{f} - \frac{1}{f_2} = \frac{1}{2.1 \text{ cm}} - \frac{1}{2.3 \text{ cm}} = 0.0414/\text{cm}$$

$f_1 = \underline{24.2 \text{ cm}}$

The corrective contact lens is converging, or positive. This is the proper correction, since incoming parallel rays are focused behind the retina of the unaided eye and must be converged further to a focus on the retina by the corrective lens.

PROBLEMS

1. A man in a clothing store is standing 3 m from a plane mirror, and a salesman is located half-way between the man and the mirror. How far from the man is the image of the salesman?

2. A light ray has an angle of incidence of 20° and an angle of refraction of 40° at the boundary between two materials. If the angle of incidence is increased, at what angle will the beam first undergo total internal reflection?

3. The image of an object located 2 cm in front of a concave spherical mirror is formed 3 cm behind the mirror. What is a) the focal length of the mirror and b) the character of the image?

4. Find a) the position and b) the character of the image of an object located 12 cm from a converging lens of focal length 4 cm.

5. For a diverging lens with focal length -10 cm, find a) the image distance and b) the character of the image for an object located 15 cm from the lens.

6. The radius of the first surface of a converging thin lens made from glass ($n = 1.5$) is 6 cm, and the radius of the second surface is -9 cm. Find the focal length of the lens in air.

7. A nearsighted person has a focal point that is 1 mm in front of the retina. What is the focal length of the contact lens required to restore normal vision, assuming an effective lens-retina distance of 2.1 cm?

Chapter 40
Physical Optics: Interference

PREVIEW

This chapter introduces the branch of optics that is based on the wave nature of light and known as physical optics. The phenomenon of interference is discussed in terms of the concept of coherence, and Young's classic double-slit interference experiment is presented. The chapter also considers optical interference in thin films and the operation of optical interferometers, including the instrument developed by Michelson.

SUMMARY

40.1 INTERFERENCE

We have discussed the principle of superposition and interference previously in Chapter 17 and have stated that constructive (destructive) interference occurs in general where superposition causes waves to reinforce (oppose) one another. We now define constructive (destructive) interference in terms of wave intensity.

<u>Definition</u>. <u>Constructive interference</u>: the phenomenon in which the intensity of superimposed waves is greater than the sum of the intensities of the separate waves.

<u>Definition</u>. <u>Destructive interference</u>: the phenomenon in which the intensity of superposed waves is less than the sum of the intensities of the separate waves.

The term complete constructive interference (complete destructive interference) is used to denote constructive (destructive) interference when the intensity of the superposed waves has its maximum (minimum or zero) value.

40.2 THE INTERFERENCE OF LIGHT: THOMAS YOUNG'S EXPERIMENT

The interference of light was first demonstrated by Thomas Young in a classic experiment in the early 1800's that established the wave theory of light. In this experiment, monochromatic light passes through an aperture and illuminates a double-slit system, from which secondary wavelets spread outward in phase according to Huygens' principle. These wavelets interfere constructively and destructively and produce a pattern of bright and dark bands on a screen placed beyond the slits.

<u>Definition</u>. <u>Interference fringes</u>: an alternating series of bright and dark bands produced by the interference of light waves.

The nature of the interference of waves from the two slits at any point on the screen depends on the difference in their path lengths, which corresponds to a phase difference according to $\Delta\phi = 2\pi(\Delta x/\lambda)$ (Eq. 17.17). A point is an intensity maximum (complete constructive interference) if the path difference is an integral number of wavelengths or if the phase difference is an integral multiple of 2π (waves in phase or crest on crest). An intensity minimum (complete destructive interference for equal amplitude waves) results if the path difference is an odd number of half wavelengths or if the phase difference is an odd-integral multiple of π (waves out of phase or crest on trough).

These conditions lead to expressions for the angular and linear position of the point on the screen (Eqs. 40.14, 40.19, 40.20, 40.21):

	Angular Position	Linear Position	Order
Intensity Maxima:	$\sin\theta = m\lambda/a$	$y = Rm\lambda/a$	$m = 0, 1, 2...$
Intensity Minima:	$\sin\theta = (m - 1/2)\lambda/a$	$y = R(m - 1/2)\lambda/a$	$m = 1, 2...$

where a is the slit separation, R is the horizontal distance to the screen, θ and y are the angular position and linear displacement (along the screen) measured from straight ahead up to the point, and m is the order of either the intensity maximum or minimum. The maxima are evenly spaced along the screen at intervals of $(R\lambda/a)$ on either side of the central maximum ($m = 0$) at $\theta = 0$. The minima are located between the maxima, giving a spacing between adjacent bright and dark fringes of $(R\lambda/2a)$. The expressions for the linear position have been obtained from those for angular position using the small-angle approximation for the case in which the slit-screen distance is much larger than the linear position.

It was the observation of an intensity maximum at the center position in Young's experiment that confirmed the wave nature of light, since the particle theory predicts a shadow at this position. Young's experiment can be used to measure wavelength, because the expressions for the intensity maxima and minima are proportional to the wavelength for a given slit system.

40.3 COHERENCE

Definition. Coherence: a property possessed by waves that have equal frequencies and maintain a constant phase difference that enables them to exhibit interference phenomena.

In order to be coherent, waves must meet both the frequency and the phase requirements. Otherwise, they are incoherent and do not produce interference effects. Ordinary light waves are incoherent, since they originate in atomic transitions that occur randomly and thus have a random rather than a constant phase difference. In Young's experiment, the waves originate from a single aperture that is equidistant from the two slits and are therefore in phase and coherent.

The intensity of two waves that are coherent is obtained by superposing the two wave amplitudes and squaring the resultant amplitude to obtain the resultant intensity. For incoherent waves, the procedure is reversed. The individual amplitudes are squared to obtain the individual intensities and the resultant intensity is the sum of the individual intensities.

For two coherent sinusoidal waves of the same frequency and amplitude, the intensity is given by

$$I = 2I_1\{1 + \cos[2\pi(x_1 - x_2)/\lambda]\} \tag{40.48}$$

where x_1 and x_2 are the positions of the waves and I_1 and λ are the intensity and wavelength of either wave. The value of $2I_1$ for incoherent waves is altered by the cosine term, whose argument is the phase difference for the two waves. If the phase difference is an odd multiple of π, the resultant intensity is zero and the result is known as complete destructive interference.

Complete constructive interference corresponds to a phase difference equal to an integral multiple of 2π and an intensity of $4I_1$. Conditions of incomplete destructive and constructive interference give intensities between these two extreme values of zero and four times the individual intensity.

> Definition. <u>Coherence length</u>: the distance along the direction of propagation of a beam of light over which phase coherence exists.

If an atom radiates for a time t, the emitted wave train has a length ct, where c is the speed of light, and the coherence length is given by L_c = ct (Eq. 40.49). For ordinary light, the coherence length is the average length of the wave trains emitted by the individual atoms. If a beam of light is split and travels along two paths, such as in Young's experiment and thin film interference, there can be interference effects in the reunited beam only if the path difference is less than the coherence length.

40.4 THIN-FILM INTERFERENCE

> Definition. <u>Optical path length</u>: the product of the physical path length and the index of refraction.

In the previous section, we have seen that the phase difference between two coherent waves that have traveled distances x_1 and x_2 in a medium of refractive index n is given by $\delta = 2\pi(x_1 - x_2)/\lambda_n$ (Eq. 40.51), where λ_n is the wavelength in the medium.

When a wave goes from one medium into another, its speed and wavelength change, but its frequency is constant, because wave crests are neither generated or consumed at the boundary. The wavelength in vacuum λ is thus given in terms of λ_n by $\lambda_n = \lambda/n$ (Eq. 40.53), and the phase difference can be expressed in terms of λ and the optical path length nx:

$$\delta = 2\pi n(x_1 - x_2)/\lambda \qquad (40.54)$$

Differences in optical paths are not the only source of phase changes. They can also occur as the result of reflections: a reflected wave undergoes a phase change of π if it strikes a surface of higher index of refraction (and thus travels faster than the transmitted wave), and there is no phase change if the wave strikes a surface of lower index of refraction (and thus travels slower than the transmitted wave).

In the use of thin films to minimize reflection (antireflection coating) or maximize reflection (reflective coating), we must take into account both the difference in optical path length and the difference in π phase changes upon reflection for the light reflected from the 1st and 2nd surfaces of the film. For example, let us assume that the 1st surface is in air (or a material whose index of refraction is less than that of the film). There will be a phase change of π for the light reflected from this surface.

For the light reflected from the 2nd surface, there is no reflection phase change if the film is in air, and the difference in reflection phase changes for light reflected from the two surfaces is π. If the film is coated on a material whose index of refraction is larger than that of the film, there is a phase change of π upon reflection, giving a difference in reflection phase changes for light reflected from the two surfaces of zero. To each of these reflection phase change differences must be added the phase change due to the difference in optical path, given for a film by Eq. 40.51 or Eq. 40.54 as $\delta = 2\pi n(2d)/\lambda$.

The net phase change difference for light reflected from the two surfaces is therefore $\delta = 4\pi nd/\lambda$ if the film is coated on a material whose index of refraction is larger than that of the film and is $\delta = 4\pi nd/\lambda + \pi$ if the film is in air. This net phase change must be $\delta = \pi, 3\pi, 5\pi\ldots$ for an antireflection coating and $\delta = 2\pi, 4\pi\ldots$ for a reflective coating.

Applying this example to an antireflection coating on a material having an index of refraction larger than that of the film, the phase difference is given by

$$\delta = 4\pi nd/\lambda = \pi, 3\pi, 5\pi\ldots$$

The minimum thickness of the coating has a phase difference of π and is given by $d = \lambda/4n$ (Eq. 40.61). A reflective coating for this case would require a phase difference of 2π and a thickness of $d = \lambda/2n$.

40.5 OPTICAL INTERFEROMETERS

<u>Definition</u>. <u>Interferometer</u>: an instrument capable of making precise measurements of optical path differences by splitting a beam of light, sending the two beams along different paths, and recombining the beams to produce an interference pattern.

The Michelson interferometer is important historically because it led to a definition of the standard meter in terms of the wavelength of light and because its use in the Michelson-Morley experiment to obtain a negative result for the velocity of the earth relative to the ether helped establish an experimental basis for Einstein's special theory of relativity.

In the Michelson interferometer, the relationship between the distance d that the interferometer is moved and the wavelength of the light is given by

$$\lambda = 2d/m \qquad (40.63)$$

where m is the number of fringe shifts (return of interference pattern to original form). The interferometer can be used to determine either wavelength or difference in optical paths. As usual, it is necessary that these differences be less than the coherence length of the light.

EXAMPLE PROBLEMS
Example 1

In a Young's interference experiment with the slits separated by 0.16 mm, for what wavelength is the separation of the first-order maximum and minimum 5 mm on a screen 2.8 m from the slits?

<u>Solution</u>

Given: $a = 0.16$ mm $= 1.6 \times 10^{-4}$ m
$\Delta y = 5$ mm $= 5 \times 10^{-3}$ m
$R = 2.8$ m

Required: λ

The order of the intensity maxima and minima in an interference pattern is specified by the integer m in the expressions for the linear positions given in Eqs. 40.19 and 40.21. Since the first-order minimum lies between the central maximum (m = 0) and the first-order maximum, the maximum always lies outside the minimum for a given order. Their separation Δy can be obtained from these equations by subtracting the position of the mth-order minimum from that of the mth-order maximum:

$$\Delta y = R\lambda m/a - R\lambda[m - (1/2)]/a = R\lambda/2a$$

The fact that this separation is constant for a given slit separation and slit-screen distance indicates that the interference fringes in a double-slit experiment are evenly spaced.

Solving this equation for λ and substituting for the slit separation a, the slit-screen distance R, and the separation Δy:

$$\lambda = 2a\, \Delta y/R = 2(1.6 \times 10^{-4}\text{ m})(5 \times 10^{-3}\text{ m})/2.8\text{ m} = 5.71 \times 10^{-7}\text{ m} = \underline{571\text{ nm}}$$

Since optical wavelengths are approximately 400-700 nm and small angles are often involved, a variety of length units often appears in interference calculations and should be converted prior to use.

Example 2

In a double-slit experiment, light of wavelength 600 nm illuminates slits separated by 0.27 mm, and an interference pattern is formed on a screen 2.4 m away. a) What is the phase difference between waves arriving at a point on the screen 8 mm from the central maximum? b) Is this position an intensity maximum or minimum and, if so, what is the order?

Solution

Given: $a = 0.27\text{ mm} = 2.7 \times 10^{-4}\text{ m}$
 $y = 8\text{ mm} = 8 \times 10^{-3}\text{ m}$
 $R = 2.4\text{ m}$
 $\lambda = 600\text{ nm} = 6 \times 10^{-7}\text{ m}$

Required: a) $\Delta\phi$, b) m

a) According to Eq. 40.13, the path difference is given by

$$\Delta x = a\sin\theta$$

where a is the slit separation and θ is the angle up to the point in question.

Substituting this equation into the expression for the corresponding phase shift given by Eq. 40.7:

$$\Delta\phi = 2\pi\, \Delta x/\lambda = 2\pi a\sin\theta/\lambda$$

Since the slit-screen distance is much larger than the linear position along the screen, the small-angle approximation can be used:

$$\sin\theta \approx \tan\theta \approx y/R$$

Substituting this equation into the expression for the phase shift along with the slit separation a, the position along the screen y, the slit-screen distance R, and the wavelength :

$$\Delta\phi = 2\pi a\sin\theta/\lambda = 2\pi a y/\lambda R$$
$$\Delta\phi = 2\pi(8 \times 10^{-3}\text{ m})(2.7 \times 10^{-4}\text{ m})/(6 \times 10^{-7}\text{ m})(2.4\text{ m}) = \underline{3\pi}$$

b) Complete destructive interference corresponds to a phase difference of an odd-integral multiple of π. Since the phase difference in this case is 3π, the point at 8 mm is an intensity minimum. The order m can be obtained from an expression for the phase shift:

$$\Delta\phi = 2\pi[m - (1/2)] = 3\pi$$
$$m = \underline{2}$$

The intensity minimum at 8 mm is therefore the 2nd order minimum of the two-slit interference pattern.

Example 3

Light with wavelength 500 nm is incident normally on a thin film of index of refraction 1.2 and thickness 200 nm. What is a) the wavelength of the light in the film, b) the optical thickness of the film, c) the phase change for light that passes through the film, d) the net phase change for light that is reflected from the second surface when the film is in air, and e) the net phase change for light that is reflected from the second surface when the film is in water?

Solution

Given: $n = 1.2$
$\lambda = 500$ nm
$d = 200$ nm

Required: a) n, b) $n \Delta x$, c) δ_1, d) δ_2, e) δ_3

a) The wavelength of light in a medium of index of refraction n is given in terms of its wavelength in air by Eq. 40.53:

$\lambda_n = \lambda/n = 500$ nm$/(1.2) = \underline{417 \text{ nm}}$

b) The optical path length in a medium of index of refraction n is given in terms of its physical path length Δx by $n \Delta x$ (Eq. 40.50). Here, the physical path length is the thickness d of the film:

$n \Delta x = (1.2)(200 \text{ nm}) = \underline{240 \text{ nm}}$

c) The phase change δ_1 is given in terms of the optical path length $n \Delta x$ and the wavelength in vacuum λ by Eq. 40.54:

$\delta_1 = 2\pi(n \Delta x)/\lambda = 2\pi(240 \text{ nm})/500 \text{ nm} = \underline{3.02 \text{ rad}}$

d) When the film is in air, the light reflected from the second surface does not undergo an additional reflection phase shift, and the net phase change is due only to the optical path length of twice the optical thickness of the film:

$\delta_2 = 2\pi(n \Delta x)/\lambda = 2\pi(2)(240 \text{ nm})/500 \text{ nm} = \underline{6.03 \text{ rad}}$

e) Since the index of refraction of water (n = 1.33) is larger than that of the film, there is a phase shift of π upon reflection from the second surface, and the net phase change is the sum of the reflection phase shift and the phase change due to the optical path length found in d):

$\delta_3 = 6.03 \text{ rad} + \pi \text{ rad} = \underline{9.17 \text{ rad}}$

Example 4

Find the minimum thickness (>0) of a film located in air having an index of refraction of 1.5 that will minimize the reflection of light of wavelength 400 nm.

Solution

Given: $n = 1.5$
$\lambda = 400$ nm

Required: d

For a film in air, the net phase change difference for light reflected from both surfaces of the film, taking into account phase changes due to both optical path difference ($4\pi nd/\lambda$) and reflection phase shift difference (π) is given by $\delta = 4\pi nd/\lambda + \pi$. To minimize reflection the light from the two surfaces must be out of phase at the first surface and the phase difference must be equal to an odd integral multiple of π:

$\delta = 4\pi nd/\lambda + \pi = \pi, 3\pi, 5\pi...$

The minimum thickness of the film is given by a phase difference of 3π, since a value of π would correspond to zero thickness:

$4\pi nd/\lambda + \pi = 3\pi$

$d = \lambda/2n$

Substituting in this equation for the wavelength in air λ and the index of refraction n:

    d = 400 nm/(2)(1.5) = <u>133 nm</u>

The thickness of the film that minimizes the reflection of light of wavelength 400 nm is one-half the wavelength of the light in the film.

Example 5

What is the minimum thickness of a film of index of refraction 1.5 that maximizes the reflection of light of wavelength 400 nm, assuming that the film is located in air?

Solution

    Given: n = 1.5
                  λ = 400 nm

    Required: d

Since the film is located in air, the net phase change difference for the light reflected from the two surfaces of the film is the sum of 4πnd/λ for the optical path difference and π for the reflection phase shift difference, as in the previous example. However, in this case, the light from the two surfaces must be in phase at the first surface to maximize the reflection, and the phase difference must be an integral multiple of 2π:

    δ = 4πnd/λ + π = 2π, 4π ...

The minimum thickness is given by a phase difference of 2π:

    4πnd/λ + π = 2π

    d = λ/4n

Substituting in this equation for the wavelength λ and the index of refraction n:

    d = 400 nm/(4)(1.5) = <u>66.7 nm</u>

The thickness of the film that maximizes the reflection of light of wavelength 400 nm in air is one-quarter the wavelength of the light in the film.

PROBLEMS

1. In a Young's interference experiment with a slit separation of 0.2 mm and a distance to the screen of 2.3 m, find the wavelength of the light required for two adjacent bright fringes in the interference pattern on the screen to be separated by 7 mm.
2. In a double-slit experiment, slits separated by 0.15 mm are illuminated with light of wavelength 592 nm. At what angle will there be a 0.5 rad phase difference between waves from the two slits, assuming that the slit-screen distance is large compared to the separation of fringes on the screen?
3. Light of wavelength 200 nm is incident normally on a thin film having an index of refraction 1.4 and a thickness 100 nm. Find a) the wavelength of the light in the film, b) the optical thickness of the film, c) the phase change of the light that passes through the film, d) the net phase change of light that passes through the film and is reflected back through the film from the second surface when the film is in water, and e) repeat d) for the film in benzene.
4. What is the minimum thickness (>0) of a film in air that has an index of refraction of 1.2 and will minimize the reflection of light of wavelength 650 nm?
5. Find the minimum thickness (>0) of a thin film of index of refraction 1.3 that will serve as a reflective coating on glass for light of wavelength 550 nm.

Chapter 41
Physical Optics: Diffraction

PREVIEW

This chapter continues the study of physical optics with the consideration of another aspect of interference known as diffraction. Diffraction occurs when waves encounter an obstacle or aperture, and a diffraction pattern results from the interference of scattered waves generated in the process. The chapter investigates the diffraction patterns formed by a single narrow slit and by a system of many slits called a diffraction grating. The use of diffraction patterns called holograms to form three-dimensional images is also discussed.

SUMMARY

41.1 DIFFRACTION

In Chapter 40 we found that the interference pattern produced by two narrow slits indicates that the light does not travel along straight lines but bends and enters the region of the geometric shadow. Since diffraction is the result of interference and this spreading of the light is typical, diffraction is sometimes referred to as the deviation of light from a straight-line path.

In general, diffraction occurs whenever light encounters an obstacle or aperture that blocks a portion of the wave and scatters the remainder. The importance of diffraction is indicated by the fact that practically every optical system uses only portions of wave fronts and therefore experiences these effects. Diffraction is characteristic of waves of all types, including sound waves and water waves.

Diffraction can be analyzed according to Huygens' principle by dividing the portion of the wave front that is not obscured into many small regions that serve as the sources of coherent secondary waves. These secondary waves are superposed in the usual manner by taking their amplitudes and phases into account, and the resultant intensity distribution is known as a diffraction pattern.

Diffraction is therefore a consequence of the interference of secondary waves from <u>many coherent sources</u>. The system of two narrow slits is a special case in which we assume only two sources of secondary waves, and we usually do not speak of a two-slit diffraction pattern. All diffraction is the result of interference, but not all interference phenomena produce diffraction.

The general case of diffraction is known as Fresnel diffraction and occurs when the distances involved are large but finite and the wave fronts are curved. The special case of Fraunhofer diffraction occurs when the incident waves can be treated as plane waves and the distances traveled by the outgoing waves are large enough that the corresponding rays can be considered approximately parallel. We consider only Fraunhofer diffraction.

For a single narrow slit, the many secondary sources are narrow strips on the wave front within the slit parallel to the length of the slit. The contributions from all sources are superposed to give the resultant wave amplitude, and the intensity on the screen, or diffraction pattern, is proportional to the square of the wave amplitude. The single-slit diffraction pattern is given by

$$I/I_0 = [\sin(\pi D \sin\theta/\lambda)]^2/(\pi D \sin\theta/\lambda)^2 \quad (41.17)$$

where I is the intensity on the screen at angle θ, I_0 is the intensity at $\theta = 0$, and D is the slit width.

The relative intensity (I/I_0) has the form of $(\sin x/x)^2$ and is characterized by three significant features: 1) a sharp forward maximum as the result of complete constructive interference at $\theta = 0$, 2) minima of zero intensity as the result of complete destructive interference located symmetrically on either side of the central maximum at angles given by $\sin\theta_m = m(\lambda/D)$, $m = \pm 1, \pm 2 \ldots$ (Eq. 41.18), and 3) weaker intensity peaks as the result of incomplete destructive interference at angles between the minima.

The first minimum on either side of the central maximum occurs at $\sin\theta_1 = \lambda/D$ (Eq. 41.19). For $\lambda \ll D$, the small-angle approximation gives $\theta_1 = \lambda/D$ (Eq. 41.20). This relation shows that most of the wave intensity is scattered into a narrow range of angle $2\theta_1 = 2\lambda/D$ about the forward direction, that for a fixed slit width longer wavelengths undergo more diffraction, and that for a given wavelength the angular width of the central maximum increases as the slit width decreases.

It should be noted that expressions obtained for angular position in the two-slit and single-slit systems are very similar: the intensity minima of the single slit are given by $\sin\theta_m = m(\lambda/D)$, where D is the slit width, and the intensity maxima for two slits are given by $\sin\theta = m\lambda/a$, where a is the distance between the slits.

41.3 DIFFRACTION AND ANGULAR RESOLUTION

The analysis of the single-slit diffraction pattern illustrates two general features of diffraction that occur when waves of wavelength λ encounter an obstruction or aperture of characteristic dimension $D \gg \lambda$: 1) destructive interference causes the intensity to fall to zero at an angle given approximately by λ/D, and 2) most of the diffracted intensity appears within a narrow angular range given by $2\theta = 2\lambda/D$ (Eq. 41.21).

The exact angle that defines the edge of the central maximum depends on the shape of the aperture or obstacle, and we have seen that it is given by $\sin\theta = \lambda/D$ for slit of width D. For a circular aperture of diameter D, this angle is given by $\sin\theta = 1.22 \lambda/D$, and the circular area of maximum intensity defined by this expression is called the Airy disk.

> Definition. Angular resolution: the smallest angular separation that can be distinguished (resolved) by an optical system.

> Definition. Rayleigh's criterion: a criterion used to establish the angular resolution of an optical device that states that the images of two point sources are resolved when the central intensity peak of the diffraction pattern of one source falls on the first intensity zero of the other source and is given by $\sin\phi = 1.22(\lambda/D)$ (Eq. 41.23) for a device of diameter D collecting waves of wavelength λ, where ϕ is the angular separation of the two sources as viewed from the device.

41.4 DIFFRACTION GRATINGS

  <u>Definition</u>. <u>Diffraction grating</u>: any optical device that is equivalent to an array of parallel slits.

For a diffraction grating, the coherent sources are the slits. As usual, the wave amplitude at a point on the screen is the superposition of the amplitudes of the waves from the individual slits, and the intensity is proportional to the square of the wave amplitude. The N-slit diffraction pattern is given by

$$I/I_o = [\sin(\pi D \sin\theta/\lambda)]^2 [\sin(N\delta/2)]^2/(\pi D \sin\theta/\lambda)^2 \sin^2(\delta/2) \qquad (41.34)$$

where I_o is the intensity at $\theta = 0$ for one slit, $\delta = 2\pi a \sin\theta/\lambda$ is the phase difference between waves from adjacent slits, and a is the center-to-center distance of the slits.

The intensity is given by the product of the intensity of a single slit (Eq. 41.16), which describes interference arising from the width D of the slits, and an interference factor that describes interference arising from the separation a of the slits and contains the dependence on the number of slits N.

The interference factor produces large peaks in the diffraction pattern called principal maxima that correspond to complete constructive interference and are given by $\sin\theta = m\lambda/a$, where $m = 0, 1, 2...$ is called the order of the maximum. Positions of zero intensity result from complete destructive interference and are given by $\Delta\theta = \lambda/Na \cos\theta$ (Eq. 41.40), where $\Delta\theta$ is the difference in the angular positions of a principal maximum and an adjacent zero and thus measures the angular radius of the maxima. As N increases the angular radius decreases, and light of a given wavelength goes into a smaller angular range.

Again, the similarity of the expression for the angular position of the principal maxima of a diffraction grating to those for the intensity minima of the single slit and the intensity maxima for two slits should be kept in mind.

The resolving power is a measure of the precision with which a grating can measure wavelengths and is given by $R = \lambda/\Delta\lambda = mN$ (Eqs. 41.43 and 41.44), where $\Delta\lambda$ is the smallest resolvable difference in wavelength for two wavelengths of average wavelength λ. The larger the resolving power the smaller the wavelength difference that can be resolved. The dependence of the resolving power on the order of the maxima indicates that two wavelengths whose $m = 1$ maxima overlap may have their $m = 2$ maxima resolved. The maxima of order m for different wavelengths are referred to as the mth-order spectrum. The resolving power is zero for the central or $m = 0$ principal maximum.

41.5 HOLOGRAPHY

  <u>Definition</u>. <u>Hologram</u>: a photographic record of the phase and amplitude distributions of the interference pattern formed by the superposition of two coherent light waves.

In the simplest type of hologram, coherent light illuminates an object, and a photographic film is located so that it intercepts light scattered from the object and direct light from the source. The developed film, or hologram, can be used to produce a three-dimensional image by using coherent light from the original source. The light diffracted by the stored interference pattern reconstructs the wave fronts that originally produced the hologram and creates the holographic images.

In order to produce any type of interference pattern, different portions of the same wave train must be superposed. In holography, one portion of a wave train becomes the reference wave, and the other becomes the wave scattered by the object. If the optical path lengths of the two portions differ by more than the coherence length of the light, the superposed waves will be incoherent and will not interfere. The development of the laser increased the coherence length of available sources from the millimeter range to the order of meters and thus increased dramatically the size of objects that can be holographed.

EXAMPLE PROBLEMS

Example 1

Light of wavelength 450 nm illuminates a slit 0.3 mm wide and forms a diffraction pattern on a screen 2 m away. Find a) the angular radius of the central maximum of the diffraction pattern and b) the separation on the screen of the positions of zero intensity on either side of the central maximum.

Solution

Given: $\lambda = 450$ nm $= 4.5 \times 10^{-7}$ m
         $D = 0.3$ mm $= 3 \times 10^{-4}$ m
         $L = 2$ m

Required: a) θ_1,   b) d

a) The angular radius of the central peak is given by Eq. 41.19:
$$\sin\theta_1 = \lambda/D$$
Since $\lambda \ll D$, the small-angle approximation (Eq. 41.20) can be used, and upon substitution for the wavelength λ and the slit width D yields
$$\theta_1 = \lambda/D = (4.5 \times 10^{-7})/(3 \times 10^{-4}) = \underline{1.5 \times 10^{-3} \text{ rad}}$$

b) The angular separation on the screen of the positions of zero intensity define the diameter $d = 2r$ of the central maximum, where
$$\tan\theta_1 = r/L = \sin\theta_1 = \theta_1$$
Solving this equation for r and substituting for the slit-screen distance L and θ_1 from a):
$$d = 2r = 2(2 \text{ m})(1.5 \times 10^{-3} \text{ rad}) = 6 \times 10^{-3} \text{ m} = \underline{6 \text{ mm}}$$

The small-angle approximation is useful in situations such as this involving very small ratios, including ratios of wavelength and slit or slit-system dimensions, source separation and source distance, and intensity maxima and minima separation and screen distance. In these cases, $\tan\theta = \sin\theta = \theta$ (in rad).

Example 2

A telescope has an objective lens of diameter 5 cm. a) For a wavelength of 500 nm, what is the angular resolution of the telescope? b) What is the maximum distance at which the telescope can just resolve points separated by 5 mm, assuming that the distance is determined by diffraction effects?

Solution

Given: $D = 5$ cm $= 5 \times 10^{-2}$ m
         $\lambda = 500$ nm $= 5 \times 10^{-7}$ m
         $d = 5$ mm $= 5 \times 10^{-3}$ m

Required: a) ϕ, b) L

a) The angular resolution, or smallest angular separation ϕ that an instrument of diameter D collecting light of wavelength λ can resolve, is given by Rayleigh's criterion (Eq. 41.23):
$$\sin \phi = 1.22(\lambda/D)$$
Using the small-angle approximation (Eq. 41.24) and substituting for the wavelength λ and the diameter D:
$$\phi = 1.22(\lambda/D) = 1.22(5 \times 10^{-7} \text{ m})/(5 \times 10^{-2} \text{ m}) = \underline{1.22 \times 10^{-5} \text{ rad}}$$
b) In order for two points separated by d at a distance L to be just resolved, their angular separation d/L must be equal to the angular resolution of the telescope ϕ:
$$d/L = \phi = 1.22(\lambda/D)$$
Solving for the distance L and substituting for the separation d and the angular resolution ϕ:
$$L = d/\phi = (5 \times 10^{-3} \text{ m})/(1.22 \times 10^{-5}) = \underline{410 \text{ m}}$$

Example 3

Light of wavelength 694 nm illuminates a grating with 527 lines/mm. a) What is the separation on a wall 10 m away of the central and first-order principal maxima? b) If the grating is 32 mm wide, what is the width of the central principal maximum on the wall?

Solution

Given: $\lambda = 694 \text{ nm} = 6.94 \times 10^{-7}$ m

$1/a = 527$ lines/mm

$w = 32 \text{ mm} = 3.2 \times 10^{-2}$ m

$L = 10$ m

Required: a) y,   b) Δy

a) The angle of the principal maximum of order m can be obtained from Eq. 41.35 and is given by
$$\sin\theta_m = m\lambda/a \tag{1}$$
The slit spacing a is the reciprocal of the number of lines/m:
$$a = 1/(527 \text{ lines/mm}) = 10^{-3} \text{ m}/527 = 1.90 \times 10^{-6} \text{ m}$$
Substituting in Eq. (1) for the order m, the wavelength λ, and the slit spacing a:
$$\sin\theta_1 = (1)(6.94 \times 10^{-7} \text{ m})/(1.9 \times 10^{-6} \text{ m}) = 0.36526$$
$$\theta_1 = 21.4°$$
The separation on the wall can be obtained from $\tan \theta_1$ and the distance L to the wall:
$$y = L \tan \theta_1 = (10 \text{ m})(0.39237) = \underline{3.92 \text{ m}}$$
b) The angular diameter of the central principal maximum can be obtained from the angular radius given in Eq. 41.40:
$$2 \Delta\theta = 2\lambda/Na \cos\theta = 2\lambda/w \cos\theta$$
Substituting in this equation for the wavelength λ, the grating width w, and the angle of the principal maximum θ:
$$2 \Delta\theta = 2(6.94 \times 10^{-7})/(3.2 \times 10^{-2})(\cos 0°) = 4.34 \times 10^{-5} \text{ rad}$$
Using the small-angle approximation, the width of the central maximum on the wall is given by
$$y = L(2 \Delta\theta) = (10 \text{ m})(4.34 \times 10^{-5} \text{ rad}) = 4.34 \times 10^{-4} \text{ m} = \underline{0.434 \text{ mm}}$$

Example 4
A diffraction grating has 527 lines/mm and is 32 mm wide. Find a) the smallest wavelength separation that can be resolved in the third-order spectrum at 600 nm and b) the highest order spectrum that can be observed with this grating.

Solution
Given: $\lambda = 600$ nm $= 6.0 \times 10^{-7}$ m
$1/a = 527$ lines/mm
$w = 32$ mm $= 3.2 \times 10^{-2}$ m

Required: a) $\Delta\lambda$,   b) m_{max}

a) The resolving power of a grating is given by Eqs. 41.43 and 41.44:
$$R = \lambda/\Delta\lambda = mN \tag{1}$$
The number of lines is given by the product of lines per unit width of the grating and the width of the grating:
$$N = (527 \text{ lines/mm})(32 \text{ mm}) = 16{,}864 \text{ lines}$$
Solving Eq. (1) for $\Delta\lambda$ and substituting for the wavelength λ, the order m, and the number of lines N:
$$\Delta\lambda = \lambda/mN = (6.0 \times 10^{-7} \text{ m})/(3)(16{,}864) = 1.19 \times 10^{-11} \text{ m} = \underline{0.0119 \text{ nm}}$$

b) The highest order that can be observed is obtained from the expression for the angular position of the principal maximum of order m:
$$\sin\theta_m = m\lambda/a$$
where the grating spacing a is the reciprocal of the number of lines/m and the maximum angle θ possible is 90°.

Solving this equation for the order m and substituting for the wavelength λ, the grating spacing a, and the maximum angle θ:
$$m = a \sin\theta_m/\lambda = (10^{-3} \text{ m}/527)(1)/(6.0 \times 10^{-7} \text{ m}) = 3.16$$
The highest order spectrum that can be observed with this grating is therefore $\underline{m_{max} = 3}$.

PROBLEMS

1. Monochromatic light is incident on a single slit of width 0.2 mm, and the distance between the points of zero intensity on either side of the central peak of the diffraction pattern formed on a screen 1.5 m away is 4.5 mm. Find a) the angular radius of the central peak and b) the wavelength of the light.

2. Find the minimum diameter of the camera lens required to resolve the eyes of a person (separated by 7 mm) at a distance of 300 m and a wavelength of 600 nm, assuming the resolution is limited by diffraction effects.

3. Monochromatic light of wavelength 486 nm is observed at 61° in the second-order spectrum of a diffraction grating. Find the number of lines/mm in the grating.

4. A diffraction grating has 900 lines/mm and is 2 cm wide. Find a) the highest order spectrum that can be observed with this grating and b) the smallest wavelength separation that can be resolved by the grating at a wavelength of 500 nm.

Chapter 42
Quantum Physics, Lasers, and Squids

PREVIEW
   This chapter presents a summary of the discoveries and ideas involved in the development of quantum physics. These include Planck's original quantum theory of blackbody radiation and quantization of energy, Einstein's photon model of light and theory of the photoelectric effect, the alpha-particle scattering experiments of Geiger and Marsden, the Rutherford nuclear atom, Bohr's quantization of angular momentum and the planetary atom, deBroglie's wave theory of particles and its confirmation in the electron diffraction experiments of Davisson and Germer, and the wave mechanics of Heisenberg, Schrödinger, and Born. The chapter also discusses lasers and squids as applications from the field of quantum engineering.

SUMMARY

42.1 THE ORIGINS OF QUANTUM PHYSICS
   The origin of quantum physics was in 1900 when Planck quantized the energy of the oscillators emitting radiation in a blackbody, defined in Chapter 23 as a perfect absorber and emitter of thermal radiation. In Planck's theory, the energy E of an oscillator of frequency ν is given by
$$E = nh\nu \tag{42.1}$$
where n is an integer and Planck's constant is $h = 6.6262 \times 10^{-34}$ J-s.
   Einstein extended Planck's idea of energy quantization to radiant energy and introduced the photon model of light, in which the energy emitted by a light source of frequency ν is carried by a bundle of energy, a light quantum or photon, of energy
$$E = h\nu \tag{42.3}$$

   Definition. Photoelectric effect: a process in which electrons, called photoelectrons, are ejected from metal surfaces by incident ultraviolet radiation.

   In 1905, Einstein used the photon model to explain the photoelectric effect. In this theory, the incident photon disappears, and its energy is transformed into the work needed to free an electron from the metal and the kinetic energy of the photoelectron after its ejection. Einstein's photoelectric equation is given by
$$h\nu = W + K_{max} \tag{42.4}$$
where the work function W is characteristic of the metal and is the minimum energy required by an electron to escape from the surface and K_{max} is the corresponding maximum kinetic energy of the electron.
   Kinetic energies less than the maximum can result from the loss of energy by the electron in collisions enroute to the surface. The minimum frequency that ejects photoelectrons is called the threshold frequency ν_0 and is obtained by evaluating the photoelectric equation for $K_{max} = 0$:

$h\nu_0 = W$ (Eq. 42.7).

In a photoelectric effect experiment, the photoelectrons flow as a current, and the potential required to just stop the current is known as the stopping potential. With $K_{max} = eV_s$, the photoelectric equation becomes

$$V_s = h\nu/e - W/e \qquad (42.6)$$

therefore, Einstein's theory predicts a linear relationship between the stopping potential and the frequency and a universal slope h/e for a graph of V_s vs ν.

Photons exist only in motion at the speed of light c and have zero mass. Since they have energy, they possess linear momentum even though they have no mass. Their linear momentum can be obtained from the relativistic expression for total energy (Eq. 20.13): $E^2 = (pc)^2 + (mc^2)^2$. With $m = 0$, the photon linear momentum p is given by

$$p = E/c = h\nu/\nu\lambda = h/\lambda \qquad (42.10, 42.11)$$

The photon also possesses a quantized angular momentum parallel or antiparallel to its direction of motion given by $L = h/2\pi$ (Eq. 42.12).

42.2 RUTHERFORD AND THE NUCLEAR ATOM

J. J. Thomson, discoverer of the electron in 1897, developed one of the early atomic models in which the negatively charged electrons were embedded in a continuous distribution of positive charge like raisins in a pudding. The nuclear model of the atom was developed by Rutherford in 1911 on the basis of the classic alpha-particle scattering experiments carried out under his direction by Geiger and Marsden.

In these experiments, alpha particles emitted by naturally-occurring radioactive substances with energies of several MeV were incident on thin metal foils, and some were found to be deflected through large angles in the backward direction. Rutherford concluded that the positive charge and most of the mass of the atom are concentrated in a small central region which he called the nucleus. He also estimated that the nucleus is smaller than the characteristic atomic size of one angstrom by a factor of approximately 10^4.

42.3 BOHR AND THE HYDROGEN ATOM

In 1913, Bohr conceived the planetary model of the hydrogen atom, which embodies the nuclear model of Rutherford and the light quanta of Einstein. In this model, the electron revolves in a circular orbit about a fixed nucleus under the influence of their mutual electrostatic attraction. The model is based on a number of assumptions which have become known as the Bohr postulates:

1) Stationary states. There exist stationary states of the dynamic atom in which the orbiting electron does not radiate.

2) Radiation postulate. A photon is emitted when the atom goes from one stationary state to another. The energy of the photon is given by $h\nu = E_a - E_b$ (Eq. 42.17), where E_a and E_b are the energies of the initial and final states of the atom.

3) Quantization. The stationary states correspond to those orbits for which the orbital angular momentum of the electron is given by $L = nh/2\pi$ (Eq. 42.18), where $n = 1, 2, 3...$ is called a quantum number.

Since the atom can exist in only certain stationary states, these states and their corresponding radii and energies are quantized. The radii and energies are specified by the quantum number n and are given by

$$r_n = \varepsilon_0 h^2 n^2/\pi mZe^2 = 0.529177 \text{ Å} (n^2/Z) = a_0(n^2/Z) \quad (42.19, 42.21)$$
$$E_n = -mZ^2e^4/8\varepsilon_0^2 h^2 n^2 = -13.6 \text{ eV} (Z^2/n^2) \quad (42.22, 42.23)$$

The Bohr theory applies to hydrogen and hydrogen-like ions, such as singly-ionized helium, in which the electron orbits a nucleus of charge Ze, where Z is the atomic number, or number of protons in the nucleus. The results for hydrogen are therefore given by $Z = 1$. The lowest-energy state, or ground state, of hydrogen is specified by $n = 1$ and has a radius of $a_0 = 0.529$ Å and an energy of -13.6 eV. Other states are referred to as excited states. All the quantized energies are negative, and zero energy corresponds to ionization, or removal of the electron from the atom.

When the atom makes a transition from an excited state to a state of lower energy, the electron changes orbits and a photon is emitted. The energy of a state is quite different from a transition energy, or photon energy, which is given by the difference in two state energies. For a transition in hydrogen from a state specified by the quantum number n_a to a lower state specified by n_b, the energy of the emitted photon is given by

$$h\nu = E_a - E_b = -13.6 \, Z^2 \, [(1/n_a)^2 - (1/n_b)^2] \text{ eV}$$

where the photon energy is positive as it must be, since $n_a > n_b$. The ionization energy of an atom is given by the energy of a transition from $E = 0$ ($n = \infty$) to the ground state ($n = 1$) and is 13.6 eV for hydrogen.

When an atom is in an excited state, it can make a transition to any lower state, not only the ground state. Transitions that terminate on the $n = 2$ state are referred to as Balmer transitions and are said to belong to the Balmer series. Other series in hydrogen are named after the men who discovered them: Lyman ($n = 1$), Paschen ($n = 3$), Brackett ($n = 4$), and Pfund ($n = 5$). Only the Balmer series is in the visible part of the spectrum. The Lyman series is in the ultraviolet; the others are in the infrared. The transitions in any series are labeled with letters of the Greek alphabet, with α belonging to the transition of lowest energy or longest wavelength.

42.4 DE BROGLIE AND THE WAVE-PARTICLE DUALITY

The wave-particle duality of radiation refers to the fact that light behaves sometimes like a particle, as in the photoelectric effect, and other times like a wave, as in the Young interference experiment. In 1924, de Broglie suggested that duality is a universal principle of nature and that particles must also exhibit both wave-like and particle-like behavior. He postulated that for both matter and radiation, the total energy and momentum are related to the frequency and wavelength of the associated wave by

$$E = h\nu \quad (42.3)$$
$$p = h/\lambda \quad (42.11)$$

where these relations connect the particle concepts of energy E and momentum p to the wave concepts of frequency ν and wavelength λ.

The particle waves of de Broglie allowed an interpretation of the quantization postulate of the Bohr theory (Eq. 42.18): $L = nh/2\pi = mvr$. This can be rewritten using Eq. 42.11 as $2\pi r = n\lambda$, which gives a phase shift of the wave associated with the electron of $\Delta\phi = 2\pi \, \Delta x/\lambda = 2\pi n\lambda/\lambda = 2\pi n$ after each orbit. The statinary states of the Bohr theory thus correspond to orbits which contain an integral number of electron wavelengths and for which there is constructive interference of the electron waves at each traversal of the orbits.

The mathematical theory of de Broglie waves was developed during 1925-26 by Heisenberg,

Schrödinger, and Born, and the existence of these waves was confirmed in the electron diffraction experiments of Davisson and Germer in 1927. Since the wavelength of the electrons predicted by Eq. 42.11 is extremely small, Davisson and Germer utilized the atoms of a nickel crystal as a diffraction grating and were able to observe a diffraction pattern produced by the interference of the electron waves. The theories involving quantization in the early 1900's, including those of Planck, Einstein, Bohr, Compton, and de Broglie, are sometimes referred to as the old quantum theory. The new quantum theory introduced in 1925-26 is called wave mechanics and is part of a theory that is known as quantum mechanics.

42.5 SCHRÖDINGER'S EQUATION, PROBABILITY WAVES, AND QUANTUM MECHANICAL TUNNELING

In 1926, Schrödinger presented the wave equation that describes de Broglie waves. Schrödinger's equation is the fundamental equation of wave mechanics and like the classical wave equation (Eq. 18.12) is a second-order linear partial differential equation. The one-dimensional form of Schrödinger's equation is given by

$$- (h^2/8m\pi^2)(\partial^2\psi/\partial x^2) + U\psi = (ih/2\pi)(\partial\psi/\partial t) \tag{42.30}$$

where ψ is the wave function and U is the potential energy, given for the hydrogen atom by the Coulomb potential energy $-e^2/4\pi\epsilon_o r$. Solutions of Schrödinger's equation lead naturally to energy quantization as the result of boundary conditions satisfied by the wave function. The quantized energies of the hydrogen atom are the same as those of the Bohr theory. Also quantized in the Schrödinger theory are the magnitude and orientation of the orbital angular momentum.

The interpretation of the quantum mechanical wave function ψ was given by Born in 1926. The square of the magnitude of the wave function is equal to the probability per unit volume of finding the particle associated with the wave at a particular location at a particular time. Since the wave function is squared, equal positive and negative values give the same probability. The particle has some probability of being found wherever the associated wave function is non-zero and cannot be found where the wave function is zero. The results of wave mechanics reduce to those of classical mechanics when the de Broglie wavelength is much less than the dimensions of the system involved. Thus, the de Broglie wavelength determines when classical mechanics is inadequate at small distances just as the speed of light determines when it is inadequate at high speeds.

Quantum-mechanical calculations indicate that there is some probability of finding a particle within the classically-forbidden region of space beyond a classical turning point where the total energy is less than the potential energy and the kinetic energy is negative. The penetration of a particle through a potential-energy barrier region such as this between two classical-allowed regions is known as quantum-mechanical tunneling.

42.6 QUANTUM OPTICS AND LASERS

In 1917, Einstein showed that three basic processes are described by the Born radiation condition $h\nu = E_a - E_b$ (Eq. 42.17):

1) absorption - excitation of an atom by absorption of a photon of energy ($E_a - E_b$).

2) spontaneous emission - de-excitation of an atom by emission of a photon with energy ($E_a - E_b$).

3) stimulated emission - de-excitation of an atom caused by an incident photon of energy ($E_a - E_b$) with the emission of a second photon of the same energy which is coherent with the original photon.

Definition. <u>Laser</u>: derived from light amplification by stimulated emission of radiation, a device that produces a narrow, directional intense beam of coherent light.

Definition. <u>Population inversion</u>: a condition in which the population of a higher-energy state is greater than that of a lower-energy state.

A laser requires an active medium, whose atoms make the transitions that produce the laser light, and an optical resonator, a mechanism to confine the laser light to stimulate further emission, usually a pair of mirrors. Light amplification, or more stimulated emission than absorption, is achieved by creating a population inversion by pumping, or addition of energy to the active medium, and selectively populating the upper laser state.

The first laser, invented by Maiman in 1960, is a three-level laser with an active medium of ruby. Pumping is with a gaseous discharge flashtube to an excited state, with the emission of laser light of wavelength 694 nm. The He-Ne laser was invented in 1961 by Javan, Bennett, and Herriott, and is a four-level laser. An electric discharge pumps the He atoms, and they in turn pump the Ne atoms by inelastic collision to the upper laser states. Stimulated emission is to a lower excited state, with the emission of laser light in the infrared and also in the visible at 632.8 nm.

42.7 SQUIDS

Definition. <u>Squid</u>: from superconducting quantum interference device, an ultrasensitive magnetometer based on the quantization of flux in a superconducting ring.

The operation of a squid involves a number of phenomena that we have considered previously, including superconductivity, quantization, interference of coherent waves, and quantum mechanical tunneling. The quantization of magnetic flux $\Phi = BA$ (Eq. 32.2) was predicted for superconductors by Abrikosov in 1957 and verified in 1961 by Fairbank and Deaver. The quantum of magnetic flux is called the fluxon $\Phi_o = h/2e = 2.0678538 \times 10^{-15}$ T-m^2 (Eq. 42.32), and the total flux can be expressed as an integral number of fluxons: $\Phi = n\Phi_o$ (Eq. 42.34). The theory of the quantum-mechanical tunneling utilized in squids was developed in 1962 by Josephson.

Definition. <u>Josephson effect</u>: tunneling of electron pairs (Cooper pairs) through a thin insulating barrier (Josephson junction) between two superconducting materials.

The squid current varies periodically with an applied magnetic field in a manner analogous to the variation in wave amplitude produced by interfering coherent light waves. In this case the interfering waves are the de Broglie waves of the electrons of the supercurrent and the phase difference is a consequence of the applied field. The squid is called a quantum interference device because the squid current varies systematically with the passage of each quantum of magnetic flux. Counting the variations in the current thus allows the squid to measure magnetic flux changes, which can be converted to magnetic field measurements using the relationship between magnetic flux and magnetic field. The squid can be converted into a sensitive ammeter by measuring the flux changes produced by a current flowing in an adjacent superconducting coil.

EXAMPLE PROBLEMS

Example 1

An aluminum surface is irradiated with monochromatic ultraviolet light, and no photoelectrons are emitted unless the wavelength is below 300 nm. a) What is the work function of aluminum? If the incident light has a wavelength of 200 nm, what is b) the maximum kinetic energy of the ejected electrons and c) the potential required to stop the photoelectric current.

Solution

Given: $\lambda_0 = 300$ nm $= 3 \times 10^{-7}$ m
$\lambda = 200$ nm $= 2 \times 10^{-7}$ m
$h = 6.63 \times 10^{-34}$ J-s (Eq. 42.2)
$c = 3 \times 10^8$ m/s (Appendix 2)
1 eV $= 1.60 \times 10^{-19}$ J (Appendix 3)

Required: a) W, b) K_{max}, d) V_s

a) The photoelectric equation (Eq. 42.4) is given by

$$h\nu = W + K_{max}$$

The work function W is the minimum energy required to free an electron from the surface and thus corresponds to zero kinetic energy (Eq. 42.7):

$$W = h\nu_0 = hc/\lambda_0$$

where ν_0 is the threshold frequency, or lowest frequency that ejects electrons, which is related by the kinematic relation $c = \nu_0 \lambda_0$ (Eq. 17.2) to the threshold wavelength λ_0, or longest wavelength that ejects electrons.

Substituting in this equation for the threshold wavelength λ_0, the constants h and c, and the J/eV conversion:

$W = hc/\lambda_0 = (6.63 \times 10^{-34}$ J-s$)(3 \times 10^8$ m/s$)/(3 \times 10^{-7}$ m$)(1.6 \times 10^{-19}$ J/eV$) = \underline{4.14 \text{ eV}}$

b) The maximum kinetic energy of the photoelectrons is obtained by solving Eq. (1) for K_{max} and again using the kinematic relation $c = \nu\lambda$:

$$K_{max} = hc/\lambda - W$$

Substituting in this equation for the wavelength λ, the work function W from a), the constants h and c, and the J/eV conversion:

$K_{max} = (6.63 \times 10^{-34}$ J-s$)(3 \times 10^8$ m/s$)/(2 \times 10^{-7}$ m$)(1.6 \times 10^{-19}$ J/eV$) - 4.14$ eV
$= 6.21$ eV $- 4.14$ eV $= \underline{2.07 \text{ eV}}$

c) The stopping potential is the potential difference that reduces the photoelectric current to zero and is given in terms of the maximum kinetic energy by Eq. 42.5:

$$V_s = K_{max}/e$$

In this equation, if the kinetic energy has units of electron volts and the charge is expressed in units of the electronic charge, the potential difference has units of volts. Substituting for the maximum kinetic energy K_{max}:

$$V_s = 2.07 \text{ eV}/e = \underline{2.07 \text{ V}}$$

It is useful to develop an expression for the energy in electron volts (eV) in terms of the wavelength in nanometers (nm):

$E(eV) = hc/\lambda = (6.63 \times 10^{-34}$ J-s$)(3 \times 10^8$ m/s$)/(10^{-9}$ m/nm$)(1.6 \times 10^{-19}$ J/eV$)\lambda$ (nm)
$= 1240$ eV-nm$/\lambda$ (nm)

We will make use of this expression in the other examples in this chapter.

Example 2

Ultraviolet radiation is incident on a sample of hydrogen atoms in their ground states, and a number of them are excited to their third excited states. a) How many different energies are present for the photons emitted by these atoms? b) What is the longest wavelength possible?

Solution

Given: $n_{max} = 4$

Required: a) # different energies, b) λ_{max}

a) An atom in an excited state can de-excite by making a transition to any lower state, and each of these transitions can be followed by other transitions to lower states until the ground state is reached. The third excited state has quantum number n = 4. The possible transitions from states n_a to states n_b can be expressed as

$n_a \rightarrow$	n_b	Transitions
4	3, 2, 1	3
3	2, 1	2
2	1	1

The total number of transitions is 3 + 2 + 1 = 6. There is a different energy for each transition, therefore the number of different energies is also given by # = <u>6</u>.

b) The wavelength of an emitted photon can be expressed in terms of the quantum numbers of the states involved in the transition using the Bohr radiation postulate (Eq. 42.17), the kinematic relation $c = \nu\lambda$ (Eq. 17.8), and the Bohr hydrogen atom energies $E_n = -13.6$ eV/n^2 (Eq. 42.22):

$$\lambda = hc/(E_a - E_b) = (hc/13.6 \text{ eV})/[(1/n_b^2) - (1/n_a^2)]$$

The longest wavelength results from the lowest-energy transition, which is from $n_a = 4$ to $n_b = 3$. Substituting in this equation for these quantum numbers and the constant 1240 eV-nm relating energy in eV to wavelength in nm developed in Example 1:

$$\lambda_{max} = 1240 \text{ eV-nm}/(13.6 \text{ eV})[(1/9) - (1/16)] = \underline{1876 \text{ nm}}$$

Example 3

An ionized Be (Z = 4) atom has one electron in orbit around the nucleus. a) Which state of this atom has a Bohr radius equal to that of the hydrogen atom ground state and b) what is the energy and wavelength of the photon emitted in the transition to the ground state?

Solution

Given: $Z = 4$
$r_n = a_o$
$n_b = 1$

Required: a) n, b) E, λ

a) The Bohr radius for a one-electron atom of atomic number Z is given in Sec. 42.3 of the Study Guide as

$$r_n = a_o(n^2/Z)$$

Solving this equation for the quantum number n of the state and substituting for the Be radius r_n and atomic number Z:

$$n = \sqrt{Zr_n/a_o} = \sqrt{Z} = \underline{2}$$

b) The transition energy is given in Sec. 42.3 of the Study Guide as

$$h\nu = E_a - E_b = -13.6 \, Z^2 \, [(1/n_a)^2 - (1/n_b)^2] \text{ eV}$$

Substituting in this equation for the atomic number Z and the quantum numbers of the initial and final states n_a and n_b:

$$E_a - E_b = -13.6 \, (4)^2 \, (1/4 - 1/1) \text{ eV} = \underline{163 \text{ eV}}$$

The wavelength can be obtained from the relation between energy and wavelength given in Example 1:

$$\lambda = 1240 \text{ eV-nm}/E(\text{eV}) = 1240 \text{ eV-nm}/163 \text{ eV} = \underline{7.6 \text{ nm}}$$

Example 4

Find a) the velocity of an electron with a de Broglie wavelength of 0.4 nm and b) the de Broglie wavelength of another particle with a mass 207 times that of the electron that has the same kinetic energy as the electron.

Solution

Given: $\lambda_e = 0.4$ nm $= 4 \times 10^{-10}$ m
         $m = 207 \, m_e$
         $m_e = 9.1 \times 10^{-31}$ kg   (Appendix 2)
         $h = 6.63 \times 10^{-34}$ J-s   (Eq. 42.2)

Required: a) v_e,   b) λ

a) The de Broglie wavelength of a particle of mass m and velocity v can be obtained from Eq. 42.11:

$$\lambda = h/p = h/mv$$

Solving this equation for v and substituting for the electron mass and wavelength and the constant h:

$$v = h/m\lambda = (6.63 \times 10^{-34} \text{ J-s})/(9.1 \times 10^{-31} \text{ kg})(4 \times 10^{-10} \text{ m}) = \underline{1.82 \times 10^6 \text{ m/s}}$$

b) Since the electron's speed is well below relativistic speeds, we can use the classical expression for the kinetic energy (Eq. 7.8). The ratio of the velocities of the two particles can be obtained by equating their kinetic energies:

$$mv^2/2 = m_e v_e^2/2$$

$$v/v_e = \sqrt{m_e/m} = \sqrt{m_e/207 \, m_e} = 1/\sqrt{207}$$

Substituting in the expression for the de Broglie wavelength of the particle (Eq. 42.11) for the particle's mass and velocity in terms of the mass m_e of the electron:

$$\lambda = h/mv = (h/207 \, m_e)(\sqrt{207}/v_e) = \lambda_e/\sqrt{207}) = 0.4 \text{ nm}/\sqrt{207}) = \underline{0.0278 \text{ nm}}$$

PROBLEMS

1. Monochromatic light of wavelength 150 nm is incident on a copper surface, and a potential difference of 3.78 V is required to stop the photoelectric current. Find a) the maximum kinetic energy of the ejected photoelectrons, b) the work function of copper, and c) the threshold wavelength.

2. A sample of hydrogen atoms in their ground states absorbs energy from incident photons, and some of the atoms make transitions to their fourth excited states. Find a) the number of different wavelengths observed in the radiation from these atoms and b) the shortest wavelength.

3. An ionized carbon atom (Z = 6) has one electron in orbit around the nucleus. Find a) the radius of the Bohr orbit in the second excited state and b) the energy and wavelength of the first Balmer (n_b = 2) transition in this atom.

4. Assuming that classical kinematics is a good approximation for electron speeds up to 0.4 times the speed of light, find a) the de Broglie wavelength of the electron at this speed and b) the de Broglie wavelength of a proton having the same speed.

Chapter 43
Nuclear Structure and Nuclear Technology

PREVIEW

This chapter introduces the neutron-proton model and used it to consider some of the properties of the nucleus and the decay of unstable nuclei by alpha-particle, beta-particle, and gamma-photon emission. Applications of nuclear physics presented include radioactive age determination and neutron activation analysis. The chapter also discusses nuclear reactors and the use of the fission and fusion processes as sources of nuclear energy.

SUMMARY

43.1 THE NEUTRON-PROTON MODEL OF THE NUCLEUS

In the previous chapter, we found that in 1911 Rutherford identified the nucleus on the basis of alpha-particle scattering experiments and correctly estimated its size as approximately four orders of magnitude smaller than the characteristic atomic dimension of one angstrom. Rutherford's work is sometimes said to have included the first nuclear physics experiments, but the beginning of nuclear physics probably came in 1932 with Chadwick's discovery of the neutron, Anderson's discovery of the positron, the first cyclotron by Lawrence and Livingston, and the first successful production of a nuclear reaction with a high-voltage accelerator by Cockcroft and Walton.

The neutron and proton, referred to as nucleons inside the nucleus, are the constituents of the nucleus in the neutron-proton model and are held together by the strong nuclear force. This force is the strongest force in nature, has a short range of the order of 10^{-15} m = 1 fm, is attractive and independent of the nucleon charge, and saturates, which means that an individual nucleon interacts with only a limited number of other nucleons.

The number of protons in the nucleus is called the atomic number Z, and the nucleus has a positive charge Ze due to the protons. The number of neutrons is called the neutron number N, and the total number of nucleons is called the mass number A = Z + N. The notation for a particular nuclear species, or nuclide, is the same as that for neutral atoms and is given for a particular chemical symbol X and numbers Z, N, and A by $^{A}_{Z}X_{N}$. The upper right corner is reserved for the charge or ionization state. The chemical symbol and Z are redundant, as both identify the element. Different isotopes of the same element have different N and A numbers, for example, two isotopes of carbon are $^{12}_{6}C_{6}$ and $^{13}_{6}C_{7}$. Nuclear masses are denoted by m, and the mass of the neutral atom, including the Z electrons, is denoted by M. Mass tables usually give the masses of neutral atoms on a mass scale on which the mass of the ^{12}C isotope is 12 atomic mass units (u).

> Definition. Binding energy: the energy required to disassemble a system into its constituent particles.

The nuclear binding energy is the energy required to take apart the nucleus into Z protons and N neutrons and is given by the sum of the mass energies of all the nucleons reduced by the mass energy of the assembled nucleus:

$$B.E. = Z M_H c^2 + N m_n c^2 - Mc^2 \qquad (43.4)$$

where M_H is the mass of the hydrogen atom, m_n the mass of the neutron, and M the mass of the assembled nucleus. This expression neglects the binding energies of the atomic electrons, but these are negligible compared to the nuclear energies. Conversely, the binding energy is recovered if the nucleus is assembled from the Z protons and N neutrons. Using an atomic mass table in atomic mass units (u), we can find binding energy by calculating the mass difference $\Delta m = (Z M_H + N m_n - M)$ and converting it to energy using the rest mass energy of one atomic mass unit. In Sec. 25 of the Study Guide we found that $1 u = 1.66053 \times 10^{-27}$ kg, and the mass unit MeV/c^2 is thus given by $1 u = 931.48$ MeV/c^2.

As a result of the saturation of nuclear forces the binding energy of a nucleus is proportional to the number of nucleons, or mass number A, so that the binding energy per nucleon B.E./A is approximately constant for all nuclei at around 8 MeV/nucleon. The slight fall off at small and large mass numbers is actually responsible for the energy releases in nuclear fission and fusion.

The observation that the interior densities of all nuclei are approximately the same requires a spherical nuclear volume proportional to the number of nucleons and a nuclear radius proportional to the cube root of the mass number:

$$R = r_o A^{1/3} \qquad (43.7)$$

where the constant of proportionality has been determined by experiment to be approximately $r_o = 1.2 \times 10^{-15}$ m = 1.2 fm (Eq. 43.8).

43.2 NUCLEAR STABILITY

Stable light nuclei are characterized by approximately equal numbers of neutrons and protons. In heavier nuclei beyond approximately Z = 40, the saturation of nuclear forces leads to greater electrical repulsion than nuclear attraction for protons and requires a neutron excess. The extra neutrons have no electrical interaction but contribute nuclear binding as a result of the charge independence of the nuclear force. There are over 2000 known stable and unstable nuclei, and their properties are summarized on a chart of the nuclides, an arrangement of all nuclides according to atomic number and neutron number. Other information on the chart includes atomic mass, isotopic abundance of stable nuclides, and the decay mode and half life of unstable nuclides.

Definition. Nuclear reaction: any process in which a nucleus undergoes some change, including radioactive decay and interaction with another nucleus.

All nuclear reactions are characterized by conservation of energy, including mass energy, charge, and linear and angular momenta. Those reactions that involve the spontaneous decay of an unstable nucleus are referred to as radioactivity. The three types of radioactivity are called alpha (α), beta (β), and gamma (γ) decay and were named by Rutherford in increasing order of penetrating power in matter before it was known that the α particle is a helium nucleus ($^4_2He_2$), β particles are electrons or positrons, and γ rays are photons.

Since an α particle consists of two protons and two neutrons, α decay requires a decrease of two in both the atomic number and the neutron number of the parent nucleus. Alpha decay occurs primarily in nuclei with A > 200, and an example is the decay of ^{238}U to ^{234}Th:

$$^{238}_{92}U_{146} \rightarrow \,^{4}_{2}He_2 + \,^{234}_{90}Th_{144} \qquad (43.12)$$

The mass of the parent exceeds that of the daughter plus alpha particle, and the excess mass energy is shared by the two decay products in a specific way as kinetic energy and in some cases as as excitation energy of the daughter nucleus. The excess mass energy can be obtained by calculating the mass excess using an atomic mass table and converting the result using the energy equivalent of one atomic mass unit. Conservation of charge and baryon number can be applied in alpha decay as conservation of atomic number and mass number.

Beta decay is a result of the weak nuclear force and involves the decay of an unstable nucleus into three products, giving the β particle a spectrum of energies rather than a discrete energy as in α decay. There are two types of β particles, and the parent nucleus decays into a daughter nucleus and either an electron (e^-) and an antineutrino ($\bar{\nu}$) or a positron (e^+) and a neutrino (ν). Examples are the electron decay of ^{14}C and the positron decay of ^{11}C:

$$^{14}_{6}C_8 \rightarrow \,^{14}_{7}N_7 + e^- + \bar{\nu} \qquad (43.25)$$

$$^{11}_{6}C_5 \rightarrow \,^{11}_{5}B_6 + e^+ + \nu \qquad (43.17)$$

In β decay, the energy available in the reaction is given by the mass energy excess of the parent and daughter atoms, to within the difference in the binding energies of the atomic electrons in these two atoms. Charge is conserved but is not reflected in conservation of mass number, however, conservation of baryon number can be applied as conservation of mass number, with the electron and neutrino and their antiparticles having mass numbers of zero.

A free neutron undergoes β decay to a proton, but the decay of a free proton is energetically impossible because m_p (1.00727 u) < m_n (1.008665 u). Inside the nucleus, however, β decay can be viewed as the transformation of p into n (or n into p) and the simultaneous creation of the decay products at the time of transformation.

Gamma emission is entirely analogous to the emission of photons by excited atoms according to the Bohr radiation postulate: $h\nu = E_a - E_b$ (Eq. 42.7), where E_a and E_b are the quantized energies of the nuclear states. As is the case with atoms, nuclei have unique energy states and gamma photon energies that can be used for identification. The energies of the gamma photons are larger than those of atomic photons, reflecting the larger energies of the nuclear states (MeV range versus eV-keV in atoms). Gamma decay differs from α and β decays in that it cannot result from unstable nuclei in their ground states and can come only from nuclei left in excited states by other processes.

The radioactive decay law is derived by assuming a random process in which nuclei decay independently of each other and of how long a given nucleus has lived, resulting in a decay rate proportional to the number of nuclei present:

$$dn/dt = -\lambda n \qquad (43.19)$$

where the decay constant λ is the probability per unit time that any particular nucleus will decay and n is the number of nuclei present at time t.

This equation can be integrated to yield the radioactive decay law in terms of either nuclei present or instantaneous decay rate, called the activity a:

$$n = n_0 e^{-\lambda t} \tag{43.20}$$
$$a = a_0 e^{-\lambda t} \tag{43.21}$$

These relations indicate that both quantities decrease exponentially from their initial values n_0 and a_0. The activity is measured experimentally as the average decay rate $a = \Delta n/\Delta t$, where Δn is the number of disintegrations occurring in a time interval Δt. Decay rates are usually expressed in a unit called the curie (Ci), based on the radioactivity of one gram of radium (1 Ci = 3.7×10^{10} decays/s).

Definition. Half life: the time required for both the number of nuclei and the decay rate of a sample of unstable nuclei to decrease by a factor of two.

The half life $\tau_{1/2}$ is related to the decay constant λ by
$$\tau_{1/2} = \ln 2/\lambda = 0.693/\lambda \tag{43.23}$$

If the time in Eqs. 43.20 and 43.21 is expressed as a multiple of the half life, $t = n\,\tau_{1/2}$, the number of nuclei present and the activity are given in terms of their initial values by $n = n_0(1/2)^n$ and $a = a_0(1/2)^n$. These expressions are valid for any time but are most easily evaluated when n is an integer.

Nuclear reactions in which two nuclei interact to produce one or more product nuclei can be analyzed in much the same way as radioactive decay. Energy, mass number and atomic number are conserved. The mass-energy difference is known as the reaction energy (or reaction Q value) and is shared by the product nuclei as kinetic and excitation energies. The reaction energy can be calculated by converting the mass difference in atomic mass units using the energy equivalent of one atomic mass unit. Examples of these types of nuclear reactions are involved in the applications of nuclear technology given in the remainder of the chapter and include nuclear fission, nuclear fusion, the production of ^{14}C in the atmosphere, and the activation of nuclei by neutron bombardment.

43.3 RADIOACTIVE DATING

In carbon-14 dating, radioactive ^{14}C produced in the atmosphere by the interaction of neutrons with ^{14}N is ingested by living organisms at the atmospheric equilibrium ratio to total carbon of approximately 10^{-12}. When an organism dies, the ratio decreases as a result of the β decay of ^{14}C. A measurement of the decay rate per gram of total carbon allows a determination of the time since the death of the organism. The age of the sample can be calculated from
$$t = -[\ln(I/I_0)]/\lambda \tag{43.30}$$
where I_0 and I are the equilibrium and present decay rates per gram of total carbon and the decay constant is related to the half line of ^{14}C by $\lambda = \ln 2/\tau_{1/2} = 0.693/5730$ yr. The standard form of ^{14}C dating can be used for organic samples with ages up to approximately ten times the 5730-yr half life of ^{14}C.

Other dating methods involve the measurement of the concentrations in a sample of pairs of nuclei related by radioactive decay. As time goes on, the concentration of one member of the pair decreases while that of the other member increases. Examples include the rubidium-strontium, uranium-lead, and krypton-argon methods used for dating rock specimens and their fossils.

43.4 NEUTRON ACTIVATION ANALYSIS

Neutron activation analysis is used to determine the concentration in a sample of an element of interest. Nuclei in the sample are excited by neutron bombardment and rapidly decay to other nuclear species, which de-excite by gamma photon emission. Measurement of the gamma photon energies and associated activities allows unique identification of the emitting nuclei and calculation of the concentration in the sample of the nuclei of interest. The concentration can be obtained from

$$I_1/I_0 = C_1/(C_0 + C_1) \tag{43.35}$$

where C_0 and C_1 are the concentrations in the standard (known) and sample and I_0 and I are the measured specific decay rates.

43.5 NUCLEAR ENERGY

Definition. Nuclear fission: the splitting of a heavy nucleus into two lighter fragments of comparable masses and several neutrons, either spontaneously or induced by neutron absorption, with the release of binding energy.

Contemporary nuclear reactors are used at electric energy plants to convert energy from nuclear fission reactions into thermal energy that is subsequently transformed into electrical energy. The fuel is ^{235}U, and a typical fission reaction is

$$^{235}_{92}U_{143} + ^{1}_{0}n_1 \rightarrow ^{143}_{56}Ba_{87} + ^{90}_{36}Kr_{54} + 3\ ^{1}_{0}n_1 \tag{43.36}$$

The average number of neutrons released is 2.5, and these make it possible to produce a self-sustaining chain reaction of fissions. Over 100 different fission fragments with atomic number in the range $30 < Z < 60$ are possible, but the energy released is always around 200 MeV, primarily in the form of kinetic energy of the fragments.

The fission energy comes from the difference in binding energies of the fissioning nucleus and the fission fragments. Nuclei with mass numbers around that of uranium have binding energy per nucleon values of around 7.5, while those in the region of the fission fragments have values near 8.5. This difference of 1 MeV per nucleon is the energy released in the fission reaction.

Other important aspects of the physics of nuclear reactors include the requirements for moderation, enrichment, and control. The neutrons produced in the fission reaction are relatively fast and must be slowed down or moderated to enhance the probability of capture leading to fission. Moderators must have atoms of approximately the same mass as the neutron, and water is commonly employed. Absorption of neutrons by the water makes it necessary to enrich the isotopic concentration of ^{235}U from the naturally-occurring value of 0.7 percent to several percent. The rate of fission reactions in a reactor is controlled by the use of materials, such as boron and cadmium, with a high probability of capturing the neutrons before they initiate further fissions.

Definition. Nuclear fusion: a nuclear reaction in which two light nuclei combine to form a heavier nucleus and other reactions products, with the release of binding energy.

As in fission, fusion energy comes from the difference in binding energies of the reacting and product nuclei and is released primarily as kinetic energy of the products. However, the technology associated with the use of fusion in nuclear reactors is at a stage of development where fusion can be considered only as a possible long-term commercial energy source. The leading candidates for reactor fission involve hydrogen nuclei, because of the energy requirements imposed by the electrical repulsion of the reacting nuclei. One such reaction involves the interaction of the nuclei of heavy hydrogen (deuteron) and $^{3}_{1}H_{2}$ (triton) in the so-called DT reaction:

$$^{2}_{1}H_{1} + ^{3}_{1}H_{2} \rightarrow ^{4}_{2}He_{2} + ^{1}_{0}n_{1} \qquad (43.44)$$

where the difference in binding energy that is released is 17.6 MeV. The energy yield per nucleon of over 3 MeV is comparable to the 1 MeV/nucleon of fission reactions.

Other aspects of the physics of fusion reactors include energy injection and containment. The kinetic energy required to overcome the electric potential energy barrier is of the order of 10 keV, which corresponds to a so-called ignition temperature of approximately 10^8 K. At these temperatures, the reacting nuclei are ionized and form a plasma. The problems associated with achieving a workable fusion system involve containing the hot plasma and delivering sufficient energy for the plasma to reach ignition temperature and fusion reactions to be initiated. A requirement that ensures a net gain in energy from the fusion reaction called the Lawson criterion must also be achieved:

$$n\tau > 10^{14} \text{ (ions/cm}^3\text{)-s} \qquad (43.47)$$

where n is the density of the plasma and τ is the confinement time, or time during which the ions interact. Two different methods for achieving the Lawson criterion have received serious attention: magnetic confinement of the plasma and inertial confinement by laser implosion of solid fuel pellets.

EXAMPLE PROBLEMS
Example 1

For ^{17}O, find a) the binding energy per nucleon and b) the energy required to remove one neutron from the nucleus.

Solution

Given: $M(^{16}O) = 15.994915$ u   (Table 43.1)
$M(^{17}O) = 16.999133$ u
$M_H = 1.007825$ u
$m_n = 1.008665$ u

Required: a) BE/A,   b) E_s

a) The binding energy is given in terms of atomic masses by Eq. 43.4:

$$BE = (Z\, M_H + N\, m_n - M)c^2$$

where M is the mass of the assembled nucleus, here ^{17}O. For ^{17}O, the chemical symbol for oxygen identifies the atomic number Z = 8. According to the notation $^{A}_{Z}X_{N}$, A = 17 and N = A - Z = 9. Substituting in the binding energy equation for the numbers Z and N and the mass $M(^{17}O)$ of the ^{17}O atom, the mass M_H of the hydrogen atom, and the mass m_n of the neutron:

$$BE = [(8)(1.007825 \text{ u}) + (9)(1.008665 \text{ u}) - 16.999133 \text{ u}]c^2 = (0.141452 \text{ u})c^2$$
$$= (0.141452 \text{ u})c^2 \,(931.48 \text{ MeV}/c^2\text{-u}) = 131.760 \text{ MeV}$$

The binding energy per nucleon is given by the ratio of the total binding energy for the nucleus and the number of nucleons A in the nucleus:

$$BE/A = 131.760/17 = \underline{7.75 \text{ MeV/nucleon}}$$

This is the average energy per nucleon required to disassemble the ^{17}O nucleus or the energy released when 8 protons and 9 neutrons are assembled into the ^{17}O nucleus.

b) The energy required to remove one neutron from the nucleus is referred to as the binding energy of the last neutron or the neutron separation energy E_s. In this case, the assembled nucleus remains ^{17}O, but the pieces are ^{16}O and n:

$$E_s = [M(^{16}O) + m_n - M(^{17}O)]c^2 = (15.994915 \text{ u} + 1.008665 \text{ u} - 16.999133 \text{ u})c^2$$
$$= (0.004447 \text{ u})c^2 = (0.004447 \text{ u})(931.48 \text{ MeV/u}) = \underline{4.14 \text{ MeV}}$$

The fact that the neutron separation energy in ^{17}O is less than the average binding energy of a nucleon reflects the relatively strong binding (and low mass) of the ^{16}O nucleus. In calculations such as this involving small mass excesses, it may be necessary to use all significant figures given in the atomic mass table in order to obtain at least three significant figures in the result.

Example 2

A possible candidate for a reactor fusion reaction is the proton-induced fusion of ^7Li, producing an alpha particle as one of the product nuclei. Find a) the other reaction product and b) the energy released in the reaction.

Solution

Given: $M(^7Li) = 7.016005$ u   (Table 43.1)

$M_H = 1.007825$ u

$M(^4He) = 4.0026034$ u

Required: a) reaction product, b) E

a) With the unknown product nucleus denoted by $^A_Z X_N$, this fusion reaction can be written as

$$^7_3Li_4 + ^1_1H_0 \rightarrow ^4_2He_2 + ^A_Z X_N$$

The unknown product nucleus can be identified by applying conservation of mass number and atomic number in the reaction:

$A = 7 + 1 - 4 = 4$

$Z = 3 + 1 - 2 = 2$

The element with atomic number 2 is helium and $N = A - Z = 4 - 2 = 2$, therefore the product nucleus is given by

reaction product = $^4_2He_2$

b) The energy released in the reaction can be obtained by converting the mass excess calculated in atomic mass units from Table 43.1 to mass energy excess using
$E(\text{MeV}) = mc^2 = m(u)(931.48 \text{ MeV/u})$:

Reacting Nuclei	Product Nuclei
^7Li: 7.016005 u	^4He: 4.002603 u
^1H : 1.007825 u	^4He: 4.002603 u
8.023830 u	8.005206 u

The mass excess is given by

$\Delta m = 8.023830 \text{ u} - 8.005206 \text{ u} = 0.018624 \text{ u}$

The energy released in the reaction is given by the mass energy excess:

$E = \Delta mc^2 = (0.018624 \text{ u})(931.48 \text{ MeV/u}) = \underline{17.3 \text{ MeV}}$

This reaction has an energy release that compares favorably with those in the DD and DT reactions (Eqs. 43.42 - 44), however, a disadvantage for its use in a fusion reactor would be a larger electrical potential energy barrier.

Example 3

What is a) the daughter nucleus in the electron decay of ^{60}Co and b) the maximum kinetic energy of the emitted electron?

Solution

Given: $M(^{60}Co) = 59.933809$ u (Table 43.1)
$M(^{60}Ni) = 59.930778$ u

Required: a) daughter nucleus, b) K_{max}

a) With the unknown daughter nucleus denoted by $^A_Z X_N$, this β decay can be expressed as

$$^{60}_{27}Co_{33} \rightarrow e^- + ^A_Z X_N + \bar{\nu}$$

The daughter nucleus can be identified by applying conservation of charge and mass number in the reaction. The electron and the antineutrino have a mass number of zero, therefore, the mass number of the daughter nucleus is the same as the parent nucleus, and A = 60. Conservation of charge requires that the atomic number of the daughter nucleus be one higher than that of the parent nucleus, so that Z = 27 + 1 = 28. Also, N = A - Z = 60 - 28 = 32. The element with atomic number 28 is nickel, therefore, the daughter nucleus is given by

$$\text{daughter nucleus} = \underline{^{60}_{28}Ni_{32}}$$

b) The maximum kinetic energy of the electron is equal to the mass energy excess in the reaction, which is given for β decay by the mass energy difference of the parent and daughter atoms if the difference in the binding energies of 27 electrons in ^{60}Co and 28 electrons in ^{60}Ni is neglected. The mass excess can be obtained from the masses in Table 43.1:

$$m = M(^{60}Co) - M(^{60}Ni) = 59.933809 \text{ u} - 59.930778 \text{ u} = 0.003031 \text{ u}$$

The mass energy excess and the maximum kinetic energy of the electrons is given by

$$K_{max} = mc^2 = (0.003031 \text{ u})(931.48 \text{ MeV/u}) = \underline{2.82 \text{ MeV}}$$

Example 4

In 1919, Rutherford identified the proton in a nuclear reaction in which he produced the first transmutation of an element by bombarding ^{14}N with α particles from ^{214}Po. What is a) the other product nucleus in this reaction and b) the mass energy excess?

Solution

Given: $M_H = 1.007825$ u (Table 43.1)
$M(^4He) = 4.002603$ u
$M(^{14}N) = 14.003074$ u

Required: a) product nucleus, b) E

a) With the unknown product nucleus denotes by $^A_Z X_N$, this nuclear reaction can be written as

$$^{14}_7 N_7 + ^4_2 He_2 \rightarrow ^1_1 H_0 + ^A_Z X_N$$

The unknown product nucleus can be identified by applying conservation of mass number and atomic number in the reaction:

$$A = 14 + 4 - 1 = 17$$
$$Z = 7 + 2 - 1 = 8$$

The element with atomic number 8 is oxygen and $N = A - Z = 17 - 8 = 9$, therefore, the product nucleus is given by

$$\text{reaction product} = {}^{17}_{8}O_8$$

b) The mass energy excess can be obtained from the mass excess calculated from the atomic masses in Table 43.1. The atomic mass of ^{17}O is given by $M(^{17}O) = 16.999133$ u.

Reacting Nuclei	Product Nuclei
^{14}N: 14.003074 u	^{17}O: 16.999133 u
^{4}He: 4.002603 u	^{1}H: 1.007825 u
18.005677 u	18.006958 u

The mass excess is given by

$$\Delta m = 18.005677 \text{ u} - 18.006958 \text{ u} = -0.001281 \text{ u}$$

The mass energy excess is obtained by converting the mass excess using the energy equivalent of one atomic mass unit:

$$E = \Delta mc^2 = (-0.001281 \text{ u})(931.48 \text{ MeV/u}) = \underline{-1.19 \text{ MeV}}$$

The negative value of the mass energy excess indicates that this reaction will not proceed unless this energy is brought into the reaction by one of the reacting nuclei. In this case, the excess energy can be supplied by the incident α particle, because α particles from ^{214}Po have an energy of 7.69 MeV.

Example 5

Find a) the activity of a 1-gm source of ^{226}Ra and b) the time required for the activity to be reduced by 5 percent.

Solution

Given: $\tau_{1/2}(^{226}Ra) = 1.60 \times 10^3$ yr (Table 43.2)

$1 \text{ Ci} = 3.7 \times 10^{10}$ decays/s

$M(^{226}Ra) = 226.0$

$N_A = 6.02 \times 10^{23}$/mol

Required: a) a_o, b) $t(0.95 \, a_o)$

a) The activity of a radioactive sample is given by Eq. 43.19:

$$dn/dt = a = -\lambda n \tag{1}$$

where n is number of nuclei present at time t, λ is the decay constant, and the minus sign indicates that n decreases with time.

The decay constant is given by Eq. 43.23:

$$\lambda = \ln 2/\tau_{1/2} = 0.693/\tau_{1/2}$$

The number of nuclei/gm present in the source can be obtained from Avogadro's number, which is the number of atoms in a mole, and the gram molecular weight, which is the mass of a mole:

$$n = N_A/M(^{226}Ra)$$

Substituting in Eq. (1) for the decay constant λ and the number of nuclei n:

$$a_0 = (0.693/\tau_{1/2})(N_A/M(^{226}Ra))$$

Substituting in this equation for the half life $\tau_{1/2}$, Avogadro's number N_A, and the atomic mass $M(^{226}Ra)$:

$$a_0 = (0.693)(6.02 \times 10^{23}/mol)/(1.60 \times 10^3 \text{ yr})(226 \text{ gm/mol}) = 1.15 \times 10^{18} \text{ decays/gm-yr}$$

For a 1-gm source:

$$a_0 = (1.15 \times 10^{18} \text{ decays/yr})(\text{yr}/3.156 \times 10^7 \text{ s}) = \underline{3.66 \times 10^{10}} \text{ decays/s}$$

This result is not surprising, as we have seen that the disintegration rate unit called the curie is based on the radioactivity of a gram of radium and is given by 1 Ci = 3.7×10^{10} decays/s.

b) The time required for the activity to decay to a given fraction of the original activity can be obtained from Eq. 43.21:

$$a = a_0 e^{-\lambda t}$$

As we have seen in Sec. 43.2 of the Study Guide, for $t = n\,\tau_{1/2}$, this equation becomes

$$a/a_0 = 1/2^n$$

Solving this equation for n and substituting in the expression for t:

$$n = -\ln(a/a_0)/\ln 2$$

$$t = n\,\tau_{1/2} = [-\ln(a/a_0)](\tau_{1/2})/\ln 2 = [-\ln(a/a_0)](\tau_{1/2})/0.693$$

Substituting in this equation for the half life $\tau_{1/2}$ and the ratio $a/a_0 = 0.95$:

$$t(0.95\,a_0) = -\ln(0.95)(1.60 \times 10^3 \text{ yr})/0.693 = -(-0.0513)(1.60 \times 10^3 \text{ yr})/0.693 = \underline{118 \text{ yr}}$$

PROBLEMS

1. What is a) the average binding energy of a nucleon in ^{12}B and b) the energy required to remove one neutron from the ^{12}B nucleus?

2. The neutron-induced fission of ^{233}U produces a fission fragment ^{91}Kr and three neutrons.
   a) Identify the other fission fragment and b) find the energy released in this reaction.

3. Find a) the daughter nucleus in the α decay of ^{238}Pu and b) the energy available for this decay.

4. Chadwick discovered the neutron in 1932 in a nuclear reaction in which he bombarded ^9Be with α particles. a) What is the other product nucleus and b) what is the mass energy excess in this reaction?

5. What is a) the activity of a 5-μgm ^{252}Cf source and b) the time required for the activity to be reduced by a factor of 8, assuming that the half life of ^{252}Cf is 2.64 yr?

Answers to Problems

CHAPTER 1
1. M/T^2
2. L^3
3. 2.47 acres
4. 0.946 liters
5. 1.86×10^5 mi/s; 6.71×10^8 mi/hr

CHAPTER 2
1. a) 4.5° west of north
   b) 307.6 km/hr
2. a) 5.7°, b) 100.5 m
4. Area = 482.1 cm^3
5. 57.5°
6. a) Area = 275 m^2
   b) 46.1 m
7. a) $A+B = i34-j9+k41$ m
   b) $A-B = i14+j41+k31$ m
   c) $A \cdot B = 20$ m^2
   d) $A \times B = i980+j240-$

CHAPTER 3
1. a) 600 N, upward
   b) painter on earth
2. 480 N
3. a) 80 N, b) 30 N
4. a) 36 N, b) 84 N
5. 1350 N
6. 400 N, 300 N
7. 53 cm

CHAPTER 4
2. 0, -3.3 m/s^2, 0, 2.8 m/s^2
3. a) -46 m/s, -44.7 m/s, -48 m/s, -50.8 m/s, -52 m/s
   b) -53 m/s
   c) $3t^2-20t^2-25 \big|_{t=2} = -53$ m/s
4. a) 220 km/hr
   b) tangent to the track
   c) 2.2 m/s^2, d) 0, e) 0
5. 8.68 m/s^2
6. -4.17 m/s^2
7. a) 76.7 m/s, b) 7.82 s
8. 1.96 s, 10.8 m/s
9. 49 ft/s
10. 1.3°, relative velocities only
11. a) 6 knots, b) -2 knots

CHAPTER 5
2. 1.91 m
3. 50 m
4. 211 m/s, assuming height of fence is y_{max}
5. a) $\theta = \tan^{-1}(v_o/v_f)$, where $v_f = \sqrt{v_o^2+2gh}$ and $h = 2$ m
   b) yes; c) yes
6. 2.7×10^{-3} m/s^2
7. 4.47 m/s^2

CHAPTER 6
1. 22.3 N
2. 18,333 lb
3. a) 22 N, b) 11 N
4. 2.64×10^4 N, 594×10^3 lb
7. a) \$12, b) the same price
8. a) 5.55 m/s^2, b) 44 m/s
9. a) a has the same magnitude but the opposite direction.
   b) $F_{m_1,m_2} = -F_{m_2,m_1} = 30$ N

CHAPTER 7
1. 1.48×10^6 J
2. 490 J
4. $(1.5)F_o x_o$
5. $-1/6$ J
6. 47 J
7. 5.36 m/s
8. 3.16×10^2 km/hr
9. 1.125×10^5 N

CHAPTER 8
1. 1.96×10^5
2. a) A, b) 0, c) A/2
   d) $A/\sqrt{2}$
3. $v = \sqrt{gL}$
6. a) Newton's 2nd law
   b) Work-energy principle
   c) Newton's 2nd law
   d) Work-energy principle

CHAPTER 9
1. a) 16 N-s, b) 16 N-s
2. a) 0.024 N-s, b) 0.024 N-s
   c) 0.6 N, d) 0.0144 J
3. 0.8 N-s
5. 6.2 m/s
7. 250 m/s
8. 30 J for the heavy block and 60 J for the light block
9. No, $\approx 1/5$

CHAPTER 10
1. 14 cm from left
2. 2.017 m from "top"
3. 162.5 m/s^2
4. a) colliding beam
   b) twice a much
6. 1137 kg/s

CHAPTER 11
1. $-6\mathbf{k}$
2. 2.5×10^6 kg-m^2/s
3. a) 4.23×10^7 m
   b) 3×10^3 m/s
5. a) 2, b) 8
6. a) 2.7×10^{40} kg-m^2/s
   b) 2.8×10^{34} kg-m^2/s
   c) 2.73×10^{40} kg-m^2/s

CHAPTER 12
1. 13.7 rad/s^2
2. a) 454 rad/s^2
   b) 73.8 rad
3. 18,000 rpm
4. 1.54 kg-m^2/s
6. a) 2467 J, b) 2.1 N-m
   c) 11.2 s
7. a) 6.25 rad/s^2
   b) 22.4 rad/s
   c) 5.3 rad/s^2
9. $I = (1/2)M(R_1^2 + R_2^2)$
10. a) 4.1 rad/s, b) 17.5 J
    c) Internal work done by skater's arms.

CHAPTER 13
1. 1.5×10^3 W
2. 534 m-N
3. 9.55×10^5 N-m
4. $(5/7)\tan\theta$
5. 1.59 rad/s
6. 9.59×10^{-4} rad/s

CHAPTER 14
1. a) 0.0987 N/m, b) 3.15 m/s
2. 3.95×10^3 N
3. a) 120 J, b) 34.6 m/s
4. a) $x = A\cos(\omega t + \pi)$
   b) $x = A\cos(\omega t + 3\pi/2)$
5. $x = 1.37 \sin(6.12t)$
6. $\nu = \sqrt{2k/m}/2\pi$
7. 4.47×10^3 kg/s

CHAPTER 15
1. $\Delta\ell/\ell = 1.47$ mm
2. $\theta = 8.2°$
3. $\Delta\ell = .069$ mm
4. 110.25 Pa
5. $U = \dfrac{\pi\mu R^3 t}{\ell}\theta^2$

CHAPTER 16
1. 80 kg
2. a) 980 Pa, b) 7644 Pa
3. a) 20 m/s
   b) 0.0141 m^3/s
4. 16.7 m/s
5. 0.895 atm
6. a) 13,370, b) turbulent

CHAPTER 17
1. a) 4 m, 0.75 Hz, 1.57/m
   b) $\psi(x,t) = 0.8 \sin(1.57x + 4.71 t)$ m
2. a) 1.33 m/s, positive x-direction
   b) 8 m/s
3. 2 m, $5\pi/2$ m, $50/\pi$ Hz, 125 m/s
4. 0.375 m, 1.23 m, 2.08 m
5. 252 Hz
6. 40 m/s

CHAPTER 18
1. 750 N
2. a) 3.33 km/s, b) 2.0 km/s
3. 3.6×10^{-4} J
4. 90 dB
5. a) 3 m, b) 14.0 Hz
6. 60 dB

CHAPTER 19
1. a) 1.98×10^8 m/s
   b) 2.24×10^8 m/s
   c) 2.60×10^8 m/s
3. 2.65×10^7 m/s
4. .999999955 c
5. 1.67×10^{-3} s

CHAPTER 20
1. 0.417 c
2. a) 3.37×10^5 m/s
   b) 1.08×10^7 m/s
   c) No
3. 1.19 mc^2, 0.952 c
4. 5.56×10^{-6}, 5.56×10^{-4}, 6.07×10^{-2}, 2.91
5. a) 3.44×10^9 eV
   b) 0.962 c
   c) 0.756 c; 1.44×10^9 eV

CHAPTER 21
1. 23.0 F°
2. 54.4°C
3. 144 J
4. 3000 J
5. 600 J

CHAPTER 22
1. -1.17 cm
2. 22.2 liters
3. 280 K
4. a) 138 K, b) 0.0988 atm
5. 8875 kJ
6. 23.0°C

CHAPTER 23
1. 10.6 W, inward
2. 709 W, outward
3. 0.00697 m^2
4. 310 K

Answers to Problems   323

CHAPTER 24
1. a) 0.25, b) 0.333
2. -12.2 kJ/K
3. 2.52 J/K
4. a) 107 J/K, b) -104 J/K
   c) 3.38 J/K

CHAPTER 25
1. a) 44.0 gm
   b) 7.31×10^{-26} kg
2. a) $4.06 \times 10^{13}/m^3$
   b) 1.12×10^{-7} Pa
3. a) 6.24×10^4 J
   b) 1.04×10^{-19} J
   c) 7500 K
4. a) 600 m/s, b) 490 m/s
   c) -200 °C

CHAPTER 26
1. $M = 3.46 \times 10^{-18}$ Kgm
2. a) 1.09 microcoulombs
   b) the same as in Ex. Prob.
3. If $\frac{2x}{R} \ll 1$, then
   $F_T = \frac{16kQqx}{R^3}$
4. In either case, displacement in one direction yields stable equilibrium while displacement in the other direction yields unstable equilibrium.
5. $F = 7.8 \times 10^{-4}$ newtons toward the center of the arc.

CHAPTER 27
1. For $x \ll d$, $E = -Qx/(\pi\epsilon_0 d^3)$
2. $f = (1/2)\{Q/(\pi\epsilon_0 d^3 m)\}^{1/2}$

3. a) In the region between the two spheres, the field is just that of a point charge of magnitude -Q; In the region outside the larger sphere, the field is zero.
4. For $r > R$, $E = 0$. For $r < R$,
   $E = \frac{-Q}{4\pi\epsilon_0 r^2}\{1 - (\frac{r}{R})^3\}$
5. The point charge will be forced away from the center of the spherical distribution creating a field for $r > R$ identical to that of a dipole of moment
   $p = 3QE\epsilon_0/\rho$

CHAPTER 28
1. $d = 1.89 \times 10^{-10}$ M;
   $W = 7.6$ eV
2. $E = \frac{Q}{\pi\epsilon_0 a^3}\{ix + j0 + k0\}$
3. 5.8×10^{22} eV
4. $V = \frac{Q}{2\pi\epsilon_0 x} \frac{1}{1-(a/x)^2}$
5. $V = \frac{1}{2\pi\epsilon_0} \ln(r_1/r_2)$

CHAPTER 29
1. $C = 77$ pf (picofarads)
2. The dielectric causes the capacitance to increase by the fraction,
   $\Delta C/C = (\kappa - 1) y/L$
3. No significant change
4. $h = 1.224$ m
5. $V = .495$ volts

CHAPTER 30
1. $V = 0.106$ volts
   $J = \frac{\sigma V}{\ell}[1 - 4(\frac{y}{\ell})^2]$

2. $I = 2567$ amperes - Such a large current would, of course, very quickly melt the brass strip
3. $\Delta t = 5.2$ °C
4. $R_{(Static)} = 1.125 \Omega$
   $R_{(Dynamic)} = 0.05 \Omega$
5. $R = 48 \Omega$; $I = 2.5$ amperes

CHAPTER 31
1. $\mathcal{E} = 23.5$ volts
2. $I_3 = 31$ milliamperes
3. $I = 28.3$ milliamperes
4. $V = 30$ volts; $C = 100\mu f$
5. $R = \frac{VR_V}{(IR_V - V)}$

CHAPTER 32
1. 10,000 volts/m in the positive z direction
2. .033 newtons horizontally toward the west
3. $n = 0.85$ turns
4. The loop will align its magnetic moment with the field and then form itself into a circular shape
5. $\mu/J = (1/2) Q/m = .005$ C/Kgm

324 Answers to Problems

CHAPTER 33
1. $B = \mu_0 I/4\pi$
2. $B = k\mu_0 Iy/\pi[a^2+y^2]$
   Note: The perceptive student will note that this expression vanishes both for $y=0$ and for $y=\infty$ thus implying some value of y for which B is a maximum. He will be unable to eat or sleep until he finds the position of this maximum.
3. $\Phi = \dfrac{\mu_0 NIt}{2\pi} \ln(OD/ID)$
   Ignoring terms of the third and larger order in the series expansion for the logarithm, the two expressions are the same
4. Eq. 33.28 correctly gives the field for any odd number of layers; the field is zero for any even number of layers
5. $F = 5 \times 10^{-4}$ newtons, downward

CHAPTER 34
1. $V = 1.05$ microvolts
2. $V_o = .66$ volts
3. 196 volts
4. $B = .375$ gauss, 70.8° below horizontal
5. $V = 8.75$ millivolts

CHAPTER 35
1. $L = .57$ henries
2. $L = (11/16)\mu_0 N^2 a$
3. $T = (1 + \dfrac{r}{R}\dfrac{L}{r})$
4. $U = 141$ joules
5. $\nu = 1.51 \times 10^7$ Hz

CHAPTER 36
1. $x = 1.7 \times 10^{-4}$ - paramagnetic

2. $T = 0.4K$
3. $\Delta t = .0018$ s
4. a) Remanent field = 0.7 tesla
   b) Coercive force = 500 amperes/m
   c) $\kappa = 3979$
5. 257.5, i.e. 12.5° south of west

CHAPTER 37
1. $V(peak) = 60$ V;
   $V(rms) = 42.4$ V
   $\nu = 159.2$ Hz ; $\Phi = 36.9°$
2. $V_{ave} = 0$; $v_{rms} = V/\sqrt{3}$
3. $V = 120$ volts
   $I = .754$ amperes
   $P = 85.3$ watts
4. $2r\left[\sqrt{\dfrac{R^2 C}{4L} + 1} - 1\right]$
5. The heat generated in a transformer depends upon the current passing through it rather than the power transmitted through it. Because of the power factor, the product of the current and voltage can be significantly larger than the power.

CHAPTER 38
1. 2.78 m (109 inches)
2. Toward the East
3. $\kappa = 1.15$
4. 474083.4
5. $\dfrac{d^2 E_x}{dz^2} = \mu_0 \varepsilon_0 \dfrac{d^2 E_x}{dt^2}$

CHAPTER 39
1. 4.5 m
2. 32.1°
3. a) 6 cm, b) virtual, upright, enlarged

4. a) 6 cm, b) real, inverted, reduced
5. a) -6 cm, b) virtual, upright, reduced
6. 36 cm
7. -42 cm

CHAPTER 40
1. 609 nm
2. 3.14×10^{-4} rad
3. a) 143 nm, b) 140 nm
   c) 4.4 rad, d) 8.8 rad
   e) 11.9 rad
4. 271 nm
5. 212 nm

CHAPTER 41
1. a) 1.5×10^{-3} rad
   b) 300 nm
2. 31.4 mm
3. 900 lines/mm
4. a) 2, b) 0.0139 nm

CHAPTER 42
1. a) 3.78 eV, b) 4.5 eV
   c) 276 nm
2. a) 10, b) 95 nm
3. a) 0.0794 nm
   b) 68 eV, 18.2 nm
4. a) 0.0606 Å, b) 3.3 fm

CHAPTER 43
1. a) 6.63 MeV/nucleon
   b) 3.37 MeV
2. $^{140}_{56}Ba_{84}$, b) 176 MeV
3. a) $^{234}_{92}U_{142}$, b) 5.59 MeV
4. a) $^{12}_{6}C_{6}$, b) 5.70 MeV
5. a) 2.69 mCi, b) 7.92 yr